OXFORD MEDICAL PUBLICATIONS

Melatonin
Clinical Perspectives

Melatonin

Clinical Perspectives

Edited by

Andrew Miles
Department of Physiology,
University College,
Cardiff

David R. S. Philbrick
Department of Child Psychiatry,
Booth Hall Children's Hospital,
University of Manchester
School of Medicine,
Blackley, Manchester

Christopher Thompson
Academic Department of Psychiatry,
Charing Cross and Westminster
Medical School,
Charing Cross Hospital,
London

OXFORD NEW YORK TOKYO
OXFORD UNIVERSITY PRESS
1988

Oxford University Press, Walton Street, Oxford OX2 6DP
Oxford New York Toronto
Delhi Bombay Calcutta Madras Karachi
Petaling Jaya Singapore Hong Kong Tokyo
Nairobi Dar es Salaam Cape Town
Melbourne Auckland
and associated companies in
Berlin Ibadan

Oxford is a trade mark of Oxford University Press

Published in the United States
by Oxford University Press, New York

British Library Cataloguing in Publication Data
Melatonin: clinical perspectives.
1. Man. Melatonin
I. Miles, Andrew II. Philbrick, David S.
III. Thompson, Christopher
612'.492
ISBN 0-19-261652-8

Library of Congress Cataloging in Publication Data
Melatonin: clinical perspectives.
(Oxford medical publications)
Includes bibliographies and index.
1. Melatonin—Physiological effect. 2. Melatonin—
Diagnostic use. 3. Mental illness—Physiological
aspects. I. Miles, Andrew. II. Philbrick, David R. S.
III. Thompson, Chris. IV. Series.
[DNLM: 1. Melatonin. WK 350 M517]
QP572.M44M45   1988   612'.492   88-17899
ISBN 0-19-261652-8

Typeset by Cotswold Typesetting Ltd, Cheltenham
Printed in Great Britain
at the University Printing House, Oxford
by David Stanford
Printer to the University

Preface

About a year ago, in consultation with my colleagues Drs David R. S. Philbrick and Chris Thompson, I began to organize the completion of the present volume for Oxford University Press. It was my intention from the start to concentrate upon and review only those areas currently of clinical interest, my opinion being that a volume dedicated to this purpose was urgently required by those scientists engaged in this specialized area of pineal research. The response of those scientists invited to contribute was particularly enthusiastic and all authors are currently working in the areas that they have reviewed, providing major commentaries on their area of interest covered.

I join with my co-editors in extending thanks to the staff of Oxford University Press for support and guidance during the course and completion of this project, and acknowledge the interest and advice of my University colleagues and friends, particularly Professor David Wallis and Dr George Foster. I give special thanks to Professor Russel J. Reiter and Professor Richard J. Wurtman whose general encouragement in my work to date has meant a great deal to me.

Cardiff   A. M.
March, 1988

Contents

Contributors

Arendt, J.
Reader in Clinical Biochemistry, Division of Clinical Biochemistry,
Department of Biochemistry, University of Surrey, Guildford, UK.

Blask, D. E.
Associate Professor, Department of Anatomy, University of Arizona,
College of Medicine, Tucson, Arizona 85724, USA.

Checkley, S. A.
Consultant Psychiatrist, The Maudsley Hospital, Denmark Hill,
London, UK.

Hill, S. M.
Postdoctoral Fellow, Department of Medicine and Oncology, University
of Texas Health Sciences Centre at San Antonio, San Antonio, Texas
78284, USA.

Lang, U.
Research Associate, Division of Endocrinology, Hôpital Cantonal
Universitaire, 1211 Genève 4, Switzerland.

Lea, A. E.
Research Affiliate, Department of Brain and Cognitive Sciences,
Massachusetts Institute of Technology, Cambridge, Massachusetts
02139, USA.

Lewy, A. J.
Professor of Psychiatry, Ophthalmology and Pharmacology, Department
of Psychiatry, School of Medicine, The Oregon Health Sciences
University, 3181 SW Sam Jackson Park Road, Portland, Oregon 97201,
USA.

Lieberman, H. R.
Research Scientist, Department of Brain and Cognitive Sciences,
Massachusetts Institute of Technology, Cambridge, Massachusetts
02139, USA.

Miles, A.

Junior Research Fellow and Assistant Lecturer in Physiology, Department of Physiology, University College, Cardiff, UK.

Dr. E. Palazidou,

The Maudsley Hospital, Denmark Hill, London, UK.

Philbrick, D. R. S.

Senior Registrar in Psychiatry, Department of Child Psychiatry, Booth Hall Children's Hospital, University of Manchester School of Medicine, Blackley, Manchester, UK.

Reiter, R. J.

Professor of Neuroendocrinology, Department of Cellular and Structural Biology, The University of Texas Health Sciences Centre at San Antonio, 7703 Floyd Curl Drive, San Antonio, Texas 78284–7762, USA.

Sack, R. L.

Professor of Psychiatry, Department of Psychiatry, Oregon Health Sciences University, 3181, SW Sam Jackson Park Road, Portland, Oregon 97201, USA.

Sizonenko, P. C.

Professor of Paediatrics and Head of Division of Biology of Growth and Reproduction, Department of Paediatrics and Genetics, University of Geneva, Hôpital Cantonal Universitaire, 1211 Genève 4, Switzerland.

Steiner, M. A.

Professor of Psychiatry and Neuroscience and Head of Clinical Studies, McMaster Psychiatric Unit, St Joseph's Hospital, Hamilton, Ontario, Canada.

Thomas, D. R.

Consultant Psychiatrist and Lecturer in Psychiatry, Academic Department of Psychological Medicine, University of Wales College of Medicine, Whitchurch Hospital, Cardiff, UK.

Thompson, C.

Senior Lecturer in Psychiatry and Honorary Consultant in Psychiatry, Academic Department of Psychiatry, Charing Cross Hospital, Fulham Palace Road, London, UK.

Vriend, J.

Associate Professor of Anatomy, University of Manitoba Health
Science Centre, 730 William Avenue, Winnipeg, Manitoba, Canada.

Waldhauser, F.

Associate Professor, Department of Paediatrics, University of Vienna,
Waehringer Guertel 18–20, A–1090, Vienna, Austria.

Waldhauser, M.

Assistant Professor, Department of Urology, University of Vienna,
Aeserstrasse 4, A-1090, Vienna, Austria.

1. Neuroendocrinology of melatonin

Russel J. Reiter

Introduction

Although melatonin was not discovered or structurally identified until the late 1950s (Lerner *et al.* 1958, 1959), its action in reference to the aggregation of melanosomes in amphibian melanocytes was observed roughly 40 years earlier (McCord and Allen 1917). In light of these early observations, it is interesting that since its discovery melatonin has been most aggressively investigated in reference to its reproductive effects (Reiter and Sorrentino 1970; Reiter 1980; Cardinali 1981; Tamarkin *et al.* 1985; Arendt 1986) while its actions on melatonin aggregation have been less thoroughly studied. The ability of melatonin to lighten the skin of frogs has, however, been useful from the standpoint that this feature has served as the basis of a bioassay for melatonin (Mori and Lerner 1960; Bagnara and Hadley 1970).

The first proof of a 24-hour cycle of melatonin in pineal tissue was provided in 1964 when a night-time increase in rat pineal melatonin level was reported (Quay 1964). This was followed by a series of investigations over the subsequent two decades which have conclusively documented a circadian melatonin rhythm in the pineal gland of a variety of non-mammalian and mammalian vertebrates including man (Reiter 1983, 1986a; Binkley 1981; Arendt 1984). With the exception of one species in which there may be a genetic absence of the enzymatic machinery required to synthesize melatonin (Ebihara *et al.* 1986) and one species in which a melatonin rhythm may only occur when the animals are exposed to certain photoperiods (McConnell and Ellendorf 1987; Reiter *et al.* 1987a), a nocturnal increase in the conversion of serotonin to melatonin in the pineal gland of mammals is usual.

Melatonin production is not, however, confined to the pineal gland. Reports of its synthesis in the retinas (Quay 1965; Cardinali and Rosner 1971; Leino and Airaksinen 1985; Wiechmann 1986) and Harderian glands (Bubenik *et al.* 1974, 1978; Reiter *et al.* 1981; Hoffman *et al.* 1985) have been forthcoming. In these organs, as in the pineal gland, the

1

production of melatonin seems to be cyclic (Binkley *et al.* 1979; Hamm and Menaker 1980; Pang *et al.* 1980; Reiter *et al.* 1983*b*). Whereas the specific functions of melatonin in the retinas and Harderian glands remain to be identified, at least in mammals, these organs seem to contribute minimally or not at all to the 24-hour rhythm of melatonin in blood. Although less convincing data have been presented, an argument has been made that melatonin may also be produced in the gastro-intestinal tract (Quay and Ma 1976), the central nervous system (Bubenik *et al.* 1974), and in red blood cells (Rosengarten *et al.* 1972).

In this chapter, the control of melatonin production, primarily within the pineal gland, will be considered. Whereas the prevailing light:dark environment is obviously an important modulator of 24-hour melatonin rhythm, other perturbations that modify the melatonin cycle will also be reviewed. Where possible, the implications of the alteration in the metabolic conversion of serotonin to melatonin and other potential pineal hormones will be pointed out.

Melatonin synthesis

The synthetic steps in the manufacture of melatonin in the pineal gland are well known (Quay 1974; Klein *et al.* 1981; Binkley 1981; Ebadi 1984). This indoleamine is a key secretory product of the pineal gland and, as a consequence, its synthesis has been studied in detail. Melatonin, as well as a number of other potential pineal hormones, are products of tryptophan metabolism (Fig. 1.1).

The amino acid tryptophan is taken up from the circulation into the pineal gland presumably by means of an active transport mechanism. In the rat pineal tryptophan levels are reportedly highest near the end of the light period with a steady decrease during darkness (Sugden 1979). This presumed change in the pineal content of the amino acid is not in synchrony with the 24-hour rhythm in circulating tryptophan levels which are highest during the daily dark phase and the pineal tryptophan rhythm has not been confirmed by all workers (Mefford and Barchas 1980; Young and Anderson 1982).

The uptake of tryptophan into the pinealocyte is under the control of a noradrenergic receptor mechanism, but does not seem to involve cyclic AMP (Wurtman *et al.* 1969). Tryptophan loading by either its chronic or acute administration increases the pineal concentrations of serotonin, a tryptophan metabolite. The increase in serotonin levels following trypto-phan administration reportedly depends on the time during the light:dark cycle when tryptophan is given (Snyder *et al.* 1967; Zweig and Axelrod 1969). Elevated dietary tryptophan levels alter pineal serotonin values in a similar manner.

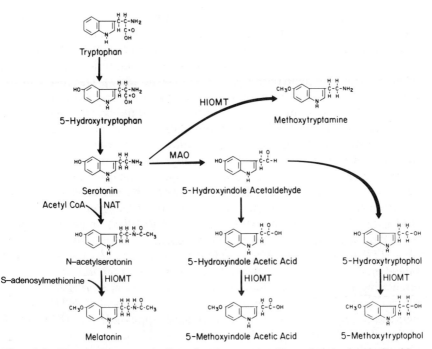

Fig. 1.1. Tryptophan metabolism in the mammalian pineal gland. Besides melatonin, methoxytryptamine and 5-methoxytryptophol have been proposed as pineal hormones. HIOMT, hydroxyindole-O-methyltransferase; MAO, mono-amine oxidase; NAT, N-acetyltransferase.

After its uptake into the pinealocytes the tryptophan destined to become serotonin is first hydroxylated to form 5-hydroxytryptophan, a conversion that depends on the enzyme tryptophan hydroxylase (Lovenberg *et al.* 1967). The enzyme requires O_2 and a reduced pterdine, tetrahydrobiopterine, which is in higher concentrations in the pineal gland than in brain tissue generally (Levine *et al.* 1978).

The activity of pineal tryptophan hydroxylase is regulated by a β-adrenergic receptor mechanism which utilizes the second messenger cyclic AMP (Shein and Wurtman 1971). Species differences have been reported in reference to the 24-hour rhythm of the tryptophan hydroxy-lase activity in mammals. Thus, in rat pineal tissue tryptophan hydroxyla-tion is reported to be greatest at night (Shibuya *et al.* 1978; Sitaram and Lees 1978) while in the Syrian hamster the activity of this enzyme reaches a peak during the daytime (Steinlechner *et al.* 1983). Few other species have been examined but findings using bovine pineal tissue suggest that, in this species, there may be no 24-hour rhythm in the rate of pineal tryptophan hydroxylation (Chan and Ebadi 1980).

Following its formation, 5-hydroxytryptophan (5-HTP) is decarboxy-lated to serotonin (5-HT) by the enzyme 5-HTP decarboxylase (L-aromatic amino acid decarboxylase) (Lovenberg *et al.* 1962). In general, the decarboxylase activity is high in pineal tissue and it does not exhibit a rhythm during the light:dark cycle (Snyder *et al.* 1967). The co-factor for this enzyme is pyridoxal-5-phosphate (Håkanson and Owman 1966). Whereas the activity of 5-HTP decarboxylase in the bovine pineal gland seems unresponsive to adrenergic stimulation (Chan and Ebadi 1980), photoperiodic and surgical perturbations which influence serotonin metabolism in the pineal gland also change the activity of this enzyme. Hence, after either constant light treatment or superior cervical gang-lionectomy, which sympathetically denervates the pineal gland, 5-HTP decarboxylase activity exhibits a significant rise (Pellegrino de Iraldi and Rodriguez de Lores Arnaiz 1964; Snyder *et al.* 1965). In general, pineal 5-HTP decarboxylase activity is high resulting in very low concentrations of 5-HTP in the gland. When 5-HTP is exogenously administered, rat pineal melatonin concentrations rise (Wurzburger *et al.* 1976) as do circulating levels of melatonin in sheep (Namboodiri *et al.* 1983). Little information is available on the variation in pineal 5-HTP over a light:dark regime; of the reports that have appeared, one (King *et al.* 1984) claimed a day:night difference in 5-HTP levels in the rat pineal gland while in Syrian hamsters (Steinlechner *et al.* 1983) no light:dark variation in pineal 5-HTP was documented.

A 5-HT rhythm in the pineal gland is documented in several mam-malian species. The 5-HT content of the pineal gland is highest during the day and lowest at night in the rat (Quay 1963; Snyder and Axelrod 1965; King *et al.* 1984), the cotton rat (Matthews *et al.* 1982), and the Syrian hamster (Steinlechner *et al.* 1983). The rhythm persists in the pineal gland of rats totally deprived of light (Snyder and Axelrod 1965) and may be a function of either 24-hour change in N-acetylation of the indole (Snyder and Axelrod 1965; Klein and Weller 1970; Brownstein *et al.* 1973) or due to a light:dark rhythm of availability of 5-HT to be oxidatively deaminated (Snyder *et al.* 1967). In the Syrian hamster it has been argued that, since the drop in pineal 5-HT levels precedes the nocturnal rise in N-acetyltransferase (NAT) activity, the reduction in pineal 5-HT content at night is not related to its metabolism via this route (Steinlechner *et al.* 1983).

Not all the 5-HT in the pineal gland is resident in the pinealocytes. It has been estimated that of the total 5-HT within the pineal gland as much as one-third may be located in the sympathetic nerve terminals which derive from the superior cervical ganglia and end among the pinealocytes (Neff *et al.* 1969). In the nerve endings 5-HT is likely to be metabolized via oxidative deamination (Falck *et al.* 1966); small amounts may also be released from the nerve endings. The serotonin present in the nerve endings is presumably synthesized in the pinealocytes and thereafter shunted into the nerve endings (Axelrod 1974).

5-HT has a variety of metabolic pathways within the pineal gland. The major pathways include oxidative deamination to form 5-hydroxyindole acetyldehyde (Keglevic *et al.* 1968; Axelrod *et al.* 1969), O-methylation with the resultant formation of 5-methoxytryptamine (Axelrod and Weissbach 1960), and the one currently of great interest to pineal scientists, N-acetylation to form N-acetylserotonin (Weissbach *et al.* 1961) (Fig. 1.2), the immediate precursor of melatonin. Melatonin has been widely investigated in reference to its endocrine properties (Reiter 1980, 1981*a*, 1982; Tamarkin *et al.* 1985; Arendt 1986) while 5-methoxytryptamine has also been touted as a pineal hormonal product (Pevet 1983).

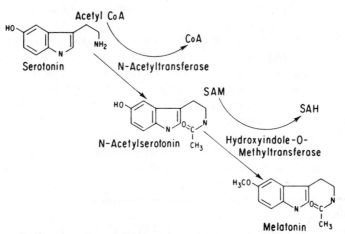

Fig. 1.2. The conversion of serotonin to melatonin the chief pineal hormone. CoA, co-enzyme A; SAH, S-adenosyl homocystein; SAM, S-adenyl methionine. (From Lewy 1983.)

NAT activity was first shown to exhibit a 24-hour rhythm in the pineal gland in 1970 (Klein and Weller 1970). In this species the nocturnal rise in the activity of the enzyme is extremely pronounced (up to 100-fold increase at night); however, this is not true for all mammals. In the gerbil and guinea-pig pineal gland, for example, the nocturnal rise in NAT activity is less than 5-fold (Rudeen *et al.* 1975). NAT has a molecular weight of roughly 39 000 daltons and transfers an acetyl group from its co-factor, acetyl coenzyme A, to an amine acceptor. Whereas the amine acceptor in the pineal gland is usually serotonin, other equally good substrates for the enzyme found in the pineal gland include 5-methoxytryptamine and tryptamine (Diguchi 1975).

The nocturnal rise in NAT activity in the pineal gland at night is initiated by stimulation of adenylate cyclase activity via an adrenergic receptor mechanism. The synthesis of mRNA is required for the induction of NAT activity, but thereafter protein synthesis maintains NAT activity at a high level (Klein *et al.* 1970; Chan and Ebadi 1980; Zatz 1981). Whether the protein being synthesized is actually NAT or some activating or inhibitory factor is unresolved. Cyclic AMP plays a multifaceted role in the regulation of NAT: it serves to induce phosphorylation of nuclear proteins thereby initiating mRNA production for NAT (Binkley 1983); it facilitates the interaction of NAT and acetyl coenzyme A by competitively displacing NAT inhibitory molecules which, in the absence of the cyclic nucleotide, depress the activity of the acetylating enzyme (Klein 1978); the drop in cyclic AMP before lights on in the morning may account for the concomitant reduction in pineal melatonin levels (Klein 1978). These interactions have recently been reviewed (Foley *et al.* 1986).

The nocturnal increase of pineal NAT is initiated by the release of noradrenalin (NA) from the postganglionic sympathetic nerves that terminate in the pineal gland; after its interaction with β-adrenoreceptors on the pinealocyte membrane it stimulates cyclic AMP production and the cascade of events summarized above (Foley *et al.* 1986). In addition to acting on β_1-adrenergic receptors, however, NA also binds to α-adrenergic receptors in the pinealocyte membrane thereby increasing phosphotidylinositol turnover (Zatz 1985); this results in the intracellular release of a number of second messengers, among them being diacylglycerol, inositol triphosphate, and phosphotidic acid. These may be involved in the α-receptor potentiation of pineal β-stimulation of NAT activity that is seen in some mammalian species (Foley *et al.* 1986).

Other neurotransmitters may be involved in the regulation of pineal NAT activity. For example, in both the cow (Chan and Ebadi 1980) and sheep (Foldes *et al.* 1983) pineal gamma-aminobutyric acid (GABA), presumably acting by means of cyclic GMP, depresses NA-enhanced NAT activity. In the ovine pineal GABA may act to inhibit protein kinase activity. Vasoactive intestinal peptide (VIP) has also been implicated in the control of pineal NAT activity. This neuromodulator, which is found in conjunction with acetylcholine in some cholinergic nerve endings, exaggerates NA-stimulated NAT activity by an undefined mechanism (Yuwiler 1983). Also, prostaglandins of the E and F series have been implicated in the control of the N-acetylation of pineal serotonin because of their interactions with the pinealocyte, thereby altering their sensitivity to NA (Cardinali *et al.* 1979, 1983; Ritta and Cardinali 1982). Binding sites for [3H]PGE$_1$, [3H]PGE$_2$, and [3H]PGF$_2$ are present in membranes obtained from bovine pineal tissue (Cardinali *et al.* 1979). The administration of indomethacin, a prostaglandin synthesis inhibitor, among other pineal changes, reduces the nighttime increase in the activity of NAT and the subsequent enzyme in this metabolic pathway, hydroxyindole-O-methyltransferase (HIOMT); however, the actual reduction in melatonin levels seems to be rather slight (Ritta and Cardinali 1980; Reiter *et al.* 1985*b*) and the physiological impact of this change remains to be identified.

The product of serotonin N-acetylation is N-acetylserotonin, the immediate precursor of melatonin. The quantity of N-acetylserotonin in the pineal gland parallels the fluctuations in NAT activity (Brownstein *et al.* 1973); thus, values are typically highest at night when the N-acetylation of serotonin is most active. The nyctohemeral increases in pineal N-acetylserotonin have been defined in at least three mammalian species (Reiter 1981*b*).

The O-methylation of N-acetylserotonin results in the formation of melatonin (Fig. 1.2). This step requires HIOMT, an enzyme involved in not only melatonin production but the synthesis of several other pineal methoxyindoles (Axelrod and Weissbach 1961). HIOMT activity is rather high and although several early reports suggested it exhibited a 24-hour rhythm in activity, this has not been routinely documented (Klein 1978; Ebadi 1984). HIOMT has two 39 000 dalton subunits and accounts for 2–4 per cent of the total soluble protein in the pineal gland. Although there are a number of potential methyl donors in pineal tissue, only S-adenosyl methionine (SAM) is involved in the O-methylation of N-acetylserotonin to melatonin (Baldessarini and Kopin 1966); the high concentrations of SAM in the pinealocyte relate to the high activity of the enzyme which produces it, i.e. ATP:L-methionine S-adenosyltransferase (Guchait and Grau 1978). A functional relationship between the activity of HIOMT and the concentrations of cyclic GMP in the pineal gland has been suggested (O'Dea and Zatz 1986; Sugden and Klein 1983), but other factors, e.g. pteridines (Ebels *et al.* 1979), may also be involved in regulation of the O-methylation of N-acetylserotonin to melatonin.

Melatonin concentrations in the pineal gland have been described for a large number of mammalian species including man (Arendt 1978; Lewy 1983; Vaughan 1984; Reiter, 1986*a*). Almost without exception, daytime values of this pineal constituent are lower than are nocturnal levels. The only species in which the actual rate of pineal melatonin production has been estimated is the Syrian hamster. Melatonin synthesis during the day in this species is about 1.6 pg/min; at night this value rises to 70.6 pg/min with the total melatonin being produced in a single 24-hour period being about 18.6 ng (Rollag *et al.* 1980).

Unlike many other hormones, melatonin in the pineal gland seems to not be stored for any appreciable amount of time; rather, once produced it quickly escapes into the systemic circulation. No active mechanism has been proposed to explain the efflux of melatonin from the pineal. Inasmuch as the molecule is highly lipid-soluble, simple diffusion is the usual process used to describe the discharge of melatonin from the pinealocyte (Illnerova *et al.* 1978; Wurtman and Moskowitz 1978; Cardinali 1981). Because of the close association of melatonin synthesis with its escape from the pineal, pineal and serum melatonin levels usually follow a parallel course (Wilkinson *et al.* 1977).

Since the alternate pathways for serotonin metabolism in the pineal gland (Fig. 1.1) have been sparingly investigated in the human, they will not be discussed here. However, the interested reader is referred to several reviews for details of these metabolic pathways (Quay 1974; Lewy 1983; Reiter 1984*a*).

Innervation and neuropharmacological regulation

The nocturnal rise in the production of pineal melatonin is under control of the sympathetic innervation to the pineal gland. The fibre connections between those nuclei in the central nervous system and pineal which regulate the function of the gland in animals are presumed to be similar to those in man (Fig. 1.3). The indoleamine rhythms in the pineal gland are believed to be initiated by neuronal activity in the suprachiasmatic nuclei (SCN) of the hypothalamus. Lesions of these nuclei prevent the nocturnal increase in pineal NAT activity (Moore and Klein 1974). The caudal projections of the SCN may include a synapse in the paraventricular nuclei (Bittman 1984) and, thereafter, a monosynaptic connection with the intermediolateral cell column of the upper thoracic cord (Swanson and Kuypers 1980); these latter preganglionic sympathetic neurons give rise to fibres which exit the spinal column in the ventral root and pass up the sympathetic trunk to synapse on postganglionic cell bodies whose axons eventually terminate in the pineal gland (Kappers 1960). During the daily dark period, action potentials in the peripheral sympathetic nervous system causes the release of the neurotransmitter noradrenalin (NA) from the terminals in the pineal gland which then drives melatonin synthesis. During the day, the neural activity within the sympathetic fibres terminating in the pineal gland is depressed by light information which is transferred to the SCN. These fibres, referred to as the retinohypothalamic tract, derive from the ganglion cells of the retina and synapse primarily on cell bodies in the contralateral SCN (Moore 1973). Besides the peripheral sympathetic innervation of the pineal gland, a central innervation via the pineal stalk

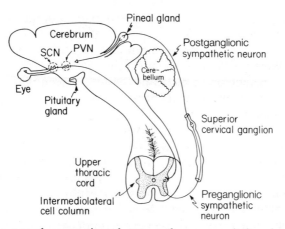

Fig. 1.3. The neural connections between the eyes and the pineal gland are believed to be similar in all mammals including man. The final common pathway in this system is the postganglionic sympathetic neuron. Besides the peripheral sympathetic innervation, in at least some mammals, there are also fibres entering the pineal via its stalk; they are also shown here. PVN, paraventricular nucleus; SCN, suprachiasmatic nucleus.

has been documented in some animals but its functional significance remains to be described (Korf and Møller 1984; Fig. 1.3).

In the human, central lesions documenting the importance of the fibre connections between the hypothalamus and the pineal gland for melatonin production have been reported. Thus, one patient with progressive supranuclear palsy (degeneration of neural structures in the brainstem) and one with Shy–Drager syndrome (central nervous system degeneration) had low melatonin values and absence of a nocturnal surge (Vaughan *et al.* 1979*b*; Vaughan *et al.* 1982). Also, the most likely explanation for the absence of a melatonin rhythm in two patients with Cushing's disease with hypothalamic involvement by tumour is that the hypothalamic lesions interfered with the transfer of the neural message from the SCN of the anterior hypothalamus to the intermediolateral cell column (Werner *et al.* 1980; Brismar *et al.* 1982). Brain lesions in the cervical cord also eliminate the 24-hour melatonin rhythm in the blood. In six quadriplegic patients with cervical cord transection urinary melatonin levels remained constant throughout the day and night (Kneisley *et al.* 1978). Control subjects included one patient with spinal cord transection in the lumbar region and neurologically normal individuals, all of whom exhibited a nyctohemeral increase in urinary melatonin excretion. Again, the most feasible explanation for these findings is that the cervical lesions prevented neural information from reaching the pineal gland. Finally, six patients with Shy–Drager syndrome

(also called multiple system atrophy) and eight with idiopathic orthostatic hypotension (with deficits in postganglionic nerves) had either low
urinary excretion of 6-hydroxymelatonin (6-OHM) or no nocturnal
increase in this urinary constituent (Tetsuo *et al.* 1981*b*). 6-OHM is the
chief hepatic metabolite of melatonin and its concentration in the urine
typically exhibits a marked nocturnal increase, paralleling the rise in
blood melatonin (Fellenberg *et al.* 1981).

The symapthetic nerve terminals within the pineal gland contain high
concentrations of NA (Morgan and Hansen 1978) with maximal
synthesis and presumed release of the neurotransmitter occurring at
night (Brownstein and Axelrod 1974; Craft *et al.* 1983). After its release
NA acts on postsynaptic adrenergic receptors in the pinealocyte
membrane. Since NA is a mixed β- and α-receptor agonist it is capable
of acting on both classes of receptors; both β- and α-receptors are
located in the pinealocyte membrane (Lewy 1983; Craft *et al.* 1985). In
most species the nocturnal rise in pineal melatonin production is
primarily the result of a β-receptor action of NA (Axelrod 1974; Zatz
1981; Lipton *et al.* 1981) (Fig. 1.4) while in others the rise in serotonin
metabolism is almost due exclusively to the action of NA on the α-
adrenergic receptors (Reiter *et al.* 1982*a*; Sugden *et al.* 1985).

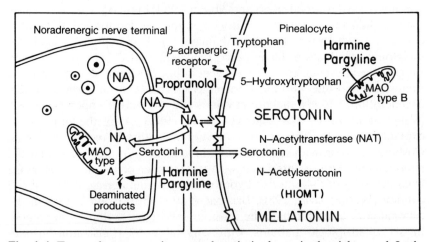

Fig. 1.4. Tryptophan conversion to melatonin is shown in the right panel. In the
left panel is a noradrenergic nerve terminal whose cell body is in the superior
cervical ganglion. In most species the release of NA from the nerve terminal acts
primarily on β-adrenergic receptors in the pinealocyte membrane. Propranolol
blocks the interaction of NA with the β-receptor. Also shown is the shunting of
serotonin (5-HT) between the pinealocyte and the nerve terminal. Monoamine
oxidase (MAO) A and B are found in the nerve terminals and pinealocytes,
respectively. (From King *et al.* 1982.)

In the human it has not been easy to prove which class of receptors mediate the nocturnal rise in serum melatonin concentrations. The infusion of β-adrenergic agents at dose levels sufficient to produce cardiovascular changes typically do not alter circulating melatonin levels when the drugs are administered during the daytime (Vaughan *et al.* 1976; Moore *et al.* 1979; Lewy 1983). In two patients with phaeo-chromocytoma and associated high circulating levels of NA, melatonin was undetectable in the blood during the day (Wetterberg 1979). Also in humans, burn injury is marked by high plasma and urinary catechol-amines and an attenuated 24-hour serum melatonin rhythm (Vaughan *et al.* 1985). These findings could, in part, be explained in terms of an active uptake of excess circulating catecholamines by the sympathetic nerve endings in the pineal gland and/or the down regulation of the adrenergic receptors on the pinealocyte membrane. Even when L-dopa is administered at dose levels that produce growth hormone release, this catecholamine precursor has no discernible influence on circulating melatonin values (Arendt 1978; Wetterberg 1978; Vaughan *et al.* 1979*b*; Lewy 1983).

Whereas it has been difficult to promote daytime pineal melatonin synthesis in the human with β-receptor agonists, the administration of β-receptor blockers does interfere with the nocturnal rises in plasma and urinary melatonin levels (Vaughan *et al.* 1976; Wetterberg 1979; Hansen *et al.* 1980; Lewy 1983). Whereas the effects of these drugs could be due to changes in neurotransmission at either central neural sites or within the pineal gland itself, the efficacy of atenolol, which does not easily cross the blood–brain barrier, in depressing nocturnal melatonin (Cowen *et al.* 1983) suggests that pinealocyte β_1-adrenergic receptors mediate the nocturnal rise of melatonin in the human. The pineal gland is generally considered to be outside the blood–brain barrier although this issue is continually debated (Møller *et al.* 1978).

The situation in reference to the pharmacological manipulation of the human pineal gland seems to be similar to that of the Syrian hamster. In both the human (Vaughan *et al.* 1976; Moore *et al.* 1979) and Syrian hamster (Lipton *et al.* 1982; Steinlechner *et al.* 1985) the pineal does not respond to exogenously administered β-receptor agonists during the day, but in both species the nocturnal rise in melatonin is prevented or attenuated by β-receptor antagonists (Hansen *et al.* 1980; Lipton *et al.* 1981). At least a partial explanation for these findings may have been recently uncovered using the Syrian hamster model. By injecting β-receptor agonists (NA or isoproterenol) during either the day or night (in animals with light-suppressed pineal melatonin levels) it was shown that the hamster pineal only responds to the drugs during the late dark phase (Steinlechner *et al.* 1985; Vaughan *et al.* 1986; Reiter *et al.* 1987*b*;

Vaughan and Reiter 1987). Indeed, the only interval during which NA or isoproterenol was effective in promoting pineal melatonin production coincided with the time of the endogenous pineal melatonin peak (Reiter *et al.* 1987*b*). The implication of these findings is the β-receptors probably mediate the normal nocturnal rise in melatonin in the Syrian hamster (likely also in the human), but the pinealocyte receptors are only sensitive (up-regulated) during the night when melatonin production is normally highest. All studies to date in which β-agonists were used in an attempt to stimulate plasma melatonin levels in humans were conducted during the day; if similar experiments were performed at night the drugs might increase the production and secretion of melatonin, as in the Syrian hamster (Reiter *et al.* 1987*b*).

In correlative studies in rats, it has been shown that pinealocyte β-receptor density is higher at night than during the day (Reiter *et al.* 1985*a*; Gonzalez-Brito *et al.* 1988). Perhaps a similar rhythm, even more exaggerated, occurs in the human and hamster pineal thereby explaining the relative unresponsiveness of the gland during the day and its potential responsiveness at night. Besides NA acting on β- and α-receptors to stimulate nocturnal melatonin production, a gamut of other neurotransmitters may modulate pineal indoleamine metabolism (Ebadi and Govitrapong 1986*a,b*).

Melatonin in body fluids

Few doubt that the usual route for the release of melatonin is into the vascular system (Hedlund *et al.* 1977; Rollag *et al.* 1978; Cardinali 1981; Reiter 1986*a*) although a direct release into the cerebrospinal fluid has also been suggested (Anton-Tay and Wurtman 1969; Collu *et al.* 1971). This latter option is a possibility in species, such as the human, where the pineal gland rests directly on the posterodorsal aspect of the third ventricle. It has been shown that, due to an interruption of the ependymal lining in the pineal recess (Hewing 1978), pinealocytes under some circumstances are directly bathed by CSF further increasing the possiblity of an intraventricular discharge of melatonin. The varied routes of melatonin release have been summarized elsewhere (Reiter *et al.* 1975). For the purposes of this report it will be accepted that pineal melatonin is first released into the blood prior to its entrance into any other body fluid.

Blood melatonin

Among the species that have been examined, a variety of different melatonin rhythms have been described. In the human, peak melatonin levels are usually measured near the middle of the dark phase with the first half of the night being associated with a gradual rise in blood concentrations of this constituent and a reduction in melatonin levels during the second half of the night (Lewy 1983; Vaughan 1984) (Fig. 1.5). From this a similar pattern of nocturnal pineal melatonin production and secretion has been deduced (Reiter 1984*a,b*); the validity of this assumption is supported by the observation that, in general, individuals who die at night have higher pineal melatonin concentrations than those who die during the day (Greiner and Chan 1978). Enzyme measurements in human pineal tissue also provide evidence of a nocturnal rise in melatonin production at night (Smith *et al.* 1981*a*). Although melatonin levels are normally highest at the time of nocturnal sleep, the pattern of melatonin release is not considered to be related to the sleep stages (Vaughan *et al.* 1979*a*; Weinberg *et al.* 1979). Dark exposure and/or sleep during the daytime is usually not accompanied by a rise in circulating melatonin levels (Vaughan *et al.* 1976; Weinberg *et al.* 1979). Likewise, remaining awake at night (without exposure to light) does not prevent the normal nocturnal melatonin surge nor does novel awakening (without light exposure) terminate melatonin secretion in either men or women (Arendt 1978; Lynch *et al.* 1978; Wetterberg 1978; Akerstedt *et al.* 1979). Blindness, without light perception, causes the melatonin rhythm to free-run (Lynch *et al.* 1978) with an estimated period of slightly greater than 24 hours (Lewy and Newsome 1983). As a result, elevated melatonin levels in the blood of blind individuals may occur anytime throughout the light:dark cycle

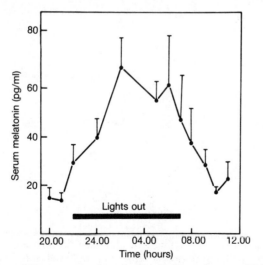

Fig. 1.5. Mean ($\pm$ SE) serum melatonin levels throughout a 24-hour period in seven healthy male volunteers. (From Waldhauser *et al.* 1984.)

(Smith *et al.* 1981*b*; Lewy and Newsome 1983; Fig. 1.6). The latter finding indicates that in humans, as in other mammals, the light:dark environment normally entrains the circadian melatonin rhythm.

Among mammals, the brightness (intensity) of light required to depress pineal or circulating melatonin levels at night varies widely (Reiter 1985). For example, melatonin production in the albino rat pineal gland can be suppressed by a light irradiance of 0.0005 μW/cm^2 (Webb *et al.* 1985); at the other extreme, it requires a light irradiance of roughly 1850 μW/cm^2 to inhibit normal melatonin production in the pineal gland of the Richardson ground squirrel (Reiter *et al.* 1983*a*). The sensitivity of the human pineal to light falls between these extremes and seems to be on the order of 150–200 μW/cm^2 ($\approx$ 2500 lux; Lewy *et al.* 1980).

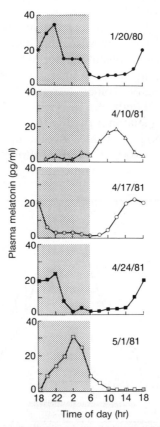

Fig. 1.6. Plasma melatonin rhythm in a blind subject measured on five different dates. Since the melatonin rhythm free-runs in blind subjects with a period slightly greater than 24-hours peak values can occur anytime during a 24-hour period. The shaded area represents darkness. (From Lewy and Newsome 1983.)

Interestingly, the induction of massive photoreceptor damage in the albino rat by constant light exposure only modestly changes the responsiveness of the pineal gland to light at night (Webb *et al.* 1985). To put these findings in perspective, the reader is reminded that on clear days the brightness of the sun can exceed 35 000 μW/cm^2 (Hughes *et al.* 1988).

Besides brightness, the wavelength (colour) of light to which mammals are exposed influences pineal melatonin production and secretion as well (Cardinali *et al.* 1972; Reiter 1985). Although the studies in this area are limited, in both the Syrian hamster (Brainard *et al.* 1984) and in the human (Brainard *et al.* 1985) wavelengths in the blue range seem to be maximally inhibitory to pineal melatonin production and serum melatonin levels. In the case of the human study, the volunteers were exposed to equal quanta of each monochromatic wavelength studied, i.e. 448, 476, 509, 542, 574, and 604 nm. Maximal plasma melatonin inhibition (64 per cent) followed the 509 nm light exposure (Fig. 1.7). Although the data reported roughly match the scotopic curve for visual sensitivity in the human, the authors cautioned, based on these data alone, about drawing any firm conclusions concerning which photo-pigment(s) may mediate the light-induced depression in melatonin.

In non-human mammals, pineal removal eliminates the nocturnal rise in circulating melatonin (Pang and Ralph 1975; Arendt *et al.* 1980; Reiter *et al.* 1982*b*; Vaughan and Reiter 1986). When the pineal gland of humans is surgically removed when it is involved with a tumour, the rise in blood melatonin normally associated with night is eliminated (Neuwelt and Lewy 1983). This is important since it suggests that melatonin is not

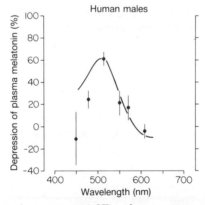

Fig. 1.7. Points are the means ($\pm$SE) of percentage depression of plasma melatonin in male human volunteers ($n=3$) exposed to different wavelengths of light; 509 nm light was most inhibitory to circulating melatonin. The solid line is the spectral sensitivity of human rod cells. (From Brainard *et al.* 1985.)

released in a circadian manner from extra-pineal sites where it is known to be produced.

Superimposed on the circadian melatonin rhythm may be an episodic cycle of melatonin release. Such episodes have been described in at least three independent laboratories (Vaughan *et al.* 1978a; Weinberg *et al.* 1979; Mullen *et al.* 1981) with the amplitude of the episode being great enough, in some cases, to raise plasma melatonin levels by 200 per cent. The frequency of the episodes varies among the reports, ranging anywhere from hourly to 2.5-minute intervals. The physiological significance of the melatonin secretory bursts remains to be identified.

The nocturnal rise in melatonin production and secretion may also be influenced by the ovarian cycle of mammals, presumably by means of an action of gonadal steroids directly on the pineal gland (Cardinali *et al.* 1983). In the rat, nocturnal pineal melatonin production is lowest on the night preceding the proestrous surge of ovulatory hormones (Johnson *et al.* 1982). Although there is one report to the contrary (Fellenberg *et al.* 1982), others claim that in the humans as well the nocturnal melatonin surge at the time of ovulation may be lower than at other stages of the menstrual cycle (Wetterberg *et al.* 1976; Hariharasubramanian *et al.* 1986; Fig. 1.8). The implication of these findings is that the lower melatonin levels at mid-cycle may be permissive to the release of ovulatory hormones and subsequent ovulation. Certainly, if melatonin is administered to rats just prior to the proestrous surge of gonadotrophins, both the release of luteinizing hormone and ovulation are depressed (Reiter and Sorrentino 1971a).

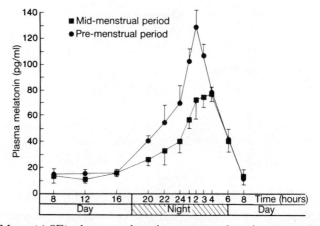

Fig. 1.8. Mean (±SE) plasma melatonin concentrations in seven adult cycling female volunteers during the mid-menstrual (days 13–17 of the menstrual cycle) and premenstrual period (days 26–31). The mid-menstrual period coincides with the time of ovulation. (From Hariharasubramanian *et al.* 1986.)

Whether there are seasonal changes in the blood melatonin rhythm in the human as in some other mammals (Goldman 1983) is subject to debate. Considering that humans closely regulate their photoperiodic environment, a procedure not available to other mammals, it might be predicted that humans would exhibit fewer seasonal changes. In view of this, marked seasonal alterations in the circadian melatonin rhythm, i.e. either the duration of nightly elevated melatonin or the amplitude of the rhythm, in humans living under regulated environments would seemingly be unusual. Yet, the duration of exposure to natural sunlight (if not to artificial light) does vary considerably, especially at the extremes of latitude (Hughes *et al.* 1987).

When blood was collected at both midnight and 0800 hours in the morning at monthly intervals throughout the year the highest melatonin levels were recorded in the winter and summer with lower values at the intermediate seasons (Arendt *et al.* 1979). This seemingly unusual pattern has been once confirmed (Birau 1981); likewise when pineal HIOMT was used as an index of melatonin production it too showed higher activity in winter and summer compared to spring and fall (Smith *et al.* 1981c). Conversely, when urinary melatonin excretion was measured as a function of the season, no annual variation was uncovered (Wetterberg *et al.* 1981). The significance of these findings is limited because of the small number of samples that were examined.

These are difficult studies to conduct in humans because of the wide individual variation in melatonin levels and the widely diverse photoperiods (both in terms of artificial and natural lighting) to which people in different countries and at different latitudes are exposed. Under controlled laboratory conditions experimental animals have been shown to vary their melatonin rhythm with the duration of the daily dark period; on the basis of this it would seem that the human melatonin rhythm has the capability, if only exposed to natural lighting conditions, of being regulated accordingly.

Stressful situations, theoretically at least, could have a major impact on the production and release of pineal melatonin by virtue of the massive discharge of catecholamines from the adrenal medulla and elsewhere under such circumstances. It has been argued, and partially experimentally documented, however, that excessive catecholamines released under specific conditions are taken up by the postganglionic sympathetic nerve endings in the pineal gland thereby protecting the pinealocytes from stimulation (Parfitt and Klein 1976). In general, this explanation is consistent with the observations in humans that neither lumbar puncture (Wetterberg 1978), insulin-induced hypoglycaemia (Vaughan *et al.* 1979b), pneumoencephalography (Vaughan *et al.* 1978b), nor strenuous exercise (Vaughan *et al.* 1979b) were found to change basal levels of melatonin in either the blood or CSF.

There is, however, one exception to the observations summarized above. When seven women exercised vigorously for 60 minutes (with a final heart rate of 85 per cent maximal), 60 minutes later their daytime basal melatonin levels had risen 2- to 3-fold (Carr *et al.* 1981; Fig. 1.9). These studies differed from the others especially in terms of the duration of the exercise (or stress) and suggests that long-term aversive demands may eventually culminate in elevated daytime melatonin levels. In view of the known reproductive inhibitory action of melatonin in experimental animals (Reiter 1980, 1982; Arendt 1986) and its potential for the same in humans, the release of melatonin after strenuous exercise has been offered as a potential explanation for menstrual cycle perturbations in long distance runners (Reiter 1986*b*).

Age seems to be a major factor in determining the amplitude of the day:night melatonin rhythm. Certainly, in the Syrian hamster, for example, the nocturnal rise in pineal melatonin is severely blunted in aged (18-month-old) versus young (2-month-old) animals (Reiter *et al.* 1980). The examination of elderly humans generally conforms to these observations. Thus, when the nocturnal melatonin rise in the plasma of healthy young men (mean age 26 years) was compared with that of old men (mean age 85 years), the nocturnal increase was much more robust (roughly 12-fold versus 3-fold in young and old men, respectively) in the younger subjects (Iguchi *et al.* 1982; Fig. 1.10). Also, when the urinary excretion of 6-hydroxymelatonin sulphate was estimated in individuals between 20 and 100 years of age a gradual reduction was seen with

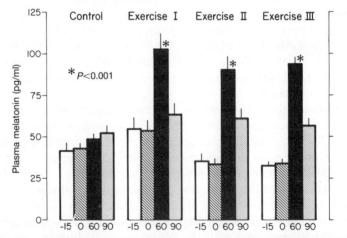

Fig. 1.9. Mean (± SE) plasma melatonin concentrations in seven healthy women before (−15 min) and after (0, 60, and 90 min) three different periods of exercise. At 60 minutes after exercise melatonin levels had risen. (Data from Carr *et al.* 1981.)

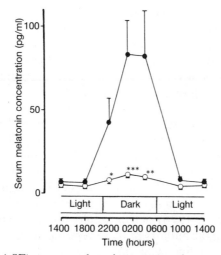

Fig. 1.10. Mean (± SE) serum melatonin concentrations over a 24-hour period in young (●) and old (○) male volunteers. Young subjects were in their mid-20s while the old subjects were in their mid-80s. The drop in melatonin with advancing age is characteristic of a number of mammals. (From Iguchi *et al.* 1982.)

increased age (Sack *et al.* 1986). A common feature of the pineal gland in elderly subjects is the variable degree of deposition of calcium deposits (Vollrath 1984); whether these deposits compromise the melatonin producing capability of the pineal gland has not been satisfactorily determined.

Melatonin has been touted as a potential sleep-inducing factor in the human (Cramer *et al.* 1974; Lieberman *et al.* 1984). Thus, the age-associated reduction in sleep duration and efficiency (Carskadon *et al.* 1980; Dement and Carskadon 1982) could be related to the commensurate drop in nocturnal melatonin secretion (Wurtman and Lieberman 1985). Indeed, the interactions of melatonin with sleep processes is a subject worthy of serious investigation.

Cerebrospinal fluid melatonin

Melatonin levels in the CSF have been carefully monitored in a number of non-human mammals. Without exception, melatonin levels in this fluid exhibit a rhythm not unlike that seen in the blood. In calves, for example, nocturnal darkness is associated with a marked rise in CSF melatonin; the CSF increase is even greater than that in the blood (Hedlund *et al.* 1977). In fact, the higher CSF values compared to those in the blood again raise the issue about the primary route of melatonin secretion, i.e.

directly into the CSF or into the blood and then secondarily into the ventricular system via the choroid plexus.

In the rhesus monkey as well, detailed studies revealed a large nocturnal rise in CSF melatonin values (Reppert *et al.* 1980). The rhythm in CSF melatonin is strongly coupled to the prevailing photoperiodic regime; the cycle is suppressed by continual light exposure and persists under prolonged darkness just as in the blood.

From the findings in these species, a similar melatonin rhythm in human CSF has been inferred although proof of this is lacking. Melatonin has been identified in the CSF of humans by both radio-immunoassay (Vaughan 1984) as well as by means of gas chromatography–mass spectrometry (Wilson *et al.* 1977). Higher than normal daytime levels of melatonin in the CSF have been reported in leukaemic children (Smith *et al.* 1976) and in adults afflicted with schizophrenia (Smith *et al.* 1979); the authors of these reports, however, made no claim that the elevated daytime CSF melatonin levels were related to the particular disease states. In humans, blood concentrations of melatonin are routinely higher than those measured in the CSF (Brown *et al.* 1979; Tan and Khoo 1981).

Salivary melatonin

In 1985, two groups identified melatonin in human saliva (Miles *et al.* 1985*a,b*; Vakkuri 1985). Vakkuri (1985) has compared salivary and serum melatonin levels throughout a 24-hour period in five male volunteers. During the daytime, serum and salivary melatonin concentrations were roughly equivalent; by comparison, at night salivary levels of the constituent were about one-third of those in the peripheral blood. The salivary melatonin secretion rate in this study was judged to be 0.5–0.9 pmol/hour during the day and between 1.0–2.4 pmol/hour at night.

The high degree of correlation between saliva and plasma melatonin levels was initially reported by Miles and colleagues (1985*a,b*, 1987) and quickly supported by McIntyre *et al.* (1987; Fig. 1.11) and Nowak and associates (1987). Besides describing the nocturnal increase in melatonin in both fluids these latter authors also observed that a 1-hour light pulse at midnight depressed both plasma and salivary melatonin (Fig. 1.12) again emphasizing the dependence of salivary melatonin concentrations on blood levels of this hormone. There are apparently some precautions that must be taken before collecting saliva for melatonin analysis. According to McIntyre and colleagues (1987) food fragments (cheese and potato chips) may negate the melatonin RIA. Likewise, the use of toothpaste and coffee too near the time of saliva collection may also interfere with the direct assessment of salivary melatonin (J. Laitinen, personal communication). Miles and colleagues

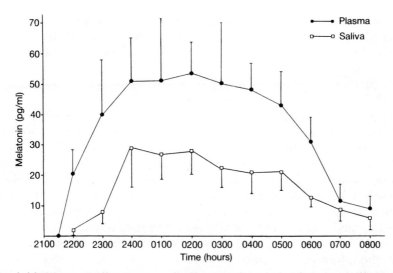

Fig. 1.11. Mean (±SE) plasma and salivary melatonin levels between 21.30 and 08.00 hours in healthy male volunteers. Salivary melatonin values are always lower than plasma concentrations, and salivary melatonin levels followed closely those in the blood. (From McIntyre *et al.* 1987; see also Miles *et al.* Chapter 13, this volume.)

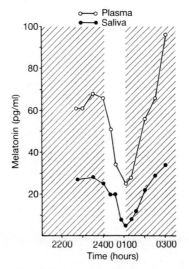

Fig. 1.12. Suppression of both nocturnal plasma and salivary melatonin with a 1-hour light pulse at night. (From McIntyre *et al.* 1987.)

have reviewed in detail the basic and clinical utility of salivary melatonin
estimation, and the reader is referred to Chapter 13 of the present
volume for more extensive discussion of this novel technique and
methodology.

Ovarian follicular fluid melatonin

Of particular interest to reproductive physiologists is the presence of
melatonin in the follicular fluid of human preovulatory follicles
(Brzezinski *et al.* 1987). The fluid samples were obtained by means of
either laprascopy or sonographically controlled follicular puncture from
infertile women who were undergoing *in vitro* fertilization and embryo
transfer; follicular development had been stimulated in these individuals
with a combination of clomiphene citrate, human menopausal gonadotrophin, and human chorionic gonadotrophin. Blood samples were
collected within 30 minutes or less from the time of follicular fluid
collection. Remarkably, mean follicular fluid concentrations of melatonin
(36.5 ± 4.8 pg/ml) were considerably higher than serum values
(10.0 ± 1.4 pg/ml) at about the same time (Fig. 1.13). In each patient
serum and follicular fluid melatonin concentrations were positively
correlated.

Since there is no evidence that the ovary has the capability of
synthesizing melatonin, it was assumed that the melatonin present in the
follicular fluid was derived from the blood (Brzezinski *et al.* 1987).

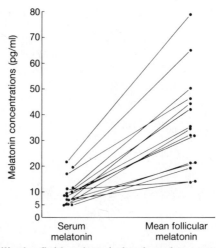

Fig. 1.13. Mean follicular fluid melatonin levels and serum melatonin concentrations in adult females. The follicular fluid and blood was collected at approximately the same time during the day. In some women fluid was collected from
several follicles. Follicular fluid melatonin levels were always as high or higher
than values in the serum. (From Brzezinski *et al.* 1987.)

Considering its role in the control of reproductive phsyiology (Reiter 1980, 1981*a*; Tamarkin *et al.* 1985) a potential site of action of melatonin at the ovarian level could be inferred from these findings. Radioactive melatonin has been shown to be concentrated in the ovaries of rats and cats after its systemic administration (Wurtman *et al.* 1964). Furthermore, melatonin receptors have been tentatively identified in ovarian tissue (Cohen *et al.* 1978) and, finally, the addition of melatonin to ovarian tissue slices *in vitro* reportedly depresses steroidogenesis (Macphee *et al.* 1974; Younglai 1979). These findings are also of interest because melatonin reportedly directly inhibits the motility of sperm collected from human males (Oosthuizen *et al.* 1986). If melatonin is also secreted into the seminal fluid in males, in both males and females it could have important peripheral effects on reproduction.

Urinary melatonin and metabolites

Melatonin is rapidly metabolized primarily in the liver and to a lesser extent in the central nervous system with the metabolites being excreted into the urine (Kopin *et al.* 1961; Kveder and McIsaac 1961). Only about 1 per cent of the melatonin secreted into the blood escapes into the urine in the unchanged form. The chief hepatic metabolite is 6-hydroxy-melatonin sulphate (6-sulphatoxymelatonin; Matthews *et al.* 1981; Fig. 1.14) while in the brain the most common degradation product of melatonin is N-acetyl-5-methoxykynurenamine. Using a mass spectro-metric method (Fellenberg *et al.* 1980) for its measurement, the urinary excretion rate of 6-sulphatoxymelatonin in adult males and females was estimated to be 6.3–30.9 μg/24 hours (Tetsuo *et al.* 1980, 1981*a*; Fellenberg *et al.* 1981). In one healthy male volunteer, where a detailed 24-hour analysis was perfomed, 80 per cent of the total urinary excretion of 6-sulphatoxymelatonin was confined to the daily dark period (Fig. 1.15); the nocturnal excretion rate was found to be highly constant in this subject over a 5-day period, with the variation during this interval being only 3 per cent. To test the effect of pinealectomy on the urinary excretion of the melatonin metabolite sheep were used as experimental subjects. In pineal intact sheep, 6-sulphatoxymelatonin levels in the urine ranged from 3.0 to 17.3 μg/24 hours while after pinealectomy these values dropped to 1.9–3.0 μg/24 hours (Fellenberg *et al.* 1981). In sheep, as in the human, the bulk of the 6-sulphatoxymelatonin measured was recovered from nocturnal urine samples. Considering the ease with which urine samples can be collected and the nature of the assay, the assessment of 6-sulphatoxymelatonin levels in the urine would seem to be a facile way of determining 24-hour melatonin production in the human, provided nothing alters the metabolic fate of melatonin in the liver.

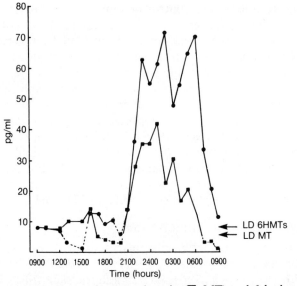

Fig. 1.14. Twenty-four-hour plasma melatonin (■, MT) and 6-hydroxymelatonin sulphate (●, 6HMTs) in a single male subject. Levels of 6-hydroxymelatonin sulphate, the chief hepatic metabolite of melatonin, follow very close to those of melatonin. LD, least detectable. (From Arendt 1981.)

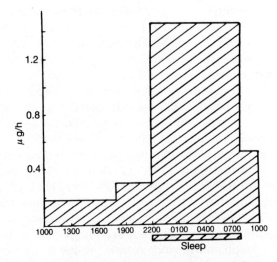

Fig. 1.15. Urinary excretion of 6-hydroxymelatonin sulphate over a 24-hour period in one male volunteer. The bulk of the melatonin metabolite is excreted at night. (From Fellenberg *et al.* 1981.)

Clinical implications and perspectives

Much research to date in experimental animals has focused on the inter-actions of the pineal gland and melatonin with the neuroendocrine-reproductive system (Fig. 1.16). However, the actions of melatonin are extremely widespread, and may perhaps influence every organ system in the body.

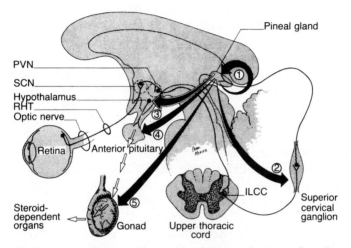

Fig. 1.16. Summary diagram illustrating the neural connections between the visual system and the pineal gland, and the humoral connection between the pineal gland and the neuroendocrine system. Potential sites of action of melatonin are labelled 1–5, but other sites may well exist. Although this figure illustrates especially the potential interactions of the pineal with the neuroendocrine-reproductive axis, this is only one of many organ systems that melatonin influences.

The specific physiological consequences of melatonin in the human have been only preliminarily defined, but all available data are discussed in the subsequent chapters of this book. Considering the profound influence of the pineal gland and its secretory products on a gamut of endocrine (Johnson 1982; Bittman 1984) and non-endocrine (Armstrong *et al.* 1982; Wetterberg 1982) functions in non-human mammals, it would seem scientifically naive to assume that melatonin does not have similar capabilities in man. There are a number of points to remember when investigating the consequences of melatonin administration in man. Certainly, one of these is the timing of the melatonin injections. It is clear from many animal studies that the efficacy of melatonin as a hormonal agent is closely linked to the time of day at which the drug is administered. Thus, for example, despite the profound inhibitory influence of daily melatonin injections on reproductive physiology in Syrian hamsters, the indoleamine must be given during very restricted portions of the 24-hour period in order to be effective in suppressing gonadotrophins and associated peripheral reproductive physiological functions; at other times during the 24-hour period melatonin administered by the same dose and route is totally inconsequential in reference to sexual physiology (Tamarkin *et al.* 1976; Reiter *et al.* 1976; Stetson and Watson-Whitmyre 1984). The Syrian hamster is not the only species in which the timing of the melatonin injection is critical to its efficacy, and to assume the human is different in this regard is probably a mistake. In fact, in the relatively few studies concerned with the endocrine consequences of melatonin in the human, timing of the melatonin injection seemed not to have been a major consideration (Fideleff *et al.* 1976; Nordlund and Lerner 1977; Shaw 1977; Weinberg *et al.* 1980). This probably influenced the outcome of the experiments and may have contributed to the essentially negative observations that were made.

In animals, as exemplified again by studies in the Syrian hamster, melatonin administration at a specific time each day must be continued for 5–10 weeks before the effects in terms of the reproductive system are obvious (Tamarkin *et al.* 1976; Reiter *et al.* 1976). In humans, melatonin has rarely been given for more than a few consecutive days and the changes induced were not particularly dramatic (Nordlund and Lerner 1977; Weinberg *et al.* 1980). As sensitive as the Syrian hamster reproductive system is to melatonin, if melatonin would only have been given for several days the results would also have been totally negative and the conclusion could well have been that the consequences of melatonin administration in this species are not great when, in fact, after long-term administration they are very substantial. Thus, in addition to the time of day the melatonin is given, the duration of treatment also may be critical to the outcome of a particular clinical investigation.

Finally, the method of melatonin administration as well as the daily dose may be important factors to consider during experimental design. In at least some species, e.g. the Syrian hamster, the actions of persistently available melatonin are dramatically different from those caused by daily melatonin administration. The ability of daily late afternoon melatonin injections to cause total involution of the neuro-endocrine-reproductive system has already been mentioned (Tamarkin *et al.* 1976; Reiter *et al.* 1976). However, it was subsequently demonstrated that a subcutaneously placed melatonin reservoir, which releases melatonin continually, actually prevented the action of daily afternoon melatonin injections and caused the animals to be reproductively competent (Reiter *et al.* 1977). The current explanation for this finding is that continually available melatonin desensitizes (down-regulates) its receptor thereby rendering the indoleamine totally ineffective (Reiter 1980). This phenomenon is not unique to the Syrian hamster but exists in the rat as well (Reiter 1980) and could certainly be operative in the human. Thus, if clinical studies were to use a method of melatonin administration that resulted in continually high levels of melatonin the consequences may be very different than after daily melatonin administration which merely causes a short-term surge of the indoleamine in the blood. Also, if the dose of melatonin is excessively high, even though only given once per day, it also physiologically down-regulates its receptor (Chen *et al.* 1980). From these studies it is obvious that the clinical use of melatonin must take into consideration a variety of parameters all of which could either maximize or, conversely, jeopardize the results of the experiment.

A number of workers have examined 24-hour melatonin patterns in patients with various diseases (Wetterberg 1978; Birau 1981; Birau *et al.* 1981; Vaughan 1984; Waldhauser *et al.* 1984; Miles *et al.* 1987; Miles and Philbrick 1988). These studies have not yet definitively identified what could be referred to as a pineal-related disease. This is not totally unexpected. In general, the pineal gland and melatonin seem to function as the 'fine-tuner' of organismal physiology and, thus, to expect remarkable changes related specifically to melatonin may be futile. Rather it seems more realistic to anticipate that this hormone may only slightly, but significantly alter (either improve or worsen) the clinical manifestation of a disease state whose development is primarily dependent upon some other malfunction.

Even if, however, melatonin rhythms in patients with a given disease are described as normal, it does not mean that the indoleamine may not be involved in the disease process, an assumption that seems to be routinely made. Again, from animal studies it is known that any number of factors can alter the sensitivity of a species to the actions of the pineal gland and melatonin; these perturbations have been referred to as pineal potentiating factors (Reiter and Sorrentino 1971*b*; Reiter 1980) and include, in rats, early androgen sterilization, anosmia, underfeeding, and cold exposure. In these animals what would usually be considered to be a normal melatonin rhythm has much greater physiology impact than it does in rats that are not sensitized due to the imposition of a potentiating factor. Thus, the usual melatonin cycle in one animal may have minimal or no effects; yet under another set of circumstances, it may be devastating to certain physiological functions. The point of this is that, just because melatonin levels or the rhythm may be within the normal range does not preclude the possibility that it is involved in the disease process. A specific example is considered. In humans with anorexia nervosa the plasma melatonin rhythm may be classified as being typical of that found in clinically normal individuals. However, if the hypothalamic sensitivity of the anorexic patients is increased, then the response to the melatonin may be greatly exaggerated. Interestingly, reduced food intake in rats is one factor that predisposes the animal to the actions of melatonin and, of course, anorexic patients typically have greatly reduced food consumption. The reader is reminded that the author is not implying that anorexia nervosa is or is not a pineal related disease; it is only used as an example of a condition in which melatonin could be involved as a causative agent of the clinical entity. The relationships of melatonin to the physiology of the organism and how these interactions may be manifested are reviewed in detail elsewhere (Reiter 1987).

In conclusion, it seems almost certain that within the next decade altered melatonin rhythms or altered sensitivity to melatonin will be found to explain several pathophysiological conditions. To assume otherwise in view of the overwhelming amount of data illustrating the wide-ranging physiological consequences of the pineal hormone would be unwise. Clearly, the subsequent chapters point out those areas which are already being exploited; it is likely, however, that the actions of melatonin are not limited to those discussed in this volume.

Acknowledgements

This work was supported by grants from the NSF, NIH, and Spanish American grants.

References

Akerstedt, T., Froberg, J. E., Friberg, Y., and Wetterberg, L. (1979). Melatonin excretion, body temperature and subjective arousal during 64 hours of sleep deprivation. *Psychoneuroendocrinology* **4**, 219–25.

Anton-Tay, F. and Wurtman, R. J. (1969). Regional uptake of ^{3}H-melatonin from blood or cerebrospinal fluid in the rat brain. *Nature* **221**, 474–5.

Arendt, J. (1978). Melatonin assays in body fluids. *Journal of Neural Transmission* (Supplement) **13**, 265–78.

Arendt, J. (1984). Mammalian pineal rhythms. *Pineal Research Reviews* **3**, 142–202.

Arendt, J. (1986). Role of the pineal gland and melatonin in seasonal reproductive function of mammals. *Oxford Reviews of Reproductive Biology* **8**, 266–320.

Arendt, J., Wirz-Justice, A., Bradtke, J., and Kornemark, M. (1979). Long-term studies on immunoreactive human melatonin. *Annals of Clinical Biochemistry* **16**, 307–12.

Arendt, J., Forbes, J. M., Brown, W. B., and Munson, A. (1980). Effect of pinealectomy on immunoassayable melatonin in sheep. *Journal of Endocrinology* **85**, 1P–2P.

Armstrong, S., Ng, K. T., and Coleman, G. J. (1982). Influence of the pineal gland on brain-behavior relationship. In *The pineal gland, Vol. III, Extra-reproductive effects* (ed. R. J. Reiter), pp. 81–106. CRC Press, Boca Raton, Florida.

Axelrod, J. (1974). The pineal gland: A neurochemical transducer. *Science* **184**, 134–48.

Axelrod, J. and Weissbach, H. (1960). Enzymatic O-methylation of N-acetyl-serotonin to melatonin. *Science* **131**, 1312.

Axelrod, J. and Weissbach, H. (1961). Purification and properties of hydroxyindole-O-methyltransferase. *Journal of Biological Chemistry* **236**, 211–13.

Axelrod, J., Shein, H. M., and Wurtman, R. J. (1969). Stimulation of C^{14}-tryptophan by noradrenaline in rat pineal in organ culture. *Proceedings of the National Academy of Science USA* **62**, 644–9.

Bagnara, J. T. and Hadley, M. E. (1970). Endocrinology of the amphibian pineal. *American Zoologist* **10**, 201–16.

Baldessarini, R. J. and Kopin, I. J. (1966). S-adenosylmethionine in brain and other tissues. *Journal of Neurochemistry* **13**, 769–77.

Binkley, S. (1981). Pineal biochemistry: Comparative aspects and circadian rhythms. In *The pineal gland, Vol. I, Anatomy and biochemistry* (ed. R. J. Reiter), pp. 155–72. CRC Press, Boca Raton, Florida.

Binkley, S. (1983). Circadian rhythms of pineal function in rats. *Endocrine Reviews* **4**, 225–70.

Binkley, S., Hryshchyshy, M., and Reilly, K. (1979). N-Acetyltransferase activity responds to environmental lighting in the eye as well as in the pineal gland. *Nature* **281**, 479–81.

Birau, N. (1981). Melatonin in human serum: progress in screening investigations and clinic. *Advances in the Biosciences* **29**, 297–326.

Birau, N., Birau, M., and Schloot, W. (1981). Melatonin rhythms in human serum. *Advances in the Biosciences* **29**, 287–95.

Bittman, E. L. (1984). Melatonin and photoperiodic time measurement: Evidence from rodents and ruminants. In *The pineal gland* (ed. R. J. Reiter), pp. 155–92. Raven Press, New York.

Brainard, G. D., Richardson, B. A., King, T. S., and Reiter, R. J. (1984). The influence of different light spectra on the suppression of pineal melatonin content in the Syrian hamster. *Brain Research* **294**, 333–9.

Brainard, G. C., *et al.* (1985). Effect of light wavelength on the suppression of nocturnal plasma melatonin in normal volunteers. *Annals of the New York Academy of Sciences* **453**, 376–8.

Brismar, K., Werner, S., and Wetterberg, L. (1982). Melatonin and corticosteroid response to metyrapone in patients with pituitary disease. In *The pineal gland and its hormones* (ed. R. J. Reiter), pp. 283–92. Alan R. Liss Publishers, New York.

Brown, G. M., Young, S. N., Gauthier, S., Tsui, H., and Grota, L. J. (1979).

Melatonin in human cerebrospinal fluid in the daytime: its origin and variation with age. *Life Science* **25**, 929–36.

Brownstein, M. and Axelrod, J. (1974). Pineal gland: 24 hour rhythm in norepinephrine turnover. *Science* **184**, 163–5.

Brownstein, M., Saavedra, J. M., and Axelrod, J. (1973). Control of pineal N-acetyltransferase activity: protection of stimulated activity by acetyl-CoA and related compounds. *Journal of Neurochemistry* **26**, 51–5.

Brzezinski, A., Seibel, M. M., Lynch, H. J., Deng, M.-H., and Wurtman, R. J. (1987). Melatonin in human preovulatory follicular fluid. *Journal of Clinical Endocrinology and Metabolism* **64**, 865–7.

Bubenik, G. A., Brown, G. M., Uhlir, T., and Grota, L. J. (1974). Immuno-histochemical localization of N-acetylindolealkylamines in pineal gland, retina and cerebellum. *Brain Research* **81**, 233–42.

Bubenik, G. A., Purtill, R. A., Brown, G. A., and Grota, L. J. (1978). Melatonin in the retina and the Harderian gland: ontogeny, diurnal variations and melatonin treatment. *Experimental Eye Research* **27**, 323–34.

Cardinali, D. P. (1981). Melatonin. A mammalian pineal hormone. *Endocrine Reviews* **2**, 237–58.

Cardinali, D. P. and Rosner. J. M. (1971). Retinal localization of the hydroxy-indole-O-methyltransferase (HIOMT) in the rat. *Endocrinology* **89**, 301–3.

Cardinali, D. P., Larin, P., and Wurtman, R. J. (1972). Control of the rat pineal gland by light spectra. *Proceedings of the National Academy of Science USA* **69**, 2003–5.

Cardinali, D. P., Ritta, N. M., Speziale, N., and Gimeno, M. F. (1979). Release and specific binding of prostaglandins in bovine pineal glands. *Prostaglandins* **18**, 577–80.

Cardinali, D. P., *et al.* (1983). Molecular aspects of neuroendocrine integrative processes in the pineal gland. In *The pineal and its endocrine role* (ed. J. Axelrod, F. Fraschini, and G. P. Velo), pp. 199–219. Plenum Publishing Company, New York.

Carr, D. B., *et al.* (1981). Plasma melatonin increases during exercise in women. *Journal of Clinical Endocrinology and Metabolism* **53**, 224–5.

Carskadon, M. A., Harvey, K., Duke, P., Anders, T. F., Litt, I. F., and Dement, W. C. (1980). Pubertal changes and daytime sleepiness. *Sleep* **2**, 453–60.

Chan, A. and Ebadi, M. (1980). The kinetics of NE-induced stimulation of SNAT in bovine pineal gland. *Neuroendocrinology* **31**, 244–51.

Chen, H. J., Brainard, G. C., III, and Reiter, R. J. (1980). Melatonin given in the morning prevents the suppressive action on the reproductive system of melatonin given in late afternoon. *Neuroendocrinology* **31**, 129–32.

Cohen, M., Roselle, D., and Chabner, B. (1978). Evidence for a cytoplasmic melatonin receptor. *Nature* **274**, 894–5.

Collu, R., Fraschini, F., and Martini, L. (1971). Blockade of ovulation by melatonin. *Experientia* **27**, 844–5.

Cowen, P. J., Fraser, S., Sammons, R., and Green, A. R. (1983). Atenolol reduces plasma melatonin concentration in man. *British Journal of Clinical Pharmacology* **15**, 579–81.

Craft, C. M., Morgan, W. W., and Reiter, R. J. (1983). 24-hour changes in catecholamine synthesis in rat and hamster pineal glands. *Neuroendocrinology* **38**, 193–8.

Craft, C. M., Morgan, W. W., Jones, D. J., and Reitner, R. J. (1985). Hamster and rat pineal gland β-receptor characterization with iodocynanopindolol and the effect of decreased catecholamine synthesis on the receptor. *Journal of Pineal Research* **2**, 51–66.

Cramer, H., Rudolph, J., Consbruch, U., and Kendel, K. (1974). On the effects of melatonin on sleep and behavior in man. *Advances in Biochemistry and Psychopharmacology* **11**, 187–91.

Dement, W. C., and Carskadon, M. A. (1982). Current perspectives on daytime sleepiness. The issues. *Sleep* **5**, S56–66.

Diguchi, T. (1975). Characteristics of serotonin-acetyl co-enzyme A-N-acetyl-transferase in the pineal gland of rat. *Journal of Neurochemistry* **24**, 1053–6.

Ebadi, M. (1984). Regulation of the synthesis of melatonin and its significance to neuroendocrinology. In *The pineal gland* (ed. R. J. Reiter), pp. 1–38. Raven Press, New York.

Ebadi, M. and Govitrapong, R. (1986*a*). Neural pathways and neurotransmitters affecting melatonin synthesis. *Journal of Neural Transmission (Supplement)* **21**, 125–35.

Ebadi, M. and Govitrapong, P. (1986*b*). Orphan transmitters and their receptor sites in the pineal gland. *Pineal Research Reviews* **4**, 1–54.

Ebels, I., de Morée, A., Hus-Citharel, A., and Moszkowska, A. (1979). A survey of some active sheep pineal fractions and a discussion on the possible significance of pteridines in those fractions in *in vitro* and *in vivo* assays. *Journal of Neural Transmission* **44**, 97–116.

Ebihara, S., Marks, T., Hudson, D. J., and Menaker, M. (1986). Genetic control of melatonin synthesis in the pineal gland of the mouse. *Science* **231**, 491–3.

Falck, B., Owman, C., and Rosengren, E. (1966). Changes in rat pineal stores of 5-hydroxytryptamine after inhibition of its synthesis or break-down. *Acta Physiologica Scandinavica* **67**, 300–5.

Fellenberg, A. J., Phillipou, G., and Seamark, R. F. (1980). Specific quantitation of urinary 6-hydroxymelatonin sulfate by GC–MS. *Biomedical Mass Spectrometry* **7**, 84–7.

Fellenberg, A. J., Phillipou, G., and Seamark, R. F. (1981). Urinary 6-sulphatoxy melatonin excretion and melatonin production rate: studies in sheep and man. In *Pineal function* (ed. C. D. Matthews and R. F. Seamark), pp. 143–50. Elsevier/North Holland Biomedical Press, Amsterdam.

Fellenberg, A. J., Phillipou, G., and Seamark, R. F. (1982). Urinary 6-sulphatoxy-melatonin excretion during the human menstrual cycle. *Clinical Endocrinology* **17**, 71–5.

Fideleff, H., Aparicio, N. J., Gutelman, A., Debeljik, L., Mancini, A., and Cramer, C. (1976). Effect of melatonin on the basal and stimulated gonadotropin levels in normal men and postmenopausal women. *Journal of Clinical Endocrinology and Metabolism.* **42**, 1014–17.

Foldes, A., Hasknson, R. M., Scaramuzzi, R. J., Hinks, N. T. and Maxwell, C. A. (1983). Modification of sheep pineal β-adrenoceptors by some gonadal steroids but not by melatonin. *Neuroendocrinology* **37**, 378–85.

Foley, P. B., Cairncross, K. D., and Foldes, A. (1986). Pineal indoles: Significance and measurement. *Neuroscience and Biobehavioral Reviews* **10**, 273–93.

Goldman, B. D. (1983). The physiology of melatonin in mammals. *Pineal Research Reviews* **1**, 145–82.

Gonzalez-Brito, A., Jones, D. J., Ademe, R. M. and Reiter, R. J. (1988). Characterization and measurement of [^{125}I]iodopindolol binding in individual rat pineal glands: existence of a 24 hour rhythm in beta-adrenergic receptor density. *Brain Research*, **438**, 108–14.

Greiner, A. C. and Chan, A. C. (1978). Melatonin content of the human pineal gland. *Science* **199**, 83–4.

Guchait, R. B. and Grau, J. E. (1978). Biosynthesis of S-adenosyl-L-methionine in the rat pineal gland. *Journal of Neurochemistry* **31**, 921–5.

Håkanson, R. and Owman, C. (1966). Pineal DOPA decarboxylase and monamine oxidase activities as related to the monoamine stores. *Journal of Neurochemistry* **13**, 597–605.

Hamm, H. E. and Menaker, M. (1980). Retinal rhythms in chicks: Circadian variation in melatonin and in serotonin N-acetyltransferase activity. *Proceedings of the National Academy of Sciences USA* **77**, 4998–5002.

Hansen, T., Heyden, T., Sundberg, I., Alfredson, G., Nyback, H., and Wetterberg, L. (1980). Propranolol in schizophrenia: clinical, metabolic, and pharmacological findings. *Archives of General Psychiatry* **37**, 685–90.

Hariharasubramanian, N., Nair, N. P. V., Pilapel, C., Thavundajil, J. X., and Quirion, R. (1986). Plasma melatonin levels during the menstrual cycle: Changes with age. In *The pineal gland during development* (ed. D. Gupta and R. J. Reiter), pp. 166–73. Croom-Helm Publishers, London.

Hedlund, L., Lischko, M. M., Rollag, M. D., and Niswender, G. D. (1977). Melatonin: Daily cycle in plasma and cerebrospinal fluid of calves. *Science* **195**, 686–7.

Hewing, M. (1978). A liquor contacting area in the pineal recess of the golden hamster (*Mesocricetus auratus*). *Anatomy and Embryology* **153**, 295–304.

Hoffman, R. A., Johnson, L. B., and Reiter, R. J. (1985). Harderian glands of golden hamsters: Temporal and sexual differences in immunoreactive melatonin. *Journal of Pineal Research* **2**, 161–8.

Hughes, P. C., Bickford, E. W., and Goldstein, E. S. (1988). Optical radiation. *Pineal Research Reviews*, **5**, 1–67.

Iguchi, H., Kato, H., and Ibayashi, H. (1982). Age-dependent reduction in serum melatonin concentrations in healthy human subjects. *Journal of Clinical Endocrinology and Metabolism* **55**, 27–9.

Illnerovà, H., Backström, M., Sääf, J., Wetterberg, L., and Vangbo, B. (1978). Melatonin in the rat pineal gland and serum; rapid parallel decline after light exposure at night. *Neuroscience Letters* **9**, 189–93.

Johnson, L. Y. (1982). The pineal gland as a modulator of the adrenal and thyroid axes. In *The pineal gland, Vol. III, Extra-reproductive effects.* (ed. R. J. Reiter), pp. 107–52. CRC Press, Boca Raton, Florida.

Johnson, L. Y., Vaughan, M. K., Richardson, B. A., Petterborg, L. J., and Reiter, R. J. (1982). Variation in pineal melatonin content during the estrous cycle of the rat. *Proceedings of the Society for Experimental Biology and Medicine* **169**,

416–19.

Kappers, J. A. (1960). The development, topographical relations and innervation of the epiphysis cerebri in the albino rat. *Zeitschrift für Zellforschung* **52**, 163–215.

Keglevïc, D., Kveder, S., and Iskrïc, S. (1968). Indoleacetaldehydes—intermediates in indolealkylamine metabolism. *Advances in Pharmacology* **6A**, 79–89.

King, T. S., Richardson, B. A., and Reiter, R. J. (1982). Regulation of rat pineal melatonin synthesis: effect of monoamine oxidase inhibition. *Molecular and Cellular Endocrinology* **35**, 327–38.

King, T. S., Steger, R. W., Steinlechner, S., and Reiter, R. J. (1984). Day-night differences in estimated rates of 5-hydroxytryptamine turnover in rat pineal gland. *Experimental Brain Research* **54**, 432–6.

Klein, D. C. (1978). Pineal gland as a model of neuroendocrine control system. In *The hypothalamus* (ed. S. Reichlin, R. J. Baldessarini, and J. B. Martin), pp. 303–29. Raven Press, New York.

Klein, D. C. and Weller, J. L. (1970). Indole metabolism in the pineal gland: A circadian rhythm in N-acetyltransferase. *Science* **169**, 1093–5.

Klein, D. C., Berg, G. R., and Weller, J. L. (1970). Melatonin synthesis: Adenosine 3′,5′-monophosphate and NE stimulated NAT. *Science* **168**, 979–80.

Klein, D. C., Auerbach, D. A., Namboodiri, M. A. A., and Wheler, G. H. T. (1981). Indole metabolism in the mammalian pineal gland. In *The pineal gland, Vol. I, Anatomy and biochemistry* (ed. R. J. Reiter), pp. 199–228. CRC Press, Boca Raton, Florida.

Kneisley, L. W., Moskowitz, M. A., and Lynch, H. J. (1978). Cervical spinal cord lesions disrupt the rhythm in human melatonin excretion. *Journal of Neural Transmission (Supplement)* **13**, 311–23.

Kopin, I. J., Pare, C. M. B., Axelrod, J., and Weissbach, H. (1961). The fate of melatonin in animals. *Journal of Biological Chemistry* **236**, 3072–5.

Korf, H.-W. and Møller, M. (1984). The innervation of the mammalian pineal gland with special reference to central pinealopetal connections. *Pineal Research Reviews* **2**, 41–86.

Kveder, S. and McIsaac, W. M. (1961). The metabolism of melatonin (N-acetyl-5-methoxytryptamine) and 5-methoxytryptamine. *Journal of Biological Chemistry* **236**, 3214–20.

Leino, M. and Airaksinen, M. M. (1985). Methoxyindoles of the retina. *Medical Biology* **63**, 160–9.

Lerner, A. B., Case, J. D., Takahashi, Y., Lee, T. H., and Mori, W. (1958). Isolation of melatonin, the pineal factor that lightens melanocytes. *Journal of the American Chemical Society* **80**, 2587.

Lerner, A. B., Case, J. D., and Heinzelman, R. V. (1959). Structure of melatonin. *Journal of the American Chemical Society* **81**, 6084–5.

Levine, R. A., Kuhn, D. M., and Lovenberg, W. (1978). The regional distribution of hydroxylase cofactor in rat brain. *Journal of Neurochemistry* **32**, 1575–8.

Lewy, A. J. (1983). Biochemistry and regulation of mammalian melatonin production. In *The pineal gland* (ed. R. Relkin) pp. 77–128. Elsevier

Biomedical Press, New York.

Lewy, A. J. and Newsome, D. A. (1983). Different types of melatonin circadian secretory rhythms in some blind subjects. *Journal of Clinical Endocrinology and Metabolism* **56**, 1103–7.

Lewy, A. J., Wehr, T. A., Goodwin, F. K., Newsome, D. A., and Markey, S. P. (1980). Light suppresses melatonin secretion in humans. *Science* **210**, 1267–9.

Lieberman, H. R., Waldhauser, F., Garfield, G., Lynch, H. J., and Wurtman, R. J. (1984). Effects of melatonin on human mood and performance. *Brain Research* **323**, 201–7.

Lipton, J. S., Petterborg, L. J., and Reiter, R. J. (1981). Influence of propanolol, phenoxybenzamine or phentolamine on the *in vivo* nocturnal rise of pineal melatonin levels in the Syrian hamster. *Life Science* **28**, 2377–82.

Lipton, J. S., Petterborg, L. J., Steinlechner, S., and Reiter, R. J. (1982). *In vivo* responses of the pineal gland of the Syrian hamster to isoproterenol or norepinephrine. In *The pineal and its hormones* (ed. R. J. Reiter), pp. 107–16. Alan R. Liss Publishers, New York.

Lovenberg, W., Weissbach, H., and Udenfriend, S. (1962). Aromatic L-amino acid decarboxylase. *Journal of Biological Chemistry* **237**, 89–92.

Lovenberg, W., Jequier, E., and Sjoerdsma, A. (1967). Tryptophan hydroxylation: measurement in pineal gland, brainstem and carcinoid tumour. *Science* **155**, 217–19.

Lynch, H. J., Jimerson, D. C., Ozaki, Y., Post, R. M., Bunney, W. E., Jr, and Wurtman, R. J. (1978). Entrainment of rhythmic melatonin secretion in man to a 12-hour phase shift in the light/dark cycle. *Life Science* **23**, 1557–63.

Macphee, A. A., Cole, F. E., and Rice, B. F. (1974). The effect of melatonin on steroidogenesis by the human ovary *in vitro*. *Journal of Clinical Endocrinology and Metabolism* **40**, 688–9.

Matthews, C. D., Kennaway, D. J., Fellenberg, A. J. G., Phillipou, G., Cox, L. W., and Seamark, R. F. (1981). Melatonin in man. *Advances in the Biosciences* **29**, 371–81.

Matthews, S. A., Evans, K. L., Morgan, W. W., Petterborg, L. J., and Reiter, R. J. (1982). Pineal indoleamine metabolism in the cotton rat, *Sigmodon hispidus*: Studies on norepinephrine, serotonin, N-acetyltransferase and melatonin. In *The pineal and its hormones* (ed. R. J. Reiter), pp. 35–44. Alan R. Liss, New York.

McConnell, S. J. and Ellendorf, F. (1987). Absence of nocturnal plasma melatonin surge under long and short artificial photoperiods in the domestic sow. *Journal of Pineal Research* **4**, 201–10.

McCord, C. P. and Allen, F. B. (1917). Evidence associating pineal gland function with alterations in pigmentation. *Journal of Experimental Zoology* **23**, 207–24.

McIntyre, I. M., Norman, T. R., Burrows, G. D., and Armstrong, S. M. (1987). Melatonin rhythm in human plasma and saliva. *Journal of Pineal Research* **4**, 177–83.

Mefford, I. N. and Barchas, J. D. (1980). Determination of tryptophan and metabolites in rat brain and pineal tissue by reverse-phase high-performance liquid chromatography with electrochemical detection. *Journal of Chromatography* **181**, 187–93.

Miles, A. and Philbrick, D. R. S. (1988). Melatonin and psychiatry, a review. *Biological Psychiatry* **23**, 405–25.

Miles, A., Philbrick, D. R. S., Tidmarsh, S. F., and Shaw, D. M. (1985*a*). Direct radioimmunoassay of melatonin in saliva. *Clinical Chemistry* **31**, 1412–13.

Miles, A., Philbrick, D. R. S., Shaw, D. M., Tidmarsh, S. F., and Pugh, A. J. (1985*b*). Salivary melatonin estimation in clinical research. *Clinical Chemistry* **31**, 2041–2.

Miles, A., Philbrick, D. R. S., Thomas, D. R., and Grey, J. E. (1987). Diagnostic and clinical implications of plasma and salivary melatonin assay. *Clinical Chemistry* **33**, 1295–7.

Møller, M., Van Deurs, B., and Westergaard, E. (1978). Vascular permeability to proteins and peptides in the mouse pineal gland. *Cell and Tissue Research* **195**, 1–15.

Moore, D. C., Paunier, L., and Sizonenko, P. C. (1979). Effect of adrenergic stimulation and blockade on melatonin secretion in the human. *Progress in Brain Research* **52**, 517–21.

Moore, R. Y. (1973). Retinohypothalamic projection in mammals: A comparative study. *Brain Research* **49**, 403–9.

Moore, R. Y. and Klein, D. C. (1974). Visual pathways and the central neural control of a circadian rhythm in pineal serotonin N-acetyltransferase activity. *Brain Research* **71**, 17–33.

Morgan, W. W. and Hansen, J. T. (1978). Time course of disappearance of pineal noradrenaline following superior cervical ganglionectomy. *Experimental Brain Research* **32**, 329–34.

Mori, W. and Lerner, A. B. (1960). A microscopic bioassay for melatonin. *Endocrinology* **67**, 443–50.

Mullen, P. E., *et al.* (1981). Melatonin and 5-methoxytryptophol, the 24-hour pattern of secretion in man. *Advances in Bioscience* **29**, 337–40.

Namboodiri, M. A. A., Sugden, D., Klein, D. C., and Mefford, I. N. (1983). 5-Hydroxytryptophan elevates serum melatonin. *Science* **221**, 659–61.

Neff, N. H., Barrett, R. E., and Costa, E. (1969). Kinetics and fluorescent histochemical analysis of the serotonin compartments in the rat pineal gland. *European Journal of Pharmacology* **6**, 348–56.

Neuwelt, E. A. and Lewy, A. J. (1983). Disappearance of plasma melatonin after removal of a neoplastic pineal gland. *New England Journal of Medicine* **308**, 1132–5.

Nordlund, J. J. and Lerner, A. B. (1977). The effects of oral melatonin on skin colour and on the release of pituitary hormones. *Journal of Clinical Endocrinology and Metabolism* **45**, 768–74.

Nowak, R., McMillen, I. C., Redman, J., and Short, R. V. (1987). The correlation between serum and salivary melatonin concentrations and urinary 6-hydroxymelatonin sulphate excretion rates. *Clinical Endocrinology* **27**, 445–52.

O'Dea, R. F. and Zatz, M. (1976). Catecholamine-stimulated cyclic GMP production in the rat pineal: Apparent presynaptic site of action. *Proceedings of the National Academy of Science USA* **73**, 3398–402.

Oosthuizen, J. M. C., Bornman, M. S., and Schulenburg, G. W. (1986). Melatonin impairs sperm motility—a novel finding. *South African Medical Journal* **70**,

566.

Pang, S. F. and Ralph, C. L. (1975). Pineal and serum melatonin at midday and midnight following pinealectomy or castration in male rats. *Journal of Experimental Zoology* **183**, 275–80.

Pang, S. F., Yip, M. K., Lui, H. W., Brown, G. M., and Tsiu, H. W. (1980). Diurnal rhythms of immunoreactive N-acetylserotonin and melatonin in the serum of male rats. *Acta Endocrinologia* **95**, 571–6.

Parfitt, A. G. and Klein, D. C. (1976). Sympathetic nerve endings in the pineal gland protect against acute stress-induced increase in N-acetyltransferase activity. *Endocrinology* **99**, 480–6.

Pellegrino de Iraldi, A. and Rodriguéz de Lores Arnaiz, G. (1964). 5-Hydroxytryptophan decarboxylase activity in normal and denervated pineal glands of rats. *Life Science* **3**, 589–93.

Pevet, P. (1983). Is 5-methyoxytryptamine a pineal hormone? *Psychoneuroendocrinology* **8**, 61–73.

Quay, W. B. (1963). Effect of dietary phenylalanine and tryptophan on pineal and hypothalamic serotonin levels. *Proceedings of the Society for Experimental Biology and Medicine* **114**, 718–21.

Quay, W. B. (1964). Circadian and estrous rhythms in pineal melatonin and 5-hydroxyindole-3-acetic acid. *Proceedings of the Society for Experimental Biology and Medicine* **115**, 710–13.

Quay, W. B. (1965). Retinal and pineal hydoxyindole-O-methyltransferase activity in vertebrates. *Life Sciences* **4**, 983–91.

Quay, W. B. (1974). *Pineal chemistry.* C. C. Thomas Press, Springfield.

Quay, W. B. and Ma, Y. H. (1976). Demonstration of gastrointestinal-O-methyltransferase. *International Reports on Clinical Sciences, Medical Science* **4**, 563.

Reiter, R. J. (1980). The pineal and its hormones in the control of reproduction. *Endocrine Reviews* **1**, 109–31.

Reiter, R. J. (1981*a*). Reproductive effects of the pineal gland and pineal indoles in the Syrian hamster and the albino rat. In *The pineal gland, Vol II, Reproductive effects* (ed. R. J. Reiter), pp. 45–82. CRC Press, Boca Raton, Florida.

Reiter, R. J. (1981*b*). The mammalian pineal gland: structure and function. *American Journal of Anatomy* **162**, 287–313.

Reiter, R. J. (1982). Neuroendocrine effects of the pineal gland and of melatonin. In *Frontiers in neuroendocrinology Vol. 7* (ed. W. F. Ganong and L. Martini), pp. 287–316. Raven Press, New York.

Reiter, R. J. (1983). The role of light and age in determining melatonin production in the pineal gland. In *The pineal and its endocrine role* (ed. J. Axelrod, F. Fraschini, and G. P. Velo), pp. 227–41. Plenum Press, New York.

Reiter, R. J. (1984*a*). Pineal indoles: Production, secretion and actions. In *Neuroendocrine perspectives, Vol. 3* (ed. R. M. MacLeod and E. E. Müller), pp. 345–77. Elsevier Press, Amsterdam.

Reiter, R. J. (1984*b*). Pineal function in mammals including man. In *Diagnosis and treatment of pineal region tumours* (ed. E. A. Neuwelt), pp. 86–107. Williams and Wilkins Publishers, Baltimore.

Reiter, R. J. (1985). Action spectra, dose-response relationship, and temporal aspects of light's effects on the pineal gland. *Annals of the New York Academy of Sciences* **453**, 215–30.

Reiter, R. J. (1986*a*). Normal patterns of melatonin levels in the pineal gland and body fluids of humans and experimental animals. *Journal of Neural Transmission (Supplement)* **21**, 35–54.

Reiter, R. J. (1986*b*). Pineal function in the human: Implications for reproductive physiology. *Journal of Obstetrics and Gynaecology (Supplement 2)* **6**, S77–81.

Reiter, R. J. (1987). The melatonin message: Duration versus coincidence hypotheses. *Life Science* **40**, 2119–31.

Reiter, R. J. and Sorrentino, S., Jr (1970). Reproductive effects of the mammalian pineal. *American Zoologist* **10**, 247–58.

Reiter, R. J. and Sorrentino, S., Jr (1971*a*). Inhibition of luteinizing hormone release and ovulation in PMS-treated rat by peripherally administered melatonin. *Contraception* **4**, 385–92.

Reiter, R. J. and Sorrentino, S., Jr (1971*b*). Factors influential in determining the gonad-inhibiting activity of the pineal gland. In *The pineal gland* (ed. G. E. W. Wolstenholme and J. Knight), pp. 329–44. J. & A. Churchill Publishers, London.

Reiter, R. J., Vaughan, M. K., and Blask, D. E. (1975). Possible role of the cerebrospinal fluid in the transport of pineal hormones in mammals. In *Brain–endocrine interaction II. The ventricular system in neuroendocrine mechanisms* (ed. K. M. Knigge, D. E. Scott, and H. Kobayashi), pp. 337–54. Karger, Basel.

Reiter, R. J., Blask, D. E., Johnson, L. Y., Rudeen, P. K., Vaughan, M. K., and Waring, P. J. (1976). Melatonin inhibition of reproduction in the male hamster: Its dependency on time of day and on an intact and sympathetically innervated pineal gland. *Neuroendocrinology* **22**, 107–16.

Reiter, R. J., Rudeen, P. K., Sackman, J. W., Vaughan, M. K., Johnson, L. Y., and Little, J. C. (1977). Subcutaneous melatonin implants inhibit reproductive atrophy in male hamsters induced by daily melatonin injections. *Endocrine Research Communications* **4**, 35–44.

Reiter, R. J., Richardson, B. A., Johnson, L. Y., Ferguson, B. N., and Dinh, D. T. (1980). Pineal melatonin rhythm: Reduction in ageing Syrian hamsters. *Science* **210**, 1372–3.

Reiter, R. J., Richardson, B. A., and Hurlbut, E. C. (1981). Pineal, retinal and Harderian gland melatonin in a diurnal species, the Richardson's ground squirrel (*Spermophilus richardsonii*). *Neuroscience Letters* **22**, 285–8.

Reiter, R. J., King, T. S., Richardson, B. A., and Hurlbut, E. C. (1982*a*). Pineal melatonin levels in a diurnal species, the eastern chipmunk (*Tamias striatus*): Effects of light at night, propranolol administration or superior cervical ganglionectomy. *Journal of Neural Transmission* **54**, 275–84.

Reiter, R. J., Vriend, J., Brainard, G. C., Matthews, S. A., and Craft, C. M. (1982*b*). Reduced pineal and plasma melatonin levels and gonadal atrophy in old hamsters kept under long photoperiods. *Experimental Aging Research* **8**, 27–30.

Reiter, R. J., Hurlbut, E. C., Brainard, G. C., Steinlechner, S., and Richardson, B. A. (1983*a*). Influence of light irradiance on hydroxyindole-O-methyltransferase activity, serotonin-N-acetyltransferase activity, and radioimmunoassayable melatonin levels in the pineal gland of the diurnally active Richardson's ground squirrel. *Brain Research* **288**, 151–7.

Reiter, R. J., Richardson, B. A., Matthews, S. A., Lane, S. J., and Ferguson, B. N. (1983*b*). Rhythms in immunoreactive melatonin in the retina and Harderian glands of rats: Persistence after pinealectomy. *Life Science* **32**, 1229–36.

Reiter, R. J., Esquifino., A. I., Champney, T. H., Craft, C. M., and Vaughan, M. K. (1985*a*). Pineal melatonin production in relation to sexual development in the male rat. In *Paediatric neuroendocrinology* (ed. D. Gupta, P. Borrelli, and A. Attanasio), pp. 190–202. Croom-Helm Publishers, London.

Reiter, R. J., Steinlechner, S., and Richardson, B. A. (1985*b*). Daytime and nighttime pineal N-acetyltransferase activity and melatonin content in male rats treated with indomethacin, a prostaglandin synthesis inhibitor. *Neuroendocrinology Letters* **7**, 281–7.

Reiter, R. J., Britt, J. H., and Armstrong, J. D. (1987*a*). Absence of a nocturnal rise in either NE, NAT, HIOMT or melatonin in the pineal gland of the domestic pig kept under natural environmental conditions. *Neuroscience Letters* **81**, 171–6.

Reiter, R. J., Vaughan, G. M., Oaknin, S., Troiani, M. E., Cozzi, B., and Li, K. (1987*b*). Norepinephrine or isoproterenol stimulation of pineal N-acetyltransferase activity and melatonin content in the Syrian hamster is restricted to the second half of the daily dark phase. *Neuroendocrinology* **45**, 249–56.

Reppert, S. M., Perlow, M. J., and Klein, D. C. (1980). Cerebrospinal fluid melatonin. In *Neurobiology of cereobrospinal fluid, vol. 1* (ed. J. H. Wood), pp. 579–89. Plenum Publishers, New York.

Ritta, M. N. and Cardinali, D. P. (1980). Effect of indomethacin on monamine metabolism and melatonin synthesis in rat pineal gland. *Hormone Research* **12**, 305–12.

Ritta, N. M. and Cardinali, D. P. (1982). Involvement of α-adrenoreceptors in norepinephrine-induced prostaglandin E_2 release by rat pineal *in vitro*. *Neuroscience Letters* **31**, 307–11.

Rollag, M. D., Morgan, R. J., and Niswender, G. D. (1978). Route of melatonin secretion in sheep. *Endocrinology* **102**, 1–8.

Rollag, M. D., Panke, E. S., Trakulrungsi, W., Trakulrungsi, C., and Reiter, R. J. (1980). Quantification of daily melatonin synthesis in the hamster pineal gland. *Endocrinology* **106**, 231–6.

Rosengarten, H., Meller, E., and Friedhoff, A. J. (1972). *In vitro* enzymatic formation of melatonin by human erythrocytes. *Research Communications in Chemical Pathology and Pharmacology* **4**, 457–65.

Rudeen, P. K., Reiter, R. J., and Vaughan, M. K. (1975). Pineal serotonin N-acetyltransferase in four mammalian species. *Neuroscience Letters* **1**, 225–9.

Sack, R. L., Lewy, A. J., Erb, D. L., Vollmer, W. M., and Singer, C. M. (1986). Human melatonin production decreases with age. *Journal of Pineal Research* **3**, 379–88.

Shaw, K. M. (1977). Hypothalamo–pituitary–adrenal function in Parkinsonism patients treated with melatonin. *Current Medical Research Opinion* **4**, 743–6.

Shein, H. M. and Wurtman, R. J. (1971). Stimulation of [^{14}C] tryptophan

5-hydroxylation by norepinephrine and dibutyryladenosine 3′,5′-monophosphate in rat pineal organ cultures. *Life Science* **10**, 935–40.

Shibuya, H., Toru, M., and Watanabe, S. (1978). A circadian rhythm of tryptophan hydroxylase in rat pineal. *Brain Research* **138**, 364–8.

Sitaram, B. R. and Lees, G. J. (1978). Diurnal rhythm and turnover of tryptophan hydroxylase in the pineal gland of the rat. *Journal of Neurochemistry* **31**, 1021–6.

Smith, J. A., Mee, J. J. X., Barnes, N. D., Thorbum, R. J., and Barnes, J. L. C. (1976). Melatonin in serum and cerebrospinal fluid. *Lancet* **ii**, 425.

Smith, J. A., Barnes, J. L. C., and Mee, T. J. X. (1979). The effect of neuroleptic drugs on serum and cerobrospinal fluid melatonin concentrations in psychiatric subjects. *Journal of Pharmacy and Pharmacology* **31**, 246–8.

Smith, J. A., Mee, T. J., Padwick, D. P., and Spokes, E. G. (1981*a*). Human post mortem pineal enzyme activity. *Clinical Endocrinology* **14**, 75–81.

Smith, J. A., O'Hara, J. and Schiff, A. A. (1981*b*). Altered diurnal serum melatonin rhythm in blind men. *Lancet* **ii**, 933.

Smith, J. A., Padwick, D. P., and Spokes, E. G. (1981*c*). Annual bimodal variation in human hydroxyindole-O-methyltransferase activity. *Advances in the Biosciences* **29**, 197–9.

Snyder, S. H. and Axelrod, J. (1965). Circadian rhythm in serotonin: Effect of monamine oxidase inhibition and reserpine. *Science* **149**, 242–4.

Snyder, S. H., Zweig, M., Axelrod, J., and Fischer, J. E. (1965). Control of the circadian rhythm in serotonin content of the rat pineal gland. *Proceedings of the National Academy of Science USA* **53**, 301–5.

Snyder, S H., Axelrod, J., and Zweig, M. (1967). Circadian rhythm in the serotonin content of the rat pineal gland: Regulating factors. *Journal of Pharmacology and Experimental Therapeutics* **158**, 206–13.

Steinlechner, S., Steger, R. W., King, T. S., and Reiter, R. J. (1983). Diurnal variation in the serotonin content and turnover in the pineal gland of the Syrian hamster. *Neuroscience Letters* **35**, 167–72.

Steinlechner, S., King, T. S., Champney, T. H., Spanel-Borowski, K., and Reiter, R. J. (1985*a*). Comparison of the effects of β-adrenergic agents on pineal serotonin N-acetyltransferase activity and melatonin content in two species of hamsters. *Journal of Pineal Research* **1**, 23–30.

Steinlechner, S., King, T. S., Champney, T. H., Richardson, B. A., and Reiter, R. J. (1985*b*). Pharmacological studies on the regulation of N-acetyltransferase activity and melatonin content of the pineal gland of the Syrian hamster. *Journal of Pineal Research* **2**, 109–20.

Stetson, M. H. and Watson-Whitmyre, M. (1984). Physiology of the pineal gland and its hormone melatonin in annual reproduction in rodents. In *The pineal gland* (ed. R. J. Reiter), pp. 109–54. Raven Press, New York.

Sugden, D. (1979). Circadian change in rat pineal tryptophan content: Lack of correlation with serum tryptophan. *Journal of Neurochemistry* **33**, 811–13.

Sugden, D. and Klein, D. C. (1983). Regulation of rat pineal hydroxyindole-O-methyltransferase in neonatal and adult rats. *Journal of Neurochemistry* **40**, 1647–53.

Sugden, D., Namboodiri, M. A. A., Klein, D. C., Pierce, J. E., Grady, R., Jr, and

Mefford, I. N. (1985). Ovine pineal α_1-adrenoceptor: Characterization and evidence for a functional role in the regulation of serum melatonin. *Endocrinology* **116**, 1960–7.

Swanson, L. W. and Kuypers, H. G. J. M. (1980). The paraventricular nucleus of the hypothalamus: Cytoarchitectonic subdivisions and organization of projections to the pituitary, dorsal vagal complex, and spinal cord as demonstrated by retrograde fluorescence double-labelling methods. *Journal of Comparative Neurology* **194**, 555–70.

Tamarkin, L., Westrom, W. K., Hamill, A. I., and Goldman, B. D. (1976). Effect of melatonin on the reproductive system of male and female hamsters: A diurnal rhythm in sensitivity to melatonin. *Endocrinology* **99**, 1534–41.

Tamarkin, L., Baird, C. J., and Almeida, O. F. X. (1985). Melatonin: A coordinating signal for mammalian reproduction? *Science* **227**, 714–20.

Tan, C. H. and Khoo, J. C. (1981). Melatonin concentrations in human serum, ventricular and lumbar cerebrospinal fluids as an index of the secretory pathway of the pineal gland. *Hormone Research* **14**, 224–33.

Tetsuo, M., Markey, S. P., and Kopin, I. J. (1980). Measurement of 6-hydroxy-melatonin in urine and its diurnal variation. *Life Science* **27**, 105–9.

Tetsuo, M., Markey, S. P., Colburn, R. W., and Kopin, I. J. (1981a). Quantitative analysis of 6-hydroxymelatonin in human urine by gas chromatography negative chemical ionization mass spectrometry. *Analytical Biochemistry* **110**, 208–15.

Tetsuo, M., Polinsky, R. J., Markey, S. P., and Kopin, T. J., (1981b). Urinary 6-hydroxymelatonin excretion in patients with orthostatic hypotension. *Journal of Clinical Endocrinology and Metabolism* **53**, 607–10.

Vakkuri, O. (1985). Diurnal rhythm of melatonin in human saliva. *Acta Physiologica Scandanavica* **23**, 151–4.

Vaughan, G. M. (1984). Melatonin in humans, *Pineal Research Reviews* **2**, 141–201.

Vaughan, G. M. and Reiter, R. J. (1986). Pineal dependence of the Syrian hamster's nocturnal serum melatonin surge. *Journal of Pineal Research* **3**, 9–14.

Vaughan, G. M. and Reiter, R. J. (1987). The Syrian hamster pineal gland responses to isoproterenol *in vivo* at night. *Endocrinology* **120**, 1682–4.

Vaughan, G. M., *et al.* (1976). Nocturnal elevation of plasma melatonin and urinary 5-hydroxyindole acetic acid in young men: Attempts at modification by brief changes in environmental lighting and sleep and by brief changes in environmental lighting and by autonomic drugs. *Journal of Clinical Endocrinology and Metabolism* **42**, 752–64.

Vaughan, G. M., Allen, J. P., Tullis, W., Siler-Khodr, T. M., de la Pena, A., and Sackman, J. W. (1978a). Overnight profiles of melatonin and certain adeno-hypophyseal hormones in men. *Journal of Clinical Endocrinology and Metabolism* **47**, 566–71.

Vaughan, G. M., *et al.* (1978b). Melatonin concentration in human blood and cerebrospinal fluid: Relationship to stress. *Journal of Clinical Endocrinology and Metabolism* **47**, 220–3.

Vaughan, G. M., Allen, J., and de la Pena, A. (1979a). Rapid melatonin transients. *Waking and Sleeping* **3**, 169–79.

Vaughan, G. M., McDonald, S. D., Jordan, R. M., Allen, J. P., Bell, R., and Stevens, E. A. (1979*b*). Melatonin, pituitary function and stress in humans. *Psychoneuroendocrinology* **4**, 351–62.

Vaughan, G. M., Taylor, T. J., Pruitt, B. A., Jr, and Mason, A. D., Jr (1985). Pineal function in burns: melatonin is not a marker for general sympathetic activity. *Journal of Pineal Research* **2**, 1–12.

Vaughan, G. M., Lasko, J., Coggins, S. H., Pruitt, B. A., Jr, and Mason, A. D., Jr (1986). Rhythmic melatonin response of the Syrian hamster gland to norepinephrine *in vitro* and *in vivo*. *Journal of Pineal Research* **3**, 235–50.

Vaughan, L. G., Harris, S., Allen, J., and Delea, C. (1982). Human immunoreactive melatonin and cortisol during acute stress and comparison of their rhythms. In *Biological markers in psychiatry and neurobiology* (ed. E. Usdin and I. Hanin), pp. 317–30. Pergamon Press, Oxford.

Vollrath, L. (1984). Functional anatomy of the human pineal gland. In *The pineal gland* (ed. R. J. Reiter), pp. 285–322. Raven Press, New York.

Waldhauser, F., Lynch, H. J., and Wurtman, R. J. (1984). Melatonin in human body fluids: Clinical significance. In *The pineal gland* (ed. R. J. Reiter), pp. 345–70. Raven Press, New York.

Webb, S. M., Champney, T. H., Lewinski, A. K., and Reiter, R. J. (1985). Photoreceptor damage and eye pigmentation: Influence on the sensitivity of rat pineal N-acetyltransferase activity and melatonin to light at night. *Neuroendocrinology* **40**, 205–9.

Weinberg, U., D'Eletto, R. D., Weitzman, E. D., Erlich, S., and Hallander, C. S. (1979). Circulating melatonin in man: episodic secretion throughout the light–dark cycle. *Journal of Clinical Endocrinology and Metabolism* **48**, 114–18.

Weinberg, U., Weitzman, E. D., Fukushima, D. K., Cancel, G. F., and Rosenfeld, R. S. (1980). Melatonin does not suppress the pituitary luteinizing hormone response to luteinizing hormone-releasing hormone in men. *Journal of Clinical Endocrinology and Metabolism* **51**, 161–2.

Weissbach, H., Redfield, B. G., and Axelrod, J. (1961). The enzymatic acteylation of serotonin and other naturally occurring amines. *Biochimica Biophysica Acta* **54**, 190–2.

Werner, S., Brismar, K., Wetterberg, L., and Everoth, P. (1980). Circadian rhythms of melatonin, prolactin, growth hormone and cortisol in patients with pituitary adenomas, empty sella syndrome and Cushing's syndrome due to adrenal tumors. In *Melatonin: current status and perspectives* (ed. N. Birau and W. Schloot), pp. 357–63. Pergamon Press, Oxford.

Wetterberg, L. (1978). Melatonin in humans: Physiological and clinical studies. *Journal of Neural Transmission (Supplement)* **13**, 289–310.

Wetterberg, L. (1979). Clinical significance of melatonin. *Progress in Brain Research* **52**, 539–47.

Wetterberg, L. (1982). Psychiatric aspects of pineal function. In *The pineal gland, Vol. III, Extra-reproductive effects* (ed. R. J. Reiter), pp. 219–28. CRC Press, Boca Raton, Florida.

Wetterberg, L., Arendt, J., Paunier, L., Sizonenko, P. C., van Donselaar, W., and Heyden, T. (1976). Human serum melatonin changes during the menstrual cycle. *Journal of Clinical Endocrinology and Metabolism* **42**, 185–8.

Wetterberg, L., *et al.* (1981). Circadian rhythmic urinary melatonin excretion in four seasons by clinically healthy Japanese subjects in Kyushu. *Chronobiologia* **8**, 188–9.

Wiechmann, A. F. (1986). Melatonin: Parallels in pineal gland and retina. *Experimental Eye Research* **42**, 507–27.

Wilkinson, M., Arendt, J., Bradtke, J., and de Ziegler, D. (1977). Determination of a dark-induced increase of pineal N-acetyltransferase activity and simultaneous radioimmunoassay of melatonin in pineal serum and pituitary tissue of the male rat. *Journal of Endocrinology* **72**, 243–4.

Wilson, B. W., Snedden, W., Silman, R. E., Smith, I., and Mullen, P. (1977). A gas chromatography-mass spectromety method for the quantitative analysis of melatonin in plasma and cerebrospinal fluid. *Analytical Biochemistry* **81**, 283–91.

Wurtman, R. J. and Lieberman, H. R. (1985). Melatonin secretion as a mediator of circadian variations in sleep and sleepiness. *Journal of Pineal Research* **2**, 301–3.

Wurtman, R. J. and Moskowitz, M. A. (1978). The pineal organ. *New England Journal of Medicine* **293**, 1329–33.

Wurtman, R. J., Axelrod, J., and Potter, L. T. (1964). The uptake of ^{3}H-melatonin in endocrine and nervous tissues and the effects of constant light exposure. *Journal of Pharmacology and Experimental Therapeutics* **143**, 314–19.

Wurtman, R. J., Shein, H. M., Axelrod, J., and Larin, F. (1969). Incorporation of ^{14}C-protein by cultured rat pineals: Stimulation by 1-norepinephrine. *Proceedings of the National Academy of Sciences USA* **62**, 749–55.

Wurzburger, R. J., Kawashima, K., Miller, R. L., and Spector, S. (1976). Determination of rat pineal gland melatonin content by radioimmunoassay. *Life Science* **18**, 867–78.

Young, S. N. and Anderson, G. M. (1982). Factors influencing melatonin, 5-hydroxytryptophan, 5-hydroxyindoleacetic acid, 5-hydroxytryptamine and tryptophan in rat pineal gland. *Neuroendocrinology* **35**, 464–8.

Younglai, E. V. (1979). *In vitro* effects of melatonin on hCG of steroid accumulation by rabbit ovarian follicles. *Journal of Steroid Biochemistry* **10**, 714–18.

Yuwiler, A. (1983). Vasoactive intestinal peptide stimulation of pineal serotonin-N-acetyl transferase activity: General characteristics. *Journal of Neurochemistry* **41**, 146–53.

Zatz, M. (1981). Pharmacology of the rat pineal gland. In *The pineal gland, Vol. I. Anatomy and biochemistry* (ed. R. J. Reiter), pp. 225–42. CRC Press, Boca Raton, Florida.

Zatz, M. (1985). Phorbol esters mimic α-adrenergic potentiation of serotonin N-acetyltransferase induction in the rat pineal. *Journal of Neurochemistry* **45**, 637–9.

Zweig, M. and Axelrod, J. (1969). Relationship between catecholamines and serotonin in sympathetic nerves of the rat pineal gland. *Journal of Neurobiology* **1**, 87–97.

2. Melatonin and the human circadian system

Josephine Arendt

Introduction

Circadian rhythms have often been regarded as an obscure area of physiology of little importance in clinical practice. The extensive investigations into the properties of human circadian rhythms that have been accomplished in a number of different centres (e.g. Aschoff 1981; Wever 1979; Minors and Waterhouse 1981) during the last 25 years are little known outside the discipline of Chronobiology. Recently, however, biological rhythms have become of increasing clinical (and commercial) interest. This is due in no small measure to our increasing knowledge of the function of the pineal gland, the rhythmic production of its hormone melatonin, and the effects of melatonin in animals and man.

Areas of clinical medicine where circadian (and in some cases seasonal) rhythms are of importance are beginning to be defined. They include affective disease, cancer, asthma, cardiovascular disease, and chronopharmacology (time-of-day effects in drug treatment and pharmacokinetics). Other chapters in this volume will enlarge on some of these considerations. This chapter reviews recent knowledge of the relationship of melatonin to the human circadian system and suggests some areas where it may prove to be of therapeutic benefit.

Melatonin and the human circadian system

The circadian nature and control of the human melatonin rhythm: effects of light

The marked 24-hour variation of human plasma and urinary melatonin, with high values during the dark phase, has been known to exist for some considerable time (Pelham *et al.* 1973; Lynch *et al.* 1975; Arendt *et al.* 1975, 1977). In normal environmental conditions peak levels are found around 02.00 to 04.00 hours (Fig. 2.1). More recently the major melatonin metabolite 6-sulphatoxymelatonin has been shown to have a marked plasma and urinary rhythm, correlating closely with melatonin values (Arendt *et al.* 1985*a*).

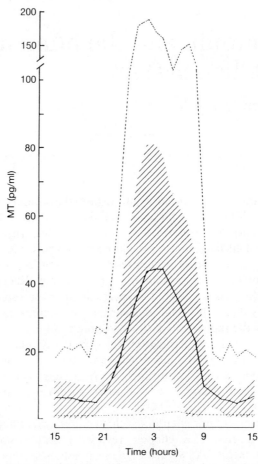

Fig. 2.1. Normal levels of human plasma melatonin in samples taken at hourly intervals throughout 24 hours. Mean levels are shown ($n=38$). The hatched area encompasses 1 standard deviation. Dotted lines indicate the range. Melatonin was measured by direct radioimmunoassay.

In animals the melatonin rhythm is known to be endogenously generated probably in the supra-chiasmatic nucleus of the hypothalamus (Moore and Klein 1974)*. It is both entrained and suppressed by light of suitable spectral composition and intensity. Cool, white fluorescent, and Vitalite light sources are commonly employed. Light wavelengths found to be most efficient in suppressing melatonin correspond both in animals and in humans to the rhodopsin absorption spectrum (Reiter 1985; Brainard *et al.* 1985).

*See Reiter, this volume, for review.

The light intensity required for suppression of human melatonin at night was originally reported to be 2500 lux (Lewy *et al.* 1980). More recently, we (Bojkowski *et al.* 1987) found that most people will respond by partial suppression of melatonin with 300 lux for 30 minutes at 00.30 hours.

The endogenous nature of the human melatonin rhythm was suggested by early work of Weitzman and colleagues (1978), who studied urinary melatonin rhythms in an environment free from time cues. The melatonin rhythm was clearly dissociated from the sleep–wake cycle in these experiments. Recently, Wever (Wever *et al.* 1988; Wever 1986) using 6-sulphatoxymelatonin as an index of melatonin secretion has made extensive investigations of the melatonin rhythm in an environment free from natural time cues but with different manipulations of the light–dark cycle. By imposing via the light–dark cycle, day lengths extending 10 minutes per day from 26 to 29 hours the circadian system undergoes fractional desynchronization (Wever 1979). The sleep–wake cycle remains entrained, but other different components lose entrainment and free-run with their own endogenous periodicity above 'day lengths' of around 27 hours. The calculated free-running periodicity of the melatonin rhythm in these circumstances is shown for one subject in Fig. 2.2. In general, in constant conditions the period of the melatonin rhythm is close to 25 hours, similar to that of other physiological variables. Even in constant bright light (≈ 3000 lux) the melatonin rhythm is still detectable as urinary 6-sulphatoxymelatonin free-running with a tau of ~ 25 hours (Wever 1986).

Wever (1986) has presented powerful evidence that melatonin is relatively unaffected by other circadian components such as the sleep–wake cycle or temperature, i.e. the rhythm is not 'masked'. Apart from the well-known suppression by bright light it is therefore an exceedingly good marker for a central circadian rhythm generating system. Coupled with the usually high-amplitude rhythm (on average 10–20-fold higher at night than during the day) and the reliability of assay methods (Arendt 1986*b*; these observations emphasize the importance of melatonin as a rhythm marker in human health and in disease. There have been a number of reports linking the melatonin and cortisol rhythms. For example, in depression Wetterberg and associates (1982) suggest that high cortisol levels are associated with low melatonin. Be that as it may, the two rhythms re-entrain at different rates after desynchronization following time-zone flight (Fevre-Montange *et al.* 1981) and thus any causal links are not likely to be very strong. The possible relationship of melatonin and cortisol secretion is reviewed later in this volume by Dr Ursula Lang.

The light–dark cycle has only recently been recognized as a major

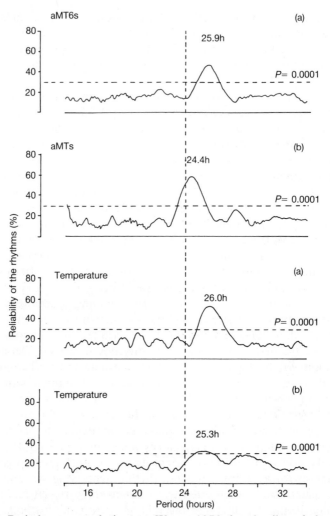

Fig. 2.2. Periodogram analysis (see Wever 1979 for details) of the urinary 6-sulphatoxymelatonin (aMT6s) rhythm and core body temperature rhythm in one volunteer in environmental isolation subjected to increasing imposed day-length (26–29 hours, 500 lux light–dark cycle). Cycle lengths up to 27 hours, (a) and from 27 hours to 29 hours, (b) are grouped separately. Note that the average period of the aMT6s rhythm (25.9 hours) in (a) indicates entrainment to the zeitgeber (26–27 hours imposed) and that the average period of the rhythm (24.4 hours) in (b) indicates desynchronization from the zeitgeber (27–29 hours imposed). The temperature rhythm behaved similarly, but in addition to the free-running period of 25.3 hours an entrained component is seen in the 27–29 hours section. Reliability of the rhythm > 30 per cent is equivalent to $P > 0.0001$ (Wever 1979). From Arendt *et al.* 1985, with acknowledgements to the CIBA Foundation.

synchronizer (or zeitgeber) for the entrainment of human circadian rhythms (Wever *et al.* 1983; Czeisler *et al.* 1986). Specifically bright light (2500 lux) is required to show strong entraining effects (Wever *et al.* 1983).

These experiments were performed subsequent to the original demonstration that only bright light would completely suppress human melatonin (Lewy *et al.* 1980) and indeed were conceived in view of the apparent lack of 'melatonin sensitivity' to dim light. It is not, therefore, unreasonable to suggest that the strong entraining effects of bright light may depend in part on this concomitant acute suppression of melatonin production. Thus, one might suggest that melatonin itself may have modulatory effects on circadian entrainment in man as will be discussed.

The precise intensity of light required to entrain melatonin itself has not, as yet, been defined. It is not necessarily the same as that required for suppression in a given individual, indeed, the two systems may use different photoreceptor mechanisms (Takahashi *et al.* 1984). Moreover, as with other circadian variables, it is likely that multiple zeitgebers (e.g. social cues, clock-time, temperature) working together with the light–dark cycle maintain synchrony under normal environmental conditions. If one, or more of these factors is weakened or absent, then others may acquire a predominant role. This statement is substantiated by experiments in environments free of natural time cues, where a simple 'gong' signal is sufficient to entrain all circadian rhythms (Wever 1979).

Antarctica provides an interesting environment in which to study light effects on melatonin and other circadian rhythms. On the British Antarctic Base at Halley (75°s) the sun does not rise for 3 months during the winter and does not set for 3 months during the summer. The only light source in winter is that of the base lighting with a maximum intensity of 500 lux. The melatonin rhythm appears entrained to 24 hours in all seasons (Fig. 2.3), but the acrophase (time of peak levels) is later with reference to clock-time in winter than at any time of the year in temperate zones (Broadway *et al.* 1987). This later peak time suggests a weakening of zeitgeber input (i.e. low light intensity). It is probable that synchronization with 24 hours is maintained by social cues—base regimes are highly structured partly in order to maintain social synchrony.

Another Antarctic observation underlining the importance of the light–dark cycle in maintaining the melatonin rhythm concerned only one subject. Following night-shift work in winter this young man had disturbed sleep and a desynchronized melatonin rhythm for at least 3 weeks, whereas in spring recovery from night shift took less than a week. In spring, access to natural sunlight is likely to be the strong entraining factor (Broadway and Arendt 1986).

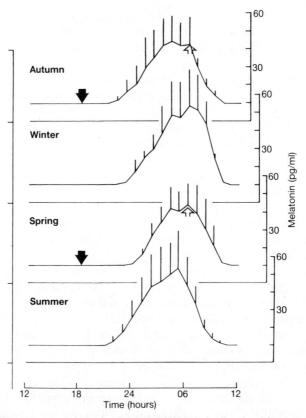

Fig. 2.3. The plasma melatonin rhythm ($\bar{x}$ SEM) in five healthy men living on a British Antarctic Base (Halley, 75°S) at the solstices and equinoxes. Light intensity does not exceed 500 lux for 3 months around the winter solstice and the sun does not set for 3 months around the summer solstice. Arrows indicate sunrise and sunset at the equinoxes. Note the delayed peak times, particularly in winter. (From Broadway *et al.* 1987, with permission.)

The application of 1 hour of bright (2500 lux) light in the morning (08.00–09.00 hours) and evening (19.30–20.30 hours) for 6 weeks following the winter solstice in Antarctica induced a marked phase advance of the melatonin rhythm, to a phase-position corresponding to that found in summer (Fig. 2.4; Arendt and Broadway 1986; Broadway *et al.* 1987). Hence, 2500 lux is sufficient to entrain melatonin in a social, but light-deprived environment and seasonal changes in the melatonin rhythm (Illnerova *et al.* 1985; Broadway *et al.* 1987) are likely to be day length dependent, as in animals (Tamarkin *et al.* 1985).

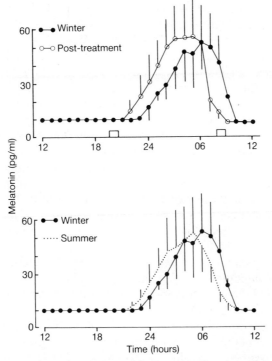

Fig. 2.4. Phase-advance of the melatonin rhythm in five men living on a British Antarctic Base (Halley, 75°S) after 1 hour of bright (>2500 lux) light morning and evening for 6 weeks following the winter solstice, and in summer. (●——●) Pretreatment levels at the winter solstice, (O——O) post-bright light treatment, (........) summer values, all mean values, error bars are left out for clarity. ANOVA gave highly significant differences in raw data and peak times between pretreatment (winter) and both post-bright light and summer values. Time of light treatment is shown by the boxes. Dim light (<500 lux) had no effect. (From Broadway *et al.* 1987, with permission.)

Effects of melatonin on circadian systems

Animal work

It has previously been noted that circadian entrainment by bright light probably suggests a modulatory role for melatonin in this function. Photoperiodic seasonal breeders read day length through the changing duration of night time melatonin and in this sense melatonin is seen as 'chemical darkness'. Suitably administered, it will act as a synchronizer of seasonal cycles, notably in reproduction (Arendt 1986a). Whether it will act as a circadian zeitgeber has been the subject of much discussion and some experimentation. In animals both light and dark pulses applied at an appropriate subjective circadian time will induce phase changes in circadian rhythms (Boulos and Rusak 1982; Pittendrigh and Daan 1976). It is possible to imagine that melatonin, administered at an appropriate time could act as a 'chemical dark pulse' on the circadian system.

In general, the evidence for an effect of melatonin on circadian rhythms concerns pharmacological dose levels. Probably, the most convincing work in rodents is that of Redman and colleagues (1983). These workers administered melatonin (1 mg/kg) by injection to free-running rats and observed entrainment of rest–activity cycles from when the injection time corresponded with activity onset. Saline injections, and injections of arginine vasotocin and MSH were in general ineffective. The dose level used in these experiments is many times the physiological production of melatonin. A most important comment by the authors was, however, that the effects of melatonin in humans are likely to be time-dependent, as in animals.

A more complex approach was described by Murakami and associates (1983) where implantation of melatonin in the supra-chiasmatic nucleus of rats hastened the re-entrainment of cortisol rhythms following a phase shift. Very much earlier, Quay (1970) reported that pinealectomized rats re-entrain faster following a phase-shift of the light–dark cycle, again suggesting an effect of pineal products on rate of re-entrainment.

Following studies on the entrainment of free-running rhythms, Armstrong and Redman (1985) have reported that melatonin will alter the direction of re-entrainment in the rat. Recently, they describe the very important finding that injections of relatively low amounts of melatonin at 9, 10, and 11 hours after onset of subjective day (but not at other times) will phase-advance the rest–activity cycle in rats, (Armstrong and Chesworth 1987). All of these experiments are of practical relevance in that, should it prove possible to manipulate human circadian rhythms in this way, a variety of conditions including jet-lag, shift work related illness, blindness and other states, associated with disturbed circadian rhythms, may be amenable to treatment.

Human work

Melatonin has been administered to humans for a variety of reasons. A concise review of the clinical pharmacology of melatonin was published by Lerner and Nordland in 1978 in which they conclude that a more extensive investigation of its pharmacological properties was desirable. Amongst the early work very large doses (up to 6.6 g/day) were used in attempts to alleviate the symptoms of Parkinson's disease (Cotzias *et al.* 1971). Although the initial report was promising, subsequent results were not consistent (Shaw *et al.* 1973). Another notable study by Carmen and colleagues (1976) concerned the administration of melatonin in large doses (up to 1.6 g/day) to depressive patients during the day. In this study the psychosis of all patients was exacerbated. Clearly, if the effects of melatonin are in part related to circadian timing systems, the administration of large amounts during the day could well induce disruption of circadian rhythms. In a depressive population where abnormal circadian rhythm relationships are common (Wehr and Goodwin 1981) it is perhaps not surprising that such treatment was counter-productive.

In view of the marked effects of melatonin on the reproductive systems of photoperiodic seasonal breeders (Tamarkin *et al.* 1985; Arendt 1986*a,b*), a number of authors have investigated the response of human reproductive hormones to both acute and chronic administration. In general, very little response has been found although a slight decrease in LH and FSH levels was found after giving 1 g per day (Nordlund and Lerner 1977) and increases in prolactin have been reported after acute dosing (Waldhauser *et al.* 1987).

The most consistent property of pharmacological doses of melatonin in both man and animals is undoubtedly that of inducing sommnolence or sleep (Cramer *et al.* 1974; Vollrath *et al.* 1981). This seems to be a time-of-day dependent effect in rats. Extensive investigations in rodents have also shown that it has very low toxicity, coupled with sedative, anti-convulsant, and analgesic properties (Sugden 1983). Bearing in mind the potential use of melatonin in the treatment of sleep disturbance and possibly other circadian rhythm problems, we have undertaken a series of investigations directed towards assessment of the chronic effects and tolerance of low-dose, melatonin administration in humans. Healthy volunteer subjects took either 2 mg melatonin in corn oil or identical vehicle daily at 17.00 hours in a double-blind cross-over design for 3–4 weeks. The experiment was performed twice, once in spring and once in autumn. A large number of variables were recorded including daily sleep logs, mood, and fatigue self-ratings several times daily, and 24-hour profiles of a number of anterior pituitary hormones, cortisol and melatonin itself, taken pre-treatment and both on the penultimate day of treatment (autumn), and the day following the last treatment (spring).

Essentially, these two studies confirmed the effects of melatonin on fatigue or sleepiness (Arendt *et al.* 1984, 1985*b*). An increase in early-evening fatigue was noted in both studies (Fig. 2.5), which was slow in onset, becoming significant in the group as a whole after 4 days, and clearly intermittent in some individuals (Fig. 2.6). Peak tiredness did not correspond with peak plasma melatonin levels, and these observations are consistent with an effect of melatonin on the timing of fatigue rather than an acute hypnotic effect. The intermittent effect in some individuals

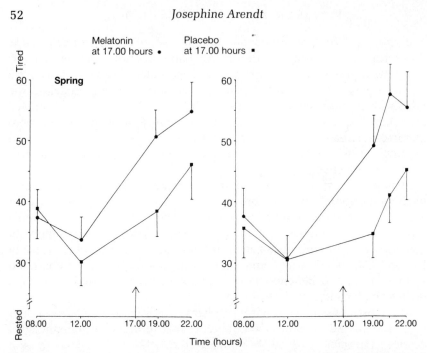

Fig. 2.5. Increase in evening 'fatigue' during chronic treatment of 12 normal volunteers (10M, 2F) with 2 mg melatonin at 1700 hours daily for 3–4 weeks compared to placebo. Data are shown for experiments in both spring (left) and autumn (right). Fatigue was rated as 'rested'=0, tired=100, using a 10 cm visual analogue scale. (From Arendt *et al.* 1985 with acknowledgements to the CIBA Foundation.)

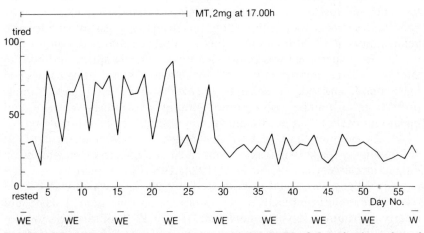

Fig. 2.6. Intermittent early evening fatigue, rated as in Fig. 2.5 during ingestion of 2 mg melatonin daily at 1700 hours compared to placebo in one subject (male). The response suggests relative co-ordination to a weak synchronizer.

could be explained on the basis of relative co-ordination to a weak zeitgeber (Wever 1979). In addition to this effect on fatigue, a modification in the timing of prolactin secretion was apparent, with an earlier decline from the night-time peak in melatonin-treated subjects (Wright *et al.* 1986). Any effects on prolactin may, of course, be subsequent to sleep changes. Most interesting of the circadian effects in these studies was the response of the endogenous melatonin rhythm itself. In the presence of an exogenous dose of melatonin at 17.00 hours, the endogenous rhythm was phase-advanced in those individuals where it was possible to differentiate exogenous and endogenous components (Fig. 2.7). As melatonin is considered an excellent marker of the circadian clock, this again indicates a probable effect on a central rhythm-generating system.

It is important to note that in spite of the previously noted observed deleterious effects of melatonin on depression (Carmen *et al.* 1976) these normal volunteers had no significant changes in mood. Likewise, LH, testosterone, growth hormone, thyroxine, and cortisol were unaffected. The effects of chronic low-dose treatment at other times of day have yet to be assessed. In general, melatonin was well-tolerated.

Although there is as yet no very striking evidence linking the pineal with the mammalian circadian clock, the experimental examples cited here cumulatively suggest that modulatory effects do exist. Indeed, it would be surprising if there were not some circadian functions related to the pineal in view of its primary importance in the generation or the synchronization of circadian rhythms in some birds and lower vertebrates (Underwood 1977; Menaker *et al.* 1981).

Studies on 'jet-lag'

Preliminary experiments in environmental isolation have attempted to investigate directly the zeitgeber properties of melatonin in man. Initial work suggested that melatonin could reinforce the light–dark cycle in the entrainment of the alertness rhythm under conditions of fractional desynchronization (Arendt *et al.* 1985*a,b*) without effect on other rhythms. Subsequent analysis of the same data but with a larger number of controls showed no significant differences (Wever 1986). Likewise, two simulated jet-lag studies (6-hour phase-shift westwards) performed both with placebo and melatonin taken at lights-off, in a blind, cross-over design, failed to show a significant effect of melatonin on the rate of re-entrainment of different circadian variables (Wever 1986).

These negative results clearly do not reinforce the hypothesis that melatonin has circadian effects in man. However, several phenomena associated with the phenomenon of jet-lag are not observed in simulation experiments. In fact subjects did not perceive the simulated 6-hour phase

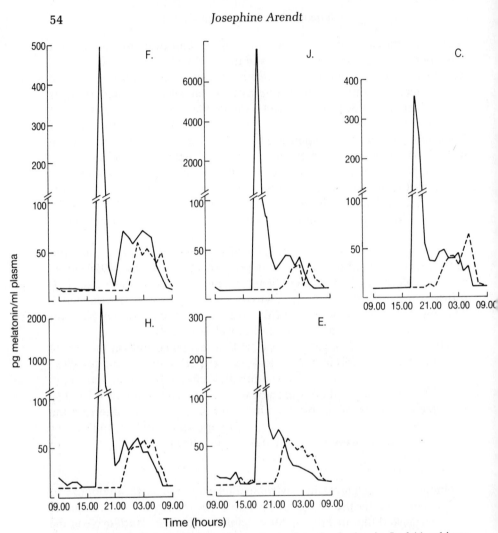

Fig. 2.7. Phase-advance of the endogenous melatonin rhythm in 5 of 11 subjects taking 2 mg melatonin daily at 1700 hours for 3 weeks (———), compared to placebo (— — — —) at 1700 hours for 3 weeks. The other six subjects had very low endogenous melatonin or slow clearance of the exogenous dose occluding the endogenous rhythm. (From Arendt *et al.* 1985, with acknowledgements to the CIBA Foundation.)

shift with or without melatonin, and decrements in performance are only observed after eastward shifts in isolation experiments (Wever 1986). Thus, it still appears worthwhile to investigate the use of melatonin in real jet-lag situations.

We have been able to conduct a small preliminary study of some of the effects of jet-lag and its treatment by melatonin (Arendt *et al.* 1986). Seventeen healthy volunteers (10 women, 7 men; age range 29–68) were flown from London to the west coast of the USA (San Francisco) and remained there for 14 days to adapt to USA time. For 3 days preceding and including the day of the flight home they took either melatonin (*n*=8) or placebo (*n*=9) at 18.00 hours local time on a random double-blind basis. For 4 days after the return flight (eight time zones eastward) the subjects took the same preparation at bedtime (22.00–24.00 hours). Seven days after their return flight they were asked to rate their jet-lag on a scale of 0 (insignificant) to 100 (very bad). Six of the nine placebo subjects rated their jet-lag at greater than 50. No melatonin subject rated their jet-lag at greater than 17 (Fig 2.8). In the absence of a double-blind cross-over study it is not possible to identify the number of melatonin subjects who would not have suffered from jet-lag anyway. Nevertheless, it is remarkable that none suffered significantly. The question then arises as to what precisely is meant by self-rated jet-lag. It can be partially answered by reference to other data from this study (Arendt *et al.* 1987). For 48 hours prior to departure from the UK and for days 1–7, 14, 15, 21, and 22 after return to the UK, subjects kept sleep-logs, recorded oral temperature and mood (18 visual analogue scales, Herbert *et al.* 1976) every 2 hours from 08.00 hours or wake up to 24.00 hours or bedtime, and performed logical reasoning and letter-cancellation tests every 4 hours.

Sequential urine samples (08.00–12.00, 12.00–18.00, 18.00–24.00,

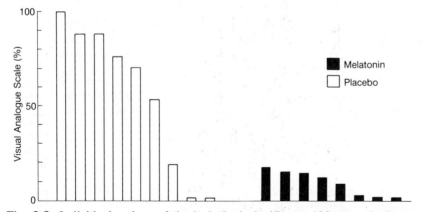

Fig. 2.8. Individual ratings of 'jet-lag' (0=insignificant, 100=very bad) on a 10-cm visual analogue scale in 17 subjects taking either melatonin (■) or placebo (□) in a precisely timed protocol (see text) before and after an 8 time-zone flight from San Francisco to London. The study design was double-blind. Subjects were in San Francisco time zone for 14 days prior to flight. (From Arendt *et al.* 1986, with permission.)

24.00–08.00 hours) were also collected during this time and for 48 hours prior to departure, from the USA. Analysis of results indicated clearly that sleep latency was greatly increased, (Fig. 2.9) and sleep quality greatly decreased in jet-lagged subjects, and both of these parameters correlated with jet-lag ratings. No important differences were found in performance measures and temperature measurements. Urine analysis showed adaptation to USA time and a more rapid re-synchronization of both melatonin (Fig. 2.10) and cortisol rhythms in melatonin subjects on return to the UK. Melatonin subjects were both less depressed and more alert than jet-lagged placebo subjects.

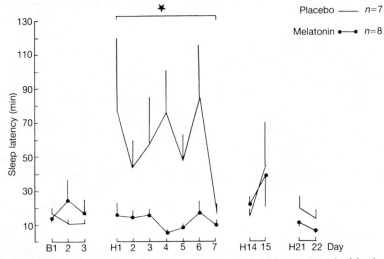

Fig. 2.9. Sleep latency ($\bar{x}$ SEM) in seven jet-lagged subjects treated with placebo and eight subjects treated with melatonin after returning from San Francisco to London. The two placebo subjects reporting no jet lag (see Fig. 2.8) were excluded from the analysis. Sleep latency was significantly decreased ($P < 0.02$) in melatonin subjects. Sleep quality (not shown) was concomitantly increased ($P < 0.05$) and both quality and latency correlated (negatively and positively, respectively) significantly with jet-lag ratings ($P < 0.001$). (From Arendt *et al.* 1987, with permission.) B=baseline days prior to leaving London. H=home days after flight from San Francisco.

Several conclusions can be drawn from this study. Subjective perception of jet-lag is primarily related to disturbed sleep and presumably the beneficial effects of melatonin are essentially sleep-related. There is possibly no need to invoke a circadian resynchronizing ability in order to explain its mode of action. Nevertheless, the more rapid resynchronization of its own endogenous rhythm, and that of cortisol once again points to an effect on the circadian clock or clocks.

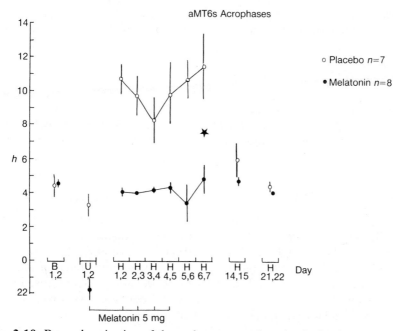

Fig. 2.10. Resynchronization of the endogenous melatonin rhythm in melatonin-treated subjects ($n=8$) compared to placebo subjects reporting jet lag ($n=7$). The calculated mean peak times (acrophases) of urinary 6-sulphatoxymelatonin are shown. B=baseline days prior to leaving London. U=the last 2 days in the USA before returning to London (14 days stay in the USA—San Francisco time zone). H=home days after flight from San Francisco to London. Note that for USA days and days H1–4 the melatonin group acrophases reflect exogenous melatonin. Endogenous aMT6s can be assessed on days H_6 and H_7 in this group: significant differences are found compared to the placebo group ($P<0.05$, ANOVA of raw data, ANOVA of acrophase shifts, $P=0.022$) with more rapid adaptation to UK time in the melatonin group.

Alternatively, the rapid adaptation of the sleep–wake cycle might entrain an accelerated adaption of other rhythms bypassing any 'clock'-effects. A direct effect of melatonin on human central rhythm-generating systems is therefore still unproven.

The apparent alleviation of jet-lag in this study is of interest whatever its explanation. It is obvious that much work remains to be done, firstly to increase the number of subjects studied, secondly to investigate different numbers of time zones and directions of flight, and thirdly to optimize the dose. A recent preliminary report (Wirz-Justice 1988) suggests the melatonin was not distinguishable from placebo in subjects after a 6-hour time-zone transition eastwards. Unfortunately, the state of circadian adaptation pre-flight, and compliance were not assessed. In uncontrolled studies only 6 of 50 flights treated with melatonin were accompanied by significant jet-lag (Arendt and Aldhous, unpublished data).

Conclusion

It is obvious that melatonin is of potential use in sleep disorders, selected depressive patients, blind subjects and old people with desynchronized rhythms, shift workers, and a number of other situations to be covered later in this volume.

Acknowledgements

This review was written during the tenure of grants from the MRC, the AFRC, and Wellcome Trust. I would like to thank my colleagues C. Bojkowski, Judie English, Christine Franey, Margaret Aldhous-Kemp, J. Broadway, Debra Skene, and J. Wright, for generating much of the data presented; and Professor V. Marks for stimulating discussions. Jet-lag studies were supported by British Caledonian Airways, Grand Metropolitan Hotels, and Horner Ltd, Montreal.

References

Arendt, J. (1986a). Role of the pineal gland and melatonin in seasonal reproductive function in mammals. *Oxford Reviews of Reproductive Biology* **8**, 266–320.

Arendt, J. (1986b). Assay of melatonin and its metabolites, results in normal and unusual environments. *Journal of Neural Transmission (Supplement)* **21**, 11–33.

Arendt, J. and Broadway, J. (1986). Phase response of human melatonin rhythms to bright light in Antarctica. *Journal of Physiology* **377**, 68.

Armstrong, S. M. and Chesworth, M. J. (1987). Melatonin phase-shifts a mammalian circadian clock. *Abstract IV Colloquium European Pineal Study Group*, Modena, Italy.

Armstrong, S. M. and Redman, J. (1985). Melatonin administration: effects on rodent circadian rhythms. In: *Photoperiodism, melatonin and the pineal* Ciba Foundation Symposium **117**, pp. 188–202.

Arendt, J., Paunier, L., and Sizonenko, P. C. (1975). Melatonin radio-immunoassay. *Journal of Clinical Endocrinology and Metabolism* **43**, 347–50.

Arendt, J., Wetterberg, L., Heyden, T., Sizonenko, P. C., and Paunier, L. (1977). Radioimmunoassay of melatonin: human serum and cerebrospinal fluid. *Hormone Research* **8**, 65–75.

Arendt, J., Borbely, A. A., Franey, C., and Wright, J. (1984). The effect of chronic, small doses of melatonin given in the late afternoon on fatigue in man: a preliminary study. *Neuroscience Letters* **45**, 317–21.

Arendt, J., Bojkowski, C., Franey, C., Wright, J., and Marks, V. (1985*a*). Immunoassay of 6-hydroxymelatonin sulphate in human plasma and urine: abolition of the urinary rhythm with atenolol. *Journal of Clinical Endocrinology and Metabolism* **60**, 1166–72.

Arendt, J., *et al.* (1985*b*). Some effects of melatonin and the control of its secretion in man. In *Photoperiodism, melatonin and the pineal, Ciba Foundation Symposium* **117**, 266–83. Pitman, London.

Arendt, J., Aldhous, M., and Marks, V. (1986). Alleviation of jet-lag by melatonin: preliminary results of controlled double-blind trial. *British Medical Journal* **292**, 1170.

Arendt, J., Aldhous, M., Marks, M., Folkard, S., English, J., Marks, V., and Arendt, J. H. (1987). Some effects of jet-lag and its treatment by melatonin. *Ergonomics.* **30**, 1379–93.

Aschoff, J. (ed.) (1981). Biological Rhythms. *Handbook of behavioural neurobiology* Vol. 4. Plenum, New York.

Bojkowski, C., Aldhous, M. E., English, J., Franey, C., Poulton, A. L., Skene, D. J., and Arendt, J. (1987). Suppression of nocturnal plasma melatonin and α-sulphatoxymelatonin by bright and dim light in man. *Hormone and Metabolic Research* **19**, 437–40.

Boulos, Z. and Rusak, B. (1982). Phase-response curves and the dual oscillator model of circadian pacemakers. In: *Vertebrate circadian systems* (ed. J. Aschoff, S. Daan, and G. Groos). pp. 215–23. Springer Verlag, Berlin.

Brainard, G. C., *et al.* (1985). Effect of light wavelength on the suppression of nocturnal plasma melatonin in normal volunteers. In: *The medical and biological effects of light* (ed. R. J. Wurtman, M. J. Baum, and J. T. Potts, Jr). *Annals of the New York Academy of Sciences* **453**, 376–8.

Broadway, J. and Arendt, J. (1986). Delayed recovery of sleep and melatonin rhythms after night shift work in the Antarctic winter. *Lancet* **ii**, 813–14.

Broadway, J., Folkard, S., and Arendt, J. (1987). Bright light phase shifts the human melatonin rhythm in Antarctica. *Neuroscience Letters*, **79**, 185–9.

Carman, J. S., Post, R. M., Buswell, R., and Goodwin, F. K. (1976). Negative effects of melatonin on depression. *American Journal of Psychiatry* **133**, 1181–6.

Cotzias, G. C., Papavasiliou, P. S., Ginos, J., Steck, A., and Duby, S. (1971). Metabolic modification of Parkinsons disease and of chronic manganese poisoning. *Annual Review of Medicine* **22** (ed. A. C. Degraff and W. P. Creger), 305–26. Annual Reviews Inc., Palo Alto, California.

Cramer, H., Rudolph, J., and Consbruch, U. (1974). On the effects of melatonin on sleep and behaviour in man. In *Serotonin: new vistas, biochemistry and behavioural and clinical studies* (ed. E. Costa) Advances in Biochemical Psychopharmacology Volume 2, pp. 187–91. Raven Press, New York.

Cramer, H. (1978). Melatonin and Sleep. *Fourth European Congress on Sleep Research* (ed. Tirgu-Mures), pp. 204–10. S. Karger, Basel.

Czeisler, C. A., *et al.* (1986). Bright light resets the human circadian pacemaker independent of the timing of the sleep-wake cycle. *Science* **233**, 667–71.

Fevre-Montange, M., Van Cauter, E., Refetoff, S., Desir, D., Tourniaire, J., and Copinschi, G. (1981). Effects of 'Jet-Lag' on hormonal patterns. II. Adaptation of melatonin circadian periodicity. *Journal of Clinical Endocrinology and Metabolism* **52**, 642–49.

Herbert, M., Johns, M. W., and Dore, C., (1976). Factor analysis of analogue scales measuring subjective feelings before and after sleep. *British Journal of Medical Psychology* **49**, 373–9.

Illnerova, H., Zvolsky, P., and Vaneck, J. (1985). The circadian rhythm in plasma melatonin concentration of the urbanised man: the effect of summer and winter time. *Brain Research* **328**, 186–9.

Lerner, A. B. and Nordland, J. J. (1978). Melatonin: clinical pharmacology. *Journal of Neural Transmission (Supplement)* **13**, 339–47.

Lewy, A. J., Wehr, T. A., Goodwin, F. K., Newsome, D. A., and Markey, S. P. (1980). Light suppresses melatonin secretion in humans. *Science* **210**, 1267–9.

Lynch, H. J., Wurtman, R. J., Moskowitz, M. A., Archer, M. C., and Ho, M. H. (1975). Daily rhythm in human urinary melatonin. *Science* **187**, 169–71.

Menaker, M., Hudson, D. J., and Takahashi, J. S. (1981). Neural and endocrine components of circadian clocks in birds. In *Biological clocks in seasonal reproductive cycles* (ed. B. K. Follett and D. E. Follett), pp. 171–83. Wright, Bristol.

Minors, D. S. and Waterhouse, J. M. (1981). *Circadian rhythms and the human.* Wright, P. S. G., Bristol.

Moore, R. Y. and Klein, D. C. (1974). Visual pathways and the central neural control of a circadian rhythm in pineal serotonin N-acetyltransferase. *Brain Research* **71**, 17–33.

Murakami, N., Hayafuji, C., Sasak, Y., Yamazaki, J., and Takahashi, K. (1983). Melatonin accelerates the reentrainment of the circadian adrenocortical rhythm in inverted illumination cycle. *Neuroendocrinology* **36**, 385–91.

Nordlund, J. J. and Lerner, A. B. (1977). The effects of oral melatonin on skin colour and on the release of pituitary hormones. *Journal of Clinical Endocrinology and Metabolism* **45**, 768–74.

Pelham, R. W. E., Vaughan, G. M., Sandcock, K. L., and Vaughan, M. K. (1973). 24 hr cycle of a melatonin-like substance in the plasma of human males. *Journal of Clinical Endocrinology and Metabolism* **37**, 341–4.

Pittendrigh, C. S. and Daan, S. (1976). A functional analysis of circadian pacemakers in nocturnal rodents. IV. Entrainment: pacemaker as clock. *Journal of Comparative Physiology* **106**, 291–331.

Quay, W. B. (1970). Precocious entrainment and associated characteristics of activity patterns following pinealectomy and reversal of photoperiod. *Physiology and Behaviour* **5**, 1281–90.

Redman, J., Armstrong, S., and Ng, K. T. (1983). Free-running activity rhythms in the rat: entrainment by melatonin. *Science* **219**, 1080–1.

Reiter, R. J. (1985). Action Spectra, dose-response relationships and temporal aspects of lights effects on the pineal gland. In *The medical and biological effects of light* (ed. R. J. Wurtman, M. J. Baum, and J. T. Potts, Jr), *Annals of the New York Academy of Sciences* **453**, 215–30.

Shaw, K. M., Stern, G. M., and Sandler, M. (1973). Melatonin and Parkinsonism. *Lancet* **1**, 271.

Sugden, D. (1983). Psychopharmacological effects of melatonin in mouse and rat. *Journal of Pharmacology and Experimental Therapeutics* **227**, 587–91.

Takahashi, J. S., De Coursey, P. J., Bauman, L., and Menaker, M. (1984). Spectral sensitivity of a novel photoreceptive system mediating entrainment of mammalian circadian rhythms. *Nature* **308**, 186–8.

Tamarkin, L., Baird, C. J., and Almeida, D. F. X. (1985). Melatonin: a coordinating signal for mammalian reproduction. *Science* **227**, 714–20.

Underwood, H. (1977). Circadian organisation in lizards: the role of the pineal organ. *Science* **195**, 587–9.

Vollrath, L., Semm, P., and Gammel, G. (1981). Sleep induction by intranasal application of melatonin. *Advances in the Biosciences* **29**, 327–9.

Waldhauser, F., *et al.* (1987). A pharmacological dose of melatonin increases prolactin levels in males without altering those of GH, LH, FSH, TSH, testosterone or cortisol. *Neuroendocrinology* **46**, 125–30.

Wehr, T. A. and Goodwin, F. K. (1981). Biological rhythms and psychiatry. *American handbook of psychiatry,* Volume VII (ed. S. Arieti and H. K. K. Brodie) 2nd edn, Basic Books, New York.

Weitzman, E. D., *et al.* (1978). Studies of the 24-hour rhythm of melatonin in man. *Journal of Neural Transmission (Supplement)* **13**, 325–37.

Wetterberg, L., *et al.* (1982). Melatonin and cortisol levels in psychiatric illness. *Lancet* **ii**, 100.

Wever, R. A. (1979). The circadian system of man: results of experiments under temporal isolation. Springer Verlag, New York.

Wever, R. A. (1986). Characteristics of circadian rhythms in human functions. In *Melatonin in humans,* Proceedings of the first international conference on melatonin in humans. (ed. R. J. Wurtman and F. Waldhauser), pp. 291–352. Vienna, Austria.

Wever, R. A., Polasek, J. and Wildgruber, C. M. (1983). Bright light affects human circadian rhythms. *Pfluger's Archives* **396**, 85–7.

Wever, R. A., Polasek, J., and Arendt, J. (1987). Control of melatonin in the human circadian system. *The pineal gland, its physiological and pharmacological role.* Erice, Sicily, Abstract.

Wirz-Justice, A. (1988). Light and dark as a drug. *Progress in Drug Research,* in press.

Wright, J., Aldhous, M., Franey, C., English, J., and Arendt, J. (1986). The effects of exogenous melatonin on endocrine function in man. *Clinical Endocrinology* **24**, 375–82.

3. Melatonin and human reproductive function

Pierre C. Sizonenko and Ursula Lang

Introduction

The concept that the pineal gland has specific effects on reproductive biology has arisen from numerous experiments in many species. This small organ located in the brain is an active neuroendocrine gland responsible for transducing environmental information such as light or season into secretory signals (Wurtman *et al.* 1964), which repress or promote reproductive functions such as behaviour, spermatogenesis, or ovulation (Reiter 1980). Wurtman and colleagues (1963) opened a new field of research linking melatonin and reproduction in observing that daily melatonin injections in female rats could delay vaginal opening, decrease the incidence of vaginal oestrous, and reduce ovarian weight. This hormone has been shown to mediate the gonadotropic effects of photoperiods in species with a seasonal reproductive cycle such as the hamster or the ewe (Reiter 1980; Arendt *et al.* 1983; Bittman *et al.* 1983; Carter and Goldman 1983; Forster *et al.* 1986). Today melatonin is known to be a possible inhibitor of sexual development in mammals (Tamarakin *et al.* 1976; Goldman *et al.* 1979; Lang 1986). Pinealectomy can prevent the gonadal regression that occurs in hamsters under long dark periods or can induce premature pubertal development which can be counteracted by melatonin administration.

Two types of secretory products have been described for the pineal gland: indoleamines and polypeptides. The active principal indoleamine is melatonin; however, other indoles like 5-methoxytryptamine have been incriminated (Pevet 1983). Among the peptides found in the pineal gland, arginine vasotocin ([Arg^8]oxytocin, AVT; Vaughan *et al.* 1974) has been attributed an antigonadotrophic role. It has been suggested that melatonin is the releasing factor for AVT (Benson *et al.* 1976; Pavel 1978).

The pineal is mostly inhibitory and the signal is carried mainly by the indoleamine melatonin (Cardinali 1981; Reiter 1980). In addition to its well-described inhibitory influence on reproduction and sexual development, melatonin can also have a stimulatory influence. This progonadal

(Turek *et al.* 1975) or 'counterantigonadotrophic' action has been described for the hamster (Reiter 1980) and the sheep (Bittman *et al.* 1983).

In the human, melatonin, as in subhuman species, is secreted during the night (Arendt 1985*a*,*b*; Lang *et al.* 1981*a*). Since studies in the human are limited due to ethical reasons, the recognition of the role of melatonin on reproductive life is particularly difficult. Plasma and urinary melatonin concentrations have been measured by several methods after extraction and purification: gas chromatography–mass spectrometry (GCMS), high-performance liquid chromatography (HPLC), thin-layer chromatography (TLC), or more commonly radio-immunoassays (RIA) (Arendt *et al.* 1977; Arendt 1986). Depending on the antimelatonin antibody and the extraction method used for the radioimmunoassay of melatonin, investigators found wide range of values. In particular, daytime plasma values were found to be very different with regard to the antimelatonin antiserum used including high values in which many investigators do not believe. These discrepancies make comparison of data between studies considerably difficult and probably explain the different results published in the human (Arendt 1985*a*,*b*).

The mechanism of action of melatonin is probably at the hypothalamic level as suggested by the decrease of the pituitary concentrations of gonadotrophin releasing-hormone (GnRH) receptors after melatonin administration (Lang *et al.* 1983), since pituitary GnRH receptor contents reflect the secretion of the hypothalamic GnRH (Clayton and Catt 1981).

This chapter will deal firstly with the effects of melatonin administered in the human on the hormones involved in reproductive function [luteinizing hormone, (LH), follicle-stimulating hormone (FSH), and prolactin]. Secondly, the levels of melatonin during adulthood will be presented in relation to the endocrinology of reproduction, in particular during menstrual cycle and during pregnancy. Thirdly, the changes in melatonin secretion and its possible role on the onset of puberty will be discussed, as well as in pathological conditions such as precocious or delayed puberty.

Effects of exogenous melatonin or of pinealectomy on reproductive endocrine functions in man and woman

No physiological role for melatonin has been yet established in man. The administration of melatonin induce somnolence and sleep and can stimulate growth hormone secretion (Smythe and Lazarus 1974; Lerner and Nordlund 1978; Vollrath *et al.* 1981; Lieberman *et al.* 1984; Wright *et al.* 1986), although the effect of melatonin on growth hormone has been debated (Weinberg *et al.* 1981). In one patient in whom the pineal gland was surgically removed, LH basal values were higher than prior to surgery (Chick *et al.* 1985). The response to GnRH was increased post-operatively, with a decrease in prolactin also being seen. Such association of low melatonin with low prolactin was also reported in a patient with a pinealoma (Kennaway *et al.* 1979). Administration of high doses of melatonin did not cause any significant changes in basal or post-GnRH-stimulation LH or FSH either in normal men, children or post-menopausal women (Fideleff *et al.* 1976; Weinberg *et al.* 1980; Aleem *et al.* 1984; Lissoni *et al.* 1986; Wright *et al.* 1986). In another report, the effects of melatonin on the plasma concentrations of prolactin depended on the time of administration; a significant increase of prolactin was observed when melatonin was given in the afternoon, in adult men and women but not in prepubertal children (Mauri *et al.* 1985). In all these studies, melatonin was administered at pharmacological doses during a short period of time. Up to now, the effects of a long-term administration of melatonin on the reproductive function in man have not been evaluted. Experimental data on the suppression of LH and FSH, and/or prolactin are based on long-term treatment of animals with melatonin which is only active in the late afternoon and therefore cannot be compared to acute experiments in the human (Reiter 1980; Lang 1986). Inhibitory effects of long-term administration of melatonin on sexual maturation were also observed in several species (Hoffmann 1979; Kenaway 1984; Lang *et al.* 1983, 1984*a*; Petterborg and Reiter 1980, Whitsett *et al.* 1984; Wurtman *et al.* 1963).

Conversely, very few studies on the effects of sex steroids on melatonin secretion have been performed in the human. In one normal man, the administration of oestradiol was associated with a decrease of testosterone and with a decrease of plasma melatonin concentrations measured during the morning (Penny and Goebelsmann 1984).

In vitro, melatonin stimulated steroidogenesis in two compartments of the human ovary (MacPhee *et al.* 1975). In a corpus luteum, melatonin has been shown to increase progesterone synthesis in a dose-related manner. In the ovarian stroma, melatonin stimulated the incorporation of C^{14}-1-acetate into androstenedione. Binding of radioactive human chorionic gonadotrophin by the corpus luteum was unaffected by melatonin (MacPhee *et al.* 1975).

Melatonin secretion during adulthood, pregnancy, and at birth

During the menstrual cycle of five healthy women, melatonin concentra-tions measured during the early morning showed elevation at the time of menstrual bleeding and had its nadir at the time of ovulation (Wetterberg *et al.* 1976). These preliminary findings were confirmed by further work (Webley and Leidenberger 1986). The circadian pattern of melatonin was studied in the follicular and the luteal phases of 10 normal women, where statistical analysis showed a significantly increased secretion of melatonin in the luteal phase compared to that in the follicular phase. The secretion of melatonin was increased in women taking the three-phase contraceptive pill in relation to the dose of progestin. These results suggest that there is a positive relationship between melatonin and progesterone, and that changes in the circadian pattern of melatonin secretion, falling before ovulation and rising during the luteal phase, may act as a modulator of menstrual cyclicity. However, these conclusions are not supported by the levels of urinary 6-sulphatoxymelatonin measured as a reliable index of melatonin production in two healthy women who showed that the amount of this metabolite excreted in the urine remained constant throughout the menstrual cycle (Fellenberg *et al.* 1982).

An inhibitory effect of the pineal gland on hormone-dependent tumorigenesis was suggested by several studies which were summarized by Lapin (1979) in his review. More recently, a striking negative correlation was found between the melatonin diurnal rhythm and the sex steroid-receptor content of the breast cancer tissue (Tamarkin *et al.* 1982b; Danforth *et al.* 1985). Women with oestrogen or progesterone receptor-positive breast tumours had a significantly lower mean plasma

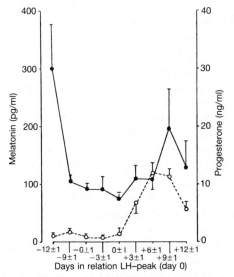

Fig. 3.1. Mean serum melatonin (solid line) and progesterone (dotted line) ±SEM during menstrual cycle in five healthy women. (Reproduced with authorization from Wetterberg *et al.* 1976.)

melatonin day–night difference than did patients with receptor-negative tumours. These data and those of Bartsch *et al.* (1981) suggest that the absence of a clear daily melatonin rhythm might be related to the presence of hormone-dependent breast cancer (see chapter by Drs Blask and Hill, this volume). Experimentally, the possible inhibitory role of melatonin on hormone-dependent tumours has been studied on anthracene-induced mammary gland tumours in the rat (Tamarkin *et al.* 1985). Melatonin protects against and pinealectomy enhances the development of such tumours. This protection supposedly exerted by melatonin occurs either through a reduction of prolactin secretion or a reduction of oestrogens receptors. Melatonin binding sites have been shown in human breast cells (Cohen *et al.* 1978), but these findings have not been completely confirmed yet. However, in another study, melatonin seemed to increase oestrogen receptor binding activity in human breast cancer cells (Danforth *et al.* 1983). In women at risk of breast cancer, the pattern of melatonin secretion was not different from that of a normal population. Therefore, the use of the daily melatonin pattern as a screening tool for breast cancer risk is not recommended.

Few studies have reported plasma levels of melatonin during pregnancy. In a recent study (Pang *et al.* 1986), it was shown that plasma melatonin concentrations measured between 10.00 and 12.30 hours were high in 1–24 weeks of gestation, low in 24 to 36 weeks, and then high again in 36–42 weeks. A nyctohemeral rhythm of melatonin has been observed in pregnant women (Kennaway *et al.* 1981). After parturition, levels of melatonin dropped rapidly. A small increase in the mean concentration of melatonin during labour was observed when compared to levels measured during late pregnancy (Mitchell *et al.* 1979).

Concentrations of melatonin have also been measured in amniotic fluid obtained during late pregnancy and labour (Mitchell *et al.* 1978). During late pregnancy, amniotic fluid contained lower concentrations of melatonin than amniotic fluid obtained by amniotomy at the time of labour. Labour seems to increase the concentrations of melatonin. The origin of the melatonin found in the amniotic fluid remains unknown, it is likely that it comes from the fetal urine. This study, however, does not clearly state the time of the day (between 08.00 and 22.30 hours) when the sampling of the amniotic fluid was performed. This is particularly important as a clear nyctohemeral rhythm has been described in pregnant women and that there is suggestion for a free maternity–fetal transfer of melatonin (see below).

It was reported that, in pregnant women, fenoterol–a β_2-agonist (Ritodrine®)—perfused in order to diminish premature uterine contractions increased plasma concentrations of melatonin during the administration of the drug (Desir *et al.* 1983). In our group we were not able to confirm this finding (Table 3.1) when perfusing the β_2-agonist either during daytime or during the night.

At birth, umbilical arterial and venous plasma contained significantly greater concentrations of melatonin than maternal plasma (Mitchell *et al.* 1979). Similar values were obtained either after spontaneous vaginal delivery or at Caesarian section. In addition, a significant difference was demonstrated for both groups of umbilical samples with raised venous levels after spontaneous vaginal delivery, but higher arterial levels at Caesarian section. In contrast, in a recent study performed in our group (Lang *et al.* 1986*a,b*), no significant difference was found between maternal, umbilical arterial, and venous concentrations (Table 3.2). There was also no difference after vaginal delivery or after Caesarian section. Very similar nyctohemeral rhythms with a nocturnal peak and a diurnal trough in melatonin levels were observed in maternal, umbilical arterial, and venous plasmas. Lowest values (23–30 pg/ml) were measured in the afternoon from 14.00 to 20.00 hours whereas highest concentrations (67–91 pg/ml) were observed late in the night from 02.00 to 08.00 hours. However, in the morning (08.00 to 14.00 hours), melatonin concentrations in mothers were found to be very high (64–86 pg/ml) compared to those of six non-pregnant women at 08.00 hours (16.6 ± 1.3 pg/ml). These results indicate, contrary to those of Mitchell *et al.* (1979), that there is no maternal–fetal gradient of melatonin suggesting free transport between the maternal and fetal compartment through the placenta (Lang *et al.* 1986*a*). The nyctohemeral variation seems to be present in mothers and fetuses at birth. These results have been confirmed in the rhesus monkey by Reppert *et al.* (1979). Radio-labelled melatonin injected in the mother near-term promptly appeared in the fetal circulation. The rates of disappearance of ^{3}H-melatonin in the maternal and fetal circulations were parallel. These findings as well as ours suggest that a daily rhythm in maternal melatonin would generate a similar rhythm in the fetus. The fetal monkey pineal gland was found to have the enzymatic system necessary for the synthesis of melatonin. It is, however, not known whether fetal melatonin synthesis is rhythmic. In a preliminary study, we have found low levels of melatonin present in the urine of newborns up to 6 weeks without difference between day and night. At 6 days of life, there is a similar plasma concentration of melatonin at noon and at midnight (Fig. 3.2). These results are questionable as these newborns were studied in a hospital nursery in which the light regimen is different from daylight and complete darkness. A diurnal and nocturnal pattern of melatonin was observed after three months of life in infants (Gupta 1986). In the rat, it has been shown that ^{3}H-melatonin injected in the mother was rapidly transferred from the maternal circulation into the lactating mammary tissue, and to the

Table 3.1. Plasma concentrations of melatonin (pg/ml) before and during the infusion of fenoterol in 10 women with premature uterine contractions

Patient No.	Time of test*	Before fenoterol	Time after beginning of infusion (min)							
			30	60	90	120	150	180	210	240
1	15.10	6.7	9.4	5.9	8.2	3.5				
2	18.15	<1.0					<1.0	<1.0	1.1	<1.0
3	19.45	18.2	5.1	15.9	37.4	18.4				
4	20.30	—					25.7	58.1	26.1	39.7
5	21.05	14.1	17.1	18.9	9.5	8.1				
6	02.30	18.6	14.0	10.2	19.4	26.2				
7	03.30	101.5					58.9	148.4	142.6	
8	03.50	38.5	22.4	20.8	25.1	27.6				
9	04.40	15.9	16.1	23.3	9.7	7.8				
10	05.40	18.4	13.9	14.4	14.6	7.7				

*Time in hours and minutes.

Table 3.2. Mean (±SEM) maternal and umbilical plasma melatonin concentrations (pg/ml) at delivery

Hours of sampling	n	Maternal blood	Umbilical blood	
			arterial	venous
08.00–12.00	4	69.1 ± 14.3	97.2 ± 28.0	85.9 ± 20.5
12.00–16.00	4	39.5 ± 14.0	56.7 ± 11.7	46.1 ± 17.2
16.00–20.00	5	20.8 ± 6.6	20.1 ± 6.0	19.8 ± 6.6
20.00–24.00	8	41.8 ± 18.3	28.2 ± 8.4	22.2 ± 6.6
24.00–04.00	8	70.7 ± 16.0	75.9 ± 15.8	61.9 ± 13.6
04.00–08.00	10	71.2 ± 15.4	114.6 ± 32.8	82.8 ± 22.5

NEWBORNS
6th Day of life
Normal at term

◆ Male n=16

○ Female n=13

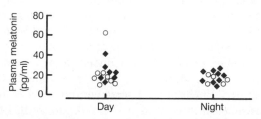

Fig. 3.2. Plasma concentrations of melatonin in newborns sampled during the day and at midnight. No nocturnal rise can be observed.

stomach of each suckling rat. These results suggest that the mother is entraining a nyctohemeral rhythm in the suckling rat which is probably unable to produce melatonin during the first 10 days of life (Reppert and Klein 1978).

Melatonin and pubertal development

As in animals, a daily rhythmic secretion of melatonin is observed in plasma, cerebrospinal fluid, and urine in humans (Arendt 1985; Lang *et al.* 1981*a*; Vaughan 1984; Waldhauser and Wurtman 1983). This pattern of secretion makes studies with multiple sampling in humans, particularly in children, difficult (see Reiter, Chapter 1; Miles *et al.* Chapter 13).

Long nights and short days of the dark winter months of the Arctic circle seem to affect the incidence of conceptions (Timonen *et al.* 1967) and increase the number of anovulatory cycles (Timonen and Carpen

1969). In blind girls it was found that the age of menarche was advanced (Zacharias and Wurtman 1964; Thomas and Pizzarello 1967; Magee *et al.* 1970). Such findings are difficult to interpret, and suggest that perhaps the pineal gland and melatonin particularly has no effect on the onset of puberty (Lehrer 1985). These studies have not taken into account the degree of light perception of all the subjects, which in turn may change the pattern of secretion of melatonin. In adult women blindness has been associated with infertility (Elden 1971), and melatonin secretion has not been assessed in these women.

In humans, pineal tumours have been associated with precocious puberty (Kitay 1954; Kitay and Altschule 1954; Waldhauser and Wurtman 1983). However, delayed puberty has been seen in boys with parenchymal pineal tumours (Kitay and Altschule 1954). It has been difficult to determine the exact role of the pineal gland in puberty as most of the tumours of the region are not pinealocytomas, but dysgerminomas secreting human chorionic gonadotrophin (Herrick 1984; Russell 1977; Sklar *et al.* 1981).

During normal pubertal development, plasma samples collected during daytime showed no differences in melatonin concentrations between prepubertal and pubertal subjects (Tamarkin *et al.* 1982*a,b*; Ehrenkranz *et al.* 1982; Lenko *et al.* 1982; Sizonenko *et al.* 1985), except in one study (Silman *et al.* 1979). Concentrations tend to be higher before puberty, but the differences were not statistically significant. Studies of nocturnal melatonin determinations appear to report also contradictory results. In three studies (Ehrenkranz *et al.* 1982; Tamarkin *et al.* 1982*a,b*; Sizonenko *et al.* 1985) night-time melatonin profiles were similar both in prepubertal and in pubertal or adult subjects. In addition, a good correlation was found between nocturnal melatonin levels and LH concentrations in four pubertal boys, suggesting an absence of decrease of melatonin during pubertal development (Fevre *et al.* 1978). Other studies reported a significant decrease of the night peak melatonin values and of the day–night increment from the prepubertal stage to the early pubertal stage (Gupta *et al.* 1983; Attanasio *et al.* 1983; Waldhauser *et al.* 1984). The latter study found differences in midnight melatonin levels between prepubertal children less than 7 years old, and prepubertal and pubertal subjects over 7 years old. These results suggest that melatonin secretion may decrease at the time of adrenarche, i.e. when the androgenic zone of the adrenal cortex matures (Sizonenko and Paunier 1975). The possible relationship between melatonin secretion and the hypothalamo–pituitary–adrenal axis during pubertal develop-ment has been studied recently (Lang *et al.* 1986*b*; Sizonenko and Aubert 1986) and is discussed in the following chapter.

Measurements of urinary melatonin can provide a better integrated picture of the total amount of melatonin secreted (Lang *et al.* 1981*a*). Urinary excretion of melatonin determined from 12-hour nocturnal or 24-hour urine collections increased with age, but was similar at all stages of puberty (Table 3.2) when expressed per square metre of body surface (Sizonenko *et al.* 1985). In contrast, Penny (1982) reported increased excretion of melatonin in the urine during puberty. Determinations of the melatonin metabolite 6-hydroxymelatonin did not show a correlation between daily excretion rates and age or pubertal stages, with the exception of an increased concentration observed at the time of onset of breast development in girls (Tetsuo *et al.* 1982). In this study, no change was observed in boys.

It is difficult from the studies and work cited to attribute a role to the pineal gland and to melatonin in the control of the onset of puberty. Results obtained in the investigations in different disorders of pubertal development also offer no definite conclusion. In boys with delayed puberty, higher daytime melatonin levels were observed than in a control population (Cohen *et al.* 1982). However, the melatonin day–night increment of subjects with delayed puberty was similar to that of prepubertal children (Attanasio *et al.* 1985). Similarly, urinary excretion of melatonin in boys with delayed puberty (Table 3.2) was not different from age-matched controls (Sizonenko *et al.* 1985). In contrast, the day–night increment of melatonin was lower in children with precocious puberty than that of age-matched prepubertal controls (Attanasio *et al.* 1983). These findings were not supported by Ehrenkranz *et al.* (1982) who reported a normal daily rhythm of plasma melatonin in precocious puberty. Tamarkin *et al.* (1982*a,b*) did not find a correlation between plasma melatonin profiles and pubertal stages or body weights of obese children and patients with Prader–Willi's syndrome. Interestingly, patients presenting with the fragile-X syndrome, enlarged testes and macrogenitosomia, showed lower melatonin profiles than age-matched controls and a significant impairment of the nocturnal rise in this hormone (O'Hare *et al.* 1986).

These discordant findings do not exclude a role of the pineal gland and of melatonin during puberty. Melatonin levels seem to decrease with age (Waldhauser and Steger 1986). Melatonin may induce subtle changes in the GnRH secretion at the arcuate nucleus level and the subsequent pulsatile secretion of gonadotrophins, such as a change in the frequency or the amplitude of LH secretion (Sizonenko and Aubert 1986). These changes are difficult to observe because our present means of investigation arc insufficient. One hypothesis is that the pineal gland interacts with a circadian pacemaker which in turn will be coupled to the arcuate nucleus through an ultradian pacemaker and will modulate the pulsatile secretion of GnRH (Lehrer 1985). In addition, melatonin is only active at a certain time during the day or the night (Reiter 1980; Lang *et al.* 1984*b*). Such action of melatonin at a certain time implies that there is a regulation of its receptors. Melatonin binding sites have been identified (Cohen *et al.* 1978; Cardinali *et al.* 1979; Lang *et al.* 1981*b*) and the sensitivity of such 'receptors' may exhibit some rhythmic and life-long modulation in particular during puberty. Thus, measurements of melatonin receptors may bring some new information on the role of melatonin on pubertal development.

General conclusions

Melatonin, the principal secretory product of the pineal gland influences reproductive function in mammals, although in the human, its role has not yet been well defined. Changes in melatonin secretion have been observed during the menstrual cycle, and also during pubertal development, but these require confirmation. However, a role for melatonin as a time-keeping hormone remains, but still has to be definitively demonstrated.

References

Aleem, F. A., Weitzman, E. D., and Weinberg, U. (1984). Suppression of basal luteinizing hormone concentrations by melatonin in post-menopausal women. *Fertility and Sterility* **42**, 923–5.

Arendt, J. (1985). Mammalian Pineal Rhythms. *Pineal Research Reviews* **3**, 161–213.

Arendt, J. (1985). General Discussion II. Melatonin assay technology. In *Photoperiodism, melatonin and the pineal.* Ciba Foundation Symposium **117**, pp. 300–4. Pitman, London.

Arendt, J. (1986). Assay of melatonin and its metabolites: Results in normal and unusual environments. *Journal of Neural Transmission* (Suppl.) **21**, 11–33.

Arendt, J., Wetterberg, L., Heyden, T., Sizonenko, P. C., and Paunier, L. (1977). Radioimmunoassay of melatonin: human serum and cerebrospinal fluid. *Hormone Research* **8**, 65–75.

Arendt, J., Symons, A. M., Land, C. A., and Pryde, S. J. (1983). Melatonin can induce early onset of the breeding season in ewes. *Journal of Endocrinology* **97**, 395–400.

Attanasio, A., Borrelli, P., Marini, R., Cambiaso, P., Cappa, M., and Gupta, D. (1983). Serum melatonin in children with early and delayed puberty. *Neuroendocrinology Letters* **5**, 387–92.

Attanasio, A., Borelli, P., and Gupta, D. (1985). Circadian rhythms in serum melatonin from infancy to adolescence. *Journal of Clinical Endocrinology and Metabolism* **61**, 388–90.

Bartsch, C., Bartsch, H., Jain, A. K., Laumas, K. R., and Wetterberg, L. (1981). Urinary melatonin levels in human breast cancer patients. *Journal of Neural Transmission* **52**, 281–94.

Benson, B., Matthews, M. J., and Hurby, V. J. (1976). Characterization and effects of a bovine pineal antigonadotrophic peptide. *American Zoology* **16**, 17–19.

Bittman, E. L., Dempsey, R. J., and Karsch, F. J. (1983). Pineal melatonin secretion drives reproductive response to daylight in the ewe. *Endocrinology* **113**, 2276–83.

Cardinali, D. P. (1981). Melatonin: a mammalian pineal hormone. *Endocrine Reviews* **2**, 327–46.

Cardinali, D. P., Vacas, M. I., and Boyer, E. E. (1979). Specific binding of melatonin in bovine brain. *Endocrinology* **105**, 437–41.

Carter, D. S. and Goldman, B. D. (1983). Progonadal role of the pineal in the Djungarian hamster: mediation by melatonin. *Endocrinology* **113**, 1268.

Chik, C. L., Talalla, A., and Brown, G. M. (1985). Effect of pinealectomy on serum melatonin, luteinizing hormone and prolactin: a case report. *Clinical Endocrinology* **23**, 367–72.

Clayton, R. N. and Catt, K. J. (1981). GnRH receptors: characterization, physiological regulation, and relationship to reproductive function. *Endocrine Reviews* **2**, 186–215.

Cohen, M., Roselle, D., Chabner, B., Schmidt, T. J., and Lippmann, M. (1978). Evidence for a cytoplasmic melatonin receptor. *Nature* **274**, 894–5.

Cohen, H. N., *et al.* (1982). Serum immunoreactive melatonin in boys with delayed puberty. *Clinical Endocrinology* **17**, 517–21.

Danforth, D. N., Jr, Tamarkin, L., and Lippman, M. E. (1983). Melatonin increases oestrogen receptor binding activity of human breast cancer cells. *Nature* **305**, 323–5.

Danforth, D. N. Jr, Tamarkin, L., Mulvihill, J. J., Bagley, C. S., and Lippman, M. E. (1985). Plasma melatonin and the hormone-dependency of human breast cancer. *Journal of Clinical Oncology* **3**, 941–8.

Desir, D., Kirkpatrick, C., Fevre-Montagne, M., and Tourniaire, J. (1983). Ritodrine increases plasma melatonin in woman. *Lancet* i, 184–5.

Elden, C. A. (1971). Sterility of blind women. *Japan Journal of Fertility and Sterility* **16**, 48–50.

Ehrenkranz, J. R., *et al.* (1982). Daily rhythm of plasma melatonin in normal and precocious puberty. *Journal of Clinical Endocrinology and Metabolism* **55**, 307–10.

Evans, A. N., and Smith, J. A. (1979). Melatonin in the maternal and umbilical circulations during human parturition (letter). *British Journal of Obstetrics and Gynaecology* **86**, 831.

Fellenberg, A. J., Phillipou, G., and Seamark, R. F. (1982). Urinary 6-sulphatoxymelatonin excretion during the human menstrual cycle. *Clinical Endocrinology* **17**, 71–5.

Fevre, M., Segel, T., Marks, J. F., and Boyar, R. M. (1978). LH and melatonin secretion patterns in pubertal boys. *Journal of Clinical Endocrinology and Metabolism* **47**, 1383–6.

Fideleff, H., Aparico, N. J., Guitelman, A., Debeljuk, L., Mancini, A., and

Cramer, C. (1976). Effect of melatonin on the basal and stimulated gonadotropin levels in normal men and postmenopausal women. *Journal of Clinical Endocrinology and Metabolism* **42**, 1014–17.

Forster, D. L., Karsch, F. J., Olster, D. H., Ryan, K. D., and Yellon, S. M. (1986). Determinants of puberty in a Seasonal Breeder. *Recent Progress in Hormone Research* **42**, 331–78.

Goldman, N. B., Hall, V., Hollister, C., Roychoudhury, P., Tamarkin, L., and Westrom, W. (1979). Effects of melatonin on the reproductive system in intact and pinealectomized male hamsters maintained under various photoperiods. *Endocrinology* **104**, 82–8.

Gupta, D. (1986). The pineal gland in relation to growth and development in children. *Journal of Neural Transmission* (Suppl.) **21**, 217–32.

Gupta, D., Riedel, L., Frick, H. J., Attanasio, A., and Ranke, M. B. (1983). Circulating melatonin in children in relation to puberty, endocrine disorders, functional tests and racial origin. *Neuroendocrinology Letters* **5**, 63–78.

Herrick, M. K. (1984). Pathology of pineal tumors. In *Diagnosis and Treatment of Pineal Region Tumors* (ed. E. A. Neuwelt), pp. 31–60. Williams & Wilkins, Baltimore, Maryland.

Hoffmann, K. (1979). Photoperiod, pineal melatonin and reproduction in hamsters. *Progress in Brain Research* **52**, 397–415.

Kennaway, D. J. (1984). Pineal functions in ungulates. *Pineal Research Reviews* **2**, 113–41.

Kennaway, D. J., McCulloch, G., Matthews, C. D., and Seamark, R. F. (1979). Plasma melatonin, luteinizing hormone, follicle-stimulating hormone, prolactin, and corticoids in two patients with pinealoma. *Journal of Clinical Endocrinology and Metabolism* **42**, 144–5.

Kennaway, D. J., Matthews, C. D., and Seamark, R. F. (1981). Pineal function in pregnancy: studies in sheep and man. *Pineal Function* (ed. C. D. Matthews, and R. F. Seamark), pp. 123–36.

Kitay, J. I. (1954). Pineal lesions and precocious puberty: a review. *Journal of Clinical Endocrinology and Metabolism* **14**, 622–5.

Kitay, J. I. and Altschule, M. D. (1954). *The pineal gland. A review of the physiologic literature*. Harvard University Press, Cambridge, Massachussetts.

Lang, U. (1986). Melatonin and Puberty. *Pineal Research Reviews* **4**, 199–243.

Lang, U., Kornemark, M., Aubert, M. L., Paunier, L., and Sizonenko, P. C. (1981*a*). Radioimmunological determination of urinary melatonin in humans: correlation with plasma levels and typical 24 hour rhythmicity. *Journal of Clinical Endocrinology and Metabolism* **53**, 645–50.

Lang, U., Aubert, M. L., and Sizonenko, P. C. (1981*b*). Location of melatonin receptors. *Paediatric Research* **15**, 80.

Lang, U., Aubert, M. L., Conne, B. S., Bradtke, J. C., and Sizonenko, P. C. (1983). Influence of exogenous melatonin on melatonin secretion and on the neuroendocrine reproductive axis of intact male rats during sexual maturation. *Endocrinology* **112**, 1578–84.

Lang, U., Aubert, M. L., Rivest, R. V. W., Vinas-Bradtke, J. C., and Sizonenko, P. C. (1984*a*). Daily afternoon administration of melatonin does not irreversibly inhibit sexual maturation in male rat. *Endocrinology* **115**, 2303–10.

Lang, U., Rivest, R. W., Schlaepfer, L., Bradtke, J., Aubert, M. L., and Sizonenko, P. C. (1984*b*). Diurnal rhythm of melatonin action on sexual maturation of male rats. *Neuroendocrinology* **38**, 261–8.

Lang, L., Beguin, F., and Sizonenko, P. C. (1986*a*). Fetal and maternal (MT) concentrations at birth in humans. *Journal of Neural Transmission* (Suppl.) **21**, 479–80.

Lang, U., Theintz, G., Rivest, R. W., and Sizonenko, P. C. (1986*b*). Nocturnal urinary melatonin excretion and plasma cortisol levels in children and adolescents after a single dose of dexamethasone. *Clinical Endocrinology* **25**, 165–72.

Lapin, V. (1979). Pineal influence on tumour. *Progress in Brain Research* **42**, 523–33.

Lehrer, S. (1985). Puberty and menopause in the human: Possible relation to gonadotropin-releasing hormone pulse frequency and the pineal gland. *Pineal Research Reviews* **3**, 237–57.

Lenko, H. L., Lang, U., Aubert, M. L., Paunier, L., and Sizonenko, P. C. (1982). Hormonal changes in puberty. VII. Lack of variation of daytime plasma melatonin. *Journal of Clinical Endocrinology and Metabolism* **54**, 1056–8.

Lerner, A. B. and Nordlund, J. J. (1978). Melatonin: clinical pharmacology. *Journal of Neural Transmission* (Suppl.) **13**, 339–47.

Lieberman, H. R., Waldhauser, F., Garfield, G., Lynch, H. J., and Wurtman, R. J. (1984). Effects of melatonin on human mood and performance. *Brain Research* **323**, 201–7.

Lissoni, P., *et al.* (1986). Effect of an acute injection of melatonin on the basal secretion of hypophyseal hormones in prepubertal and pubertal healthy subjects. *Acta Endocrinologica* **111**, 305–11.

MacPhee, A. A., Cole, F. E., and Rice, B. F. (1975). The effect of melatonin on steroidogenesis by the human ovary *in vitro*. *Journal of Clinical Endocrinology and Metabolism* **40**, 688–96.

Magee, K., Basinska, J., Quarrington, B., and Stancer, H. C. (1970). Blindness and menarche. *Life Sciences* **9**, 7–12.

Mauri, R., *et al.* (1985). Effects of melatonin on prolactin secretion during different photoperiods of the day in prepubertal and pubertal healthy subjects. *Journal of Endocrinological Investigation* **8**, 337–41.

Mitchell, M. D., Sayers, L., Keirse, M. J. N. C., Anderson, A. B. M., and Turnbull, A. C. (1978). Melatonin in amniotic fluid during parturition. *British Journal of Obstetrics and Gynaecology* **85**, 684–6.

Mitchell, M. D., Bibby, J. G., Sayers, L., Anderson, A. B. M., and Turnbull, A. C. (1979). Melatonin in the maternal and umbilical circulations during human parturition. *British Journal of Obstetrics and Gynaecology* **86**, 29–31.

O'Hare, J. P., *et al.* (1986). Does melatonin deficiency cause the enlarged genitalia of the fragile-X syndrome? *Clinical Endocrinology* **24**, 327–33.

Pang, S. F., Tang, P. L., Tang, G. W. K., and Yam, A. W. C. (1986). Melatonin and pregnancy. In *The pineal gland* (ed. G. M. Brown and S. D. Wainwright), pp. 157–62. Pergamon Press, Oxford.

Pavel, S. (1978). Arginine vasotocin as a pineal hormone. *Journal of Neural Transmission* (Suppl.) **13**, 135–55.

Penny, R. (1982). Melatonin excretion in normal males and females: increase during puberty. *Metabolism* **31**, 816–23.

Penny, R. and Goebelsmann, U. (1984). Effect of oestradiol on plasma melatonin levels. *Journal of Endocrinologcal Investigation* **7**, 55–7.

Petterborg, L. J. and Reiter, R. J. (1980). Effect of photoperiod and melatonin on testicular development in the White-footed Mouse, *Peromyscus leucopus. Journal of Reproduction and Fertility* **60**, 209–12.

Pevet, P. (1983). Is 5-methoxytryptamine a pineal hormone? *Psychoneuroendocrinology* **8**, 61–73.

Reiter, R. J. (1980). The Pineal and its Hormones in the Control of Reproduction in Mammals. *Endocrine Reviews* **1**, 109–31.

Reppert, S. M. and Klein, D. C. (1978). Transport of maternal ^{3}H-melatonin to suckling rats and the fate of ^{3}H-melatonin in the neonatal rat. *Endocrinology* **102**, 582–8.

Reppert, S. M., Chez, R. A., Anderson, A., and Klein, D. C. (1979). Maternal-fetal transfer of melatonin in the non-human primate. *Paediatrics Research* **13**, 788–91.

Russell, D. S. (1977). Pineal neoplasms. In *Pathology of tumours of the nervous system* (ed. D. S. Russell and L. J. Rubinstein), pp. 283–98. Williams & Wilkins, Baltimore, Maryland.

Silman, R. E., Leone, R. M., Hooper, R. J. L., and Preece, M. A. (1979). Melatonin, the pineal gland and human puberty. *Nature* **282**, 301–3.

Sizonenko, P. C. and Aubert, M. L. (1986). Neuroendocrine changes characteristic of sexual maturation. *Journal of Neural Transmission* (Suppl.) **21**, 159–81.

Sizonenko, P. C. and Paunier, L. (1975). Hormonal changes during puberty. III. Correlation of plasma dehydroepiandrosterone, testosterone, FSH and LH with stage of puberty and bone age in normal boys and girls and in patients with Addison's disease or hypogonadism or with premature or late adrenarche. *Journal of Clinical Endocrinology and Metabolism* **41**, 894–904.

Sizonenko, P. C., Lang, U., Rivest, R. W., and Aubert, M. L. (1985). The pineal and pubertal development. In *Photoperiodism, melatonin and the pineal.* Ciba Foundation Symposium **117**, pp. 208–25. Pitman, London.

Sklar, C. A., Conte, F. A., Kaplan, S. L., and Grumbach, M. M. (1981). Human chorionic gonadotrophin-secreting pineal tumour: Relation to pathogenesis and sex limitation of sexual precocity. *Journal of Clinical Endocrinology and Metabolism* **53**, 656–60.

Smythe, C. A. and Lazarus, L. (1974). Growth hormone response to melatonin in man. *Science* **134**, 1373–4.

Tamarkin, L., Westrom, W. K., Hamill, A. I., and Goldman, B. D. (1976). Effect of melatonin on the reproductive systems of male and female Syrian hamsters: a diurnal rhythm in sensitivity to melatonin. *Endocrinology* **99**, 1534–41.

Tamarkin, L., Abastillas, P., McNemar, A., and Sidbury, J. B. (1982*a*). The daily profile of plasma melatonin in obese and Prader-Willi syndrome children. *Journal of Clinical Endocrinology and Metabolism* **55**, 491–5.

Tamarkin, L., *et al.* (1982*b*). Decreased nocturnal melatonin peak in patients with oestrogen receptor positive breast cancer. *Science* **216**, 1003–5.

Tamarkin, L., Almeida, O. F. X., and Danforth, D. N., Jr (1985). Melatonin and malignant disease. In *Photoperiodism, melatonin and the pineal*, Ciba Foundation Symposium **117**, pp. 284–94. Pitman, London.

Tetsuo, M., Poth, M., and Markey, S. P. (1982). Melatonin metabolite excretion during childhood and puberty. *Journal of Clinical Endocrinology and Metabolism* **55**, 311–13.

Thomas, J. and Pizzarello, D. (1967). Blindness, biologic rhythms and menarche. *Obstetrics and Gynecology* **30**, 507–9.

Timonen, S. and Carpen, E. (1969). Multiple pregnancies and photoperiodicity. *Annales Chirurgiae et Gynaecologiae Fenniae* **57**, 135–8.

Timonen, S., Franzas, B., and Wichmann, K., (1967). Photosensibility of the human pituitary. *Annales Chirurgiae et Gynaecologiae Fenniae* **53**, 165–72.

Turek, F. W., Desjardins, C., and Menaker, M. (1975). Melatonin: antigonadal and progonadal effects in male golden hamsters. *Science* **190**, 280–2.

Vaughan, G. M. (1984). Melatonin in humans. *Pineal Research Review* **2**, 141–201.

Vaughan, M. K., Vaughan, G. M., and Klein, D. C. (1974). Arginine-vasotocin: effects on development of reproductive organs. *Science* **186**, 938–9.

Vollrath, L., Semm, P., and Gammel, G. (1981). Sleep induction by intranasal application of melatonin. *Advances in the Biosciences* **29**, 327–9.

Waldhauser, F. and Steger, H. (1986). Changes in melatonin secretion with age and pubescence. *Journal of Neural Transmission* (Suppl.) **21**, 183–97.

Waldhauser, F. and Wurtman, R. J. (1983). The secretion and actions of melatonin. *Biochemical Actions of Hormones* **10**, 187–225.

Waldhauser, F., Weiszenbacher, G., Frisch, U., Waldhauser, M., and Wurtman, R. J. (1984). Fall in nocturnal serum melatonin during prepuberty and pubescence. *Lancet* i, 362–5.

Webley, G. E. and Leidenberger, F. (1986). The circadian pattern of melatonin and its positive relationship with progesterone in women. *Journal of Clinical Endocrinology and Metabolism* **63**, 323–8.

Weinberg, U., Weitzman, E. D., Fukushima, D. K., Cancel, G. F., and Rosenfeld, R. S. (1980). Melatonin does not suppress the pituitary luteinizing hormone response to luteinizing hormone-releasing hormone in men. *Journal of Clinical Endocrinology and Metabolism* **51**, 161–2.

Weinberg, G., Weitzman, E. D., Horowitz, Z. D., and Burg, A. C. (1981). Lack of an effect of melatonin on the basal and the L-dopa stimulated growth hormone secretion in men. *Journal of Neural Transmission* **52**, 117–21.

Wetterberg, L., Arendt, J., Paunier, L., Sizonenko, P. C., Donselaar, W., and Heyden, T. (1976). Human serum melatonin changes during the menstrual cycle. *Journal of Clinical Endocrinology and Metabolism* **42**, 185–8.

Whitsett, J. M., Underwood, H., and Cherry, J. (1984). Influence of melatonin on pubertal development in male deer mice (*Peromyscus maniculatas*). *Journal of Reproduction and Fertility* **72**, 287–93.

Wright, J., Aldhous, M., Franey, C., English, J., and Arendt, J. (1986). The effects of exogenous melatonin on endocrine function in man. *Clinical Endocrinology* **24**, 375–82.

Wurtman, R. J., Axelrod, J., and Chu, E. W. (1963). Melatonin a pineal substance:

effect on the rat ovary. *Science* **141**, 277–8.

Wurtman, R. J., Axelrod, J., and Fischer, J. E. (1964). Melatonin synthesis in the pineal gland: effect of light mediated by the sympathetic nervous system. *Science* **143**, 1328–30.

Zacharias, L. and Wurtman, R. (1964). Blindness: its relation to age of menarche. *Obstetrics and Gynaecology* **30**, 507–9.

4. Melatonin and human adrenocortical function

Ursula Lang and Pierre C. Sizonenko

Introduction

Compared to the considerable number of studies concerning the influence of melatonin on reproduction in mammals, interactions between adrenal and pineal functions have been investigated sparingly. In addition, there exist contradictory observations in the literature, especially in early reports, as to whether or not the pineal gland or melatonin influence adrenal development and adrenal functions.

The commonest result of pinealectomy appears to be adrenal hypertrophy which can be reversed by treatment with pineal extracts or melatonin. In 1959, Wurtman *et al.* found that pinealectomized rats had an adrenal enlargement which could be reversed by pineal extracts (Wurtman *et al.* 1959). These observations have been confirmed by other investigators who found that pinealectomy increases adrenal weight and corticosteroid release in rodents (Vaughan *et al.* 1972; Relkin 1983). Melatonin injections or administration of pineal extracts, on the other hand, have been found to decrease adrenal weight and function in rodents (Motta *et al.* 1968; Vaughan *et al.* 1972; Ogle and Kitay 1977; Johnson 1982; Relkin 1983). *In vitro* studies with adrenal homogenates, and slices from rat and beef (Ogle and Kitay 1978; Giordano *et al.* 1970; Mehdi and Sandor 1977) showed that melatonin inhibited corticosteroid biosynthesis. In contrast, elevation of serum melatonin levels by administration of the hormone to normal human individuals did not influence serum cortisol concentrations (Waldhauser *et al.* 1984). Suppression of endogenous melatonin secretion by light in non-human primates also did not affect serum cortisol levels (Perlow *et al.* 1981).

The influence of the adrenal gland on pineal functions has also been studied. Reiter *et al.* (1982) reported that bilateral adrenalectomy severely limited the nocturnal rise in pineal melatonin in the rat. However, this observation was not confirmed by Champney *et al.* (1985) who found no significant alterations of melatonin levels in either adrenalectomized or corticosteroid-implanted rats. The same authors observed an increase in pineal levels of melatonin in adrenalectomized hamsters.

Likewise, contrary observations have also been reported in reference to enzymes which are necessary for the biosynthesis of melatonin, such as N-acetyltransferase (NAT) and hydroxyindole-O-methyltransferase (HIOMT) in adrenalectomized rats. Sugden and Klein (1983) found no effect of adrenalectomy or hypophysectomy on pineal HIOMT in neonatal or adult rats, whereas Deussen-Schmitter *et al.* (1976) described an increase in pineal HIOMT activity in adrenalectomized rats. Reiter and colleagues (1982) found a decreased NAT activity in pineal glands of adrenalectomized rat, but Lynch and associates (1977) observed no effect of adrenal removal on NAT activity.

The aim of the present chapter is to survey and to summarize present data on a possible interaction of pineal and adrenocortical functions in the human.

Melatonin and cortisol in psychiatric disease

The reports on alterations of melatonin and cortisol secretion in psychiatric patients are most interesting. Increased serum cortisol levels are well established in certain types of depression as well as an abnormal response to the dexamethasone suppression test (Sacher 1975; Caroll 1984). An evening dose of dexamethasone normally suppresses nocturnal cortisol surge, but in some depressed patients normal suppression does not occur.

In a summary of a series of studies by his group, Wetterberg (1983) described lower serum melatonin levels measured at 02.00 hours in 17 patients with major depression who were non-suppressors of cortisol in response to dexamethasone, compared with melatonin levels in 22 control subjects and 15 patients with major depression, but a normal suppression of cortisol in response to dexamethasone. Several other studies describe a coincidence of high serum cortisol levels and low nocturnal serum melatonin levels in depressive patients (Wetterberg *et al.* 1979, 1982, 1984; Branchey *et al.* 1982; Lewy *et al.* 1979; Claustrat *et al.* 1984). Ferrier *et al.* (1982) reported elevated serum cortisol concentrations and decreased serum melatonin levels in chronic schizophrenia. These observations were taken as evidence for an interaction in the regulation of cortisol and melatonin secretion (Wetterberg *et al.* 1981, 1982, 1984). However, Branchey *et al.* (1982) observed a dissociation of melatonin and cortisol secretory patterns in depressive patients in whom the two hormones were determined during illness and after recovery. Likewise, Thompson *et al.* (1985) reported that in six depressed patients nocturnal melatonin secretion was significantly increased after 3 weeks of treatment with desipramine, whereas cortisol values remained unchanged. When the antidepressant drug fluvoxamine was given to six healthy male volunteers, an increase in melatonin and cortisol secretion was observed, although alterations in plasma cortisol levels were less pronounced than those of melatonin levels (Demisch *et al.* 1986). For further discussion of the phenomenon, the reader is referred to chapters 10, 12, and 13 of this volume.

Melatonin in Cushing's syndrome

Studies on serum and urinary melatonin in conditions of sustained elevated free plasma cortisol (Cushing's syndrome) resulted in discordant observations.

There exist three main causes of Cushing's syndrome: adrenocortical adenoma and carcinoma, pituitary-dependent bilateral adrenocortical hyperplasia (Cushing's disease), and ectopic ACTH syndrome.

Werner *et al.* (1980) observed a normal circadian pattern of melatonin secretion in three patients with Cushing's syndrome and in four of ten patients with Cushing's disease with ACTH producing adenomas, whereas among the other six patients, four had a reduced and two had an absent nocturnal rise in serum melatonin. Young (1981) reported three cases with Cushing's disease having normal nocturnal rises in plasma melatonin. Vaughan *et al.* (1979) studied a bilaterally adrenalectomized patient suffering from Cushing's disease and found no alteration in diurnal melatonin secretion irrespective of on or off glucocorticoid replacement. In contrast, Fevre-Montagne *et al.* (1983) observed an increase of plasma melatonin levels during the day in six women with Cushing's syndrome (four adrenal adenoma, two hyperplasia) compared to values from healthy individuals.

Melatonin in congenital adrenal hyperplasia

Congenital adrenal hyperplasia (CAH) is a disease resulting from a defect in cortisol biosynthesis. The biosynthetic pathway by which cholesterol is converted to cortisol can be interrupted by various enzymes defects, the most common being the 21-hydroxylase deficiency. All those enzymic deficiencies cause an accumulation of cortisol precursor(s) and this results in reduced levels of circulating cortisol, which leads to a compensatory increase in ACTH (New and Levine 1984). Waldhauser and colleagues (1986) have studied a possible interaction of the adrenal and pineal functions in children with congenital hyperplasia both on and off corticoid replacement therapy. Thirteen children with congenital hyperplasia due to 21-hydroxylase deficiency were examined on two occasions: on one occasion 3 days after interruption of cortisol substitution (hydrocortone) and on another occasion during treatment with dexamethasone and fludrocortisone. Their serum levels of 17 OH-progesterone (an indicator for the activation of the hypothalamo–pituitary–adrenal axis in CAH) and their serum melatonin concentrations were compared to those of children suffering from minor non-endocrine diseases (control group).

Serum 17-OH progesterone levels were much higher in CAH patients after cessation of cortisol substitution treatment when compared to those of the other groups. In contrast, no difference was found in serum melatonin levels among endocrinologically normal children and children suffering from CAH in different states of stimulation of the hypothalamo–pituitary–adrenal axis. In all children, melatonin secretion showed a circadian rhythm with low serum levels during the day and markedly higher levels during the night with peak values at 02.00 hours.

The study of Waldhauser *et al.* (1984) does thus not confirm the hypothesis of an interaction in the regulation of cortisol and melatonin secretion, but rather suggests the theory of two independent mechanisms for regulating these hormones (Perlow *et al.* 1981).

Inhibition of cortisol biosynthesis and melatonin secretion

Brismar and associates (1982) showed that artificial blockage of cortisol biosynthesis by metyrapone in patients with pituitary disorders increased urinary melatonin excretion. In a second study, Brismar *et al.* (1985) confirmed this observation in two healthy volunteers and in two patients with hyperparathyroidism and prolactin-secreting microadenoma, respectively. During acute inhibition of cortisol production induced by metyrapone, urinary melatonin excretion increased whereas serum level of melatonin remained unchanged. Since metyrapone was also found to increase the metabolic rate of melatonin (Brismar *et al.* 1985) the authors concluded that melatonin secretion was stimulated during metyrapone stimulation.

Urinary melatonin excretion appears to reflect an integration of the hormone secreted, since it has been shown that unmetabolized melatonin in the nocturnal urine is in good correlation with plasma levels at midnight (Lang *et al.* 1981*a,b*). Urinary melatonin also displays the characteristic circadian rhythm which has been observed for melatonin in the blood (Ozaki *et al.* 1976; Lang *et al.* 1981*a,b*). However, it is well known that most of the circulating melatonin is rapidly metabolized in the liver by a microsomal hydroxylase to form 6-hydroxymelatonin (Kopin *et al.* 1961). Since metyrapone is an unspecific enzyme blocker

(Levine *et al.* 1978), it inhibits several other enzymes, particularly microsomal liver enzymes (Lieberman 1969) in addition to adrenal 11-hydroxylase. It is therefore possible that metyrapone increases the amount of unchanged excreted melatonin by partially blocking metabolizing enzyme(s).

Studies from our laboratory (Lang *et al.* 1986) on cortisol and melatonin secretion in children and adolescents after dexamethasone treatment showed also a coincidence of low plasma cortisol levels and high melatonin excretion.

Nocturnal urinary melatonin content and plasma cortisol levels were determined in 41 children (18 males and 23 females) with weight problems before and after a single oral dose of dexamethasone at 20.00 hours. Blood samples were taken at 08.00 and 17.00 hours before and at 08.00 hours after dexamethasone administration. Other causes of obesity than exogenous and/or familial have been ruled out. Urine was collected on two consecutive nights from 20.00 to 08.00 hours before and after dexamethasone treatment. Statistical analysis of the mean of all individuals before and after dexamethasone showed a highly significant increase in nocturnal melatonin excretion after dexamethasone administration (43.3 ± 4.0 ng/12 hr $v.$ 26.8 ± 1.8, $P < 0.001$). No correlation was found between body weight and nocturnal melatonin excretion, or between body weight and plasma levels of cortisol.

An interesting observation was made when individuals were grouped according to their cortisol cycle and cortisol response to dexamethasone, as shown in Table 4.1.

A first group of 15 patients, with plasma cortisol levels at 17.00 hours below the 50 per cent value of the morning level and with cortisol levels below 1.0 μg/100 ml after a single oral dose of dexamethasone, showed a highly significant increase of nocturnal melatonin excretion after dexamethasone treatment (Table 4.1, group 1). This increase in urinary melatonin content was significant in each pubertal stage (Fig. 4.1). In a second group of 16 patients with mean cortisol levels similar at 08.00 and 17.00 hours, but with a normal cortisol suppression after dexamethasone (< 1 μg/100 ml), nocturnal melatonin excretion also increased significantly (Table 4.1, group 2). This overall increase was due to the marked response of prepubertal children (stage PI, Fig. 4.1). A third group of 10 patients, with low-amplitude cortisol cycles and poor cortisol suppression after dexamethasone administration, showed no increase in melatonin excretion (Table 4.1, group 3).

Among these 41 subjects, six had delayed puberty. All these patients showed a marked increase in nocturnal melatonin excretion in response to oral dexamethasone, independent of cortisol values and pubertal stages (76.1 ± 7.2 ng/12 hr $v.$ 38.6 ± 6.9 for the control night). Nocturnal

Ursula Lang and Pierre C. Sizonenko

Table 4.1. Nocturnal melatonin excretion in correlation to plasma cortisol levels before and after dexamethasone administration (DEX)

Group	Plasma cortisol (μg/100 ml)			Urinary melatonin (ng/12 hr)	
	08.00 hours	17.00 hours	08.00 hours (DEX)	20.00–08.00 hours	20.00–08.00 hours (DEX)
I ($n=15$)	12.2 ± 1.4	3.1 ± 0.5	< 1	33.6 ± 3.0	63.5 ± 5.5***
II ($n=16$)	10.7 ± 1.3	9.3 ± 1.5	< 1	21.5 ± 2.2†	34.7 ± 5.6**
III ($n=10$)	16.2 ± 2.5	10.8 ± 2.4	9.1 ± 2.4	24.2 ± 2.7†	24.9 ± 3.7

*** $P < 0.001$; ** $P < 0.01$ (paired t-test: values before and after dexamethasone treatment.
† $P < 0.05$ (ANOVA) plus LSD test: control values of group II and III compared to group I. (Reproduced from Lang et al. 1986.)

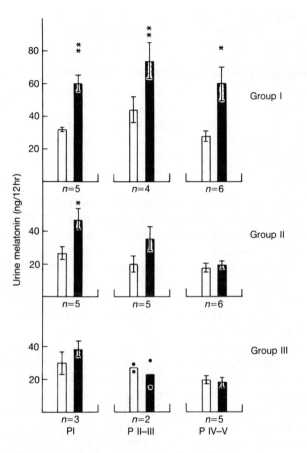

Fig. 4.1. Influence of plasma cortisol levels and pubertal stage on nocturnal melatonin excretion before (open bars) and after a single oral dose of dexamethasone (full bars).

Pubertal stages (PI–PV) are indicated at the bottom of the graph. Group I contained individuals whose plasma cortisol levels at 17.00 hours were below the 50 per cent value of the morning cortisol level and with normal cortisol suppression in response to dexamethasone. Group II represents subjects whose cortisol levels at 17.00 hours were equal or higher than the 50 per cent value at 08.00 hours, but whose response to dexamethasone was normal. Group III contained individuals with a low-amplitude cortisol cycle and an abnormal response to dexamethasone. n=number of individuals. Values are given as mean ±SEM. Statistical significance for different pubertal stages and groups were determined with paired Student's t-test and ANOVA plus LSD test (individuals and individual differences before and after dexamethasone administration). ***$P<0.001$; **$P<0.01$; *$P<0.05$.

melatonin excretion before any treatment (control night) was significantly higher in group 1 patients (normal cortisol cycle and normal cortisol suppression after dexamethasone) than in those of groups 2 and 3 (Table 4.1). Similar observations have been made by Wetterberg and associates (1979, 1983) who found that low nocturnal plasma melatonin levels were more often associated with increased cortisol secretion and an abnormal response to dexamethasone.

In contrast, dexamethasone administration at 10.00 hours had no effect on the circadian rhythm of serum melatonin levels in normal adult volunteers (Beck-Friis *et al.* 1983), except at 08.00 hours (after dexamethasone administration), when reduced melatonin levels were observed.

In this context, it may be important to take into consideration the age of the studied individuals. Recent studies in humans (Waldhauser *et al.* 1984; Gupta *et al.* 1983) reported much higher nocturnal melatonin concentrations in children younger than 7 years, and their progressive decrease during late prepuberty and puberty, indicating an interaction between the pineal gland and the pituitary–gonadal axis. Our results show that overnight cortisol suppression is more often associated with increased melatonin production during prepuberty and at the early stages of puberty than later in adolescence.

During prepubertal childhood, between 7 and 8 years of age, maturation of the androgenic zone of the adrenal cortex (adrenarche) starts and plasma levels of adrenal androgens (DHEA and DHEA-S) increase (Sizonenko 1978). The hypothalamic–pituitary factor(s) inducing this maturation remain unknown. ACTH, LH, and an unidentified pituitary factor have been suggested to stimulate adrenal androgen secretion at puberty (Sizonenko 1978). The possibility exists that melatonin might influence the hypothalamic or pituitary secretion of the factor(s) stimulating the adrenal cortex or that changes in plasma melatonin levels induce adrenocortical maturation. We observed high concentrations of specific melatonin-binding sites in adrenal glands, hypothalamus, and pituitary gland of male rats (Lang *et al.* 1981*a*). It also has been found that pineal extract and melatonin influence steroid production in adrenal glands (Ogle and Kitay 1978; Mehdi and Sandor 1977; Heiman and Porter 1980).

The opposite relationship could also exist, i.e. cortisol or/and adrenal androgens affecting the pineal gland directly or indirectly through hypothalamic or pituitary factors, inducing a change in melatonin secretion. This change could then start maturation of the neuroendocrine-reproductive axis as suggested by animal experimentation. However, further studies are necessary to confirm this hypothesis, and also to validate the observations concerning a coincidence of low cortisol secretion and high melatonin excretion (Brismar *et al.* 1982; Lang *et al.* 1986). Since most of the circulating melatonin is metabolized and excreted as the glucuronide and sulphate conjugates of 6-hydroxy-melatonin (Kopin *et al.* 1961; Jones *et al.* 1969), it would be important to compare our results of melatonin excretion and those of Brismar and colleagues (1982, 1985) with measurements of 6-hydroxymelatonin in the urine.

Such an analysis could also give information about a possible interaction of cortisol, dexamethasone, or metyrapone with melatonin-metabolizing enzymes.

Summary and concluding remarks

Several studies have reported possible interactions between adrenal and pineal functions in mammals. However, it is not yet possible to attribute a definitive influence of the pineal gland on adrenal functions, or to describe precise effect(s) of the adrenal glands on melatonin secretion.

In the animal model the most consistent observations report an inhibitory influence of the pineal gland and melatonin on adrenal glands, whereas studies on a possible adrenal influence on pineal development and melatonin secretion resulted in contradictory observations.

In man, it is even more difficult to evaluate a possible relationship in the regulation of cortisol and melatonin secretion. Here, the most con-sistent observations have described an association of both cortisol hyper-secretion and abnormal cortisol response to dexamethasone with reduced melatonin secretion in psychiatric patients. Studies on melatonin secretion with Cushing's syndrome resulted in very discordant findings, and Waldhauser and associates (1986) found no difference in melatonin secretion between endocrinologically normal children and children suffering from congenital adrenal hyperplasia on and off corticosteroid replacement therapy.

Studies on melatonin secretion and action in the animal model have shown that, in studying the role of melatonin, experimental conditions such as age, time of the day, etc., have to be very carefully defined and followed. In this context, it might be important to take into consideration the age and the pubertal stage of development when studying a possible relationship between cortisol and melatonin secretion.

In our studies we have observed that overnight cortisol suppression is more often associated with increased nocturnal melatonin production during prepuberty than later during pubertal development. Waldhauser and colleagues (1984) found considerably higher nocturnal plasma melatonin levels in very young children (1–5 years) than in adolescents. These observations suggest that in man interactions between adrenal and pineal functions are more likely to occur at a very early stage of pubertal development, possibly before or during adrenarche, rather than later during adolescence or in adulthood.

References

Beck-Friis, J., Hansen, T., Ljunggren, J.-G., Unden, F., and Wetterberg, L. (1983). Serum melatonin and cortisol in human subjects after the administration of dexamethasone and propranolol. *Psychopharmacology Bulletin* **19**, 646–8.

Branchey, L., Weinberg, U., Branchey, M., Linkowski, P., and Mendlewicz, J. (1982). Simultaneous study of 24-hour patterns of melatonin and cortisol secretion in depressed patients. *Neuropsychobiology* **8**, 225–32.

Brismar, K., Werner, S., and Wetterberg, L. (1982). Melatonin and corticosteroid response to metyrapone in patients with pituitary disease. In *The Pineal and its Hormones* (ed. R. J. Reiter), pp. 283–92. Alan R. Liss, New York.

Brismar, K., Werner, S., Thoren, M., and Wetterberg, L. (1985). Metyrapone—an agent for melatonin as well as ACTH and cortisol secretion. *Journal of Endocrinological Investigation* **2**, 91–5.

Caroll, B. J. (1984). Dexamethasone suppression test for depression. In *Frontiers in biochemical and pharmacological research in depression* (ed. E. Usdin, M. Asberg, L. Bertilsson, and F. Sjöqvist), pp. 179–88. Raven Press, New York.

Champney, T. H., Craft, C. M., Webb, S. M., and Reiter, R. J. (1985). Hormonal modulation of pineal melatonin synthesis in rats and Syrian hamsters: effects of adrenalectomy and corticosteroid implants. *Journal of Neural Transmission* **64**, 67–79.

Claustrat, B., Chazot, G., Brun, J., Jordan, J., and Sassolas, G. (1984). A chronobiological study of melatonin and cortisol secretion in depressed subjects: Plasma melatonin, a biochemical marker in major depression. *Biological Psychiatry* **19**, 1215–17.

Demisch, K., *et al.* (1986). Melatonin and cortisol increase after fluvoxamine. *British Journal of Clinical Pharmacology* **22**, 620–2.

Deussen-Schmitter, M., Garweg, G., Schwabedal, P. E., and Wartenberg, H. (1976). Simultaneous changes of the perivascular contact area and HIOMT activity in the pineal organ after bilateral adrenalectomy in the rat. *Anatomic Embriology* **149**, 297–305.

Ferrier, I. N., Johnston, E. C., Crow, T. J., and Arendt, J. (1982). Melatonin/cortisol ratio in psychiatric illness. *Lancet* **i**, 1070.

Fevre-Montagne, M., Tourniaire, J., Estour, B., and Bajard, L. (1983). 24 hour melatonin secretory pattern in Cushing's syndrome. *Clinical Endocrinology (Oxford)* **18**, 175–81.

Giordano, G., Balesteri, R., Jacopino, G. E., Foppiani, E., and Bertolini, S. (1970). L'action in vitro de la mélatonine sur l'hormonosynthèse cortico-surrénale du rat. *Annales d'Encrinologie* **31**, 1071–80.

Gupta, D., Riedel, L., Frick, H. J., Attanasio, A., and Ranke, M. B. (1983). Circulating melatonin in children in relation to puberty, endocrine disorders, functional tests and racial origin. *Neuroendocrinological Letters* **5**, 63–78.

Heiman, M. L. and Porter, J. R. (1980). Inhibitory effect of pineal extract on adrenal cortex: lack of competition with ACTH. *Hormone Research* **12**, 104–12

Johnson, L. Y. (1982). The pineal gland as a modulator of the adrenal and thyroidal axes. In *The pineal gland*, vol. 2; *Extra-reproductive effects* (ed. R. J. Reiter), pp. 107–52. CRC Press, Boca Raton, Florida.

Jones, R. T., McGeer, P. L., and Greiner, A. C. (1969). Metabolism of exogenous melatonin in schizophrenic and non-schizophrenic volunteers. *Clinica Chimica Acta* **26**, 281–3.

Kopin, I. J., Pare, C. M. P., Axelrod, J., and Weissbach, H. (1961). The fate of melatonin in animals. *Journal of Biological Chemistry* **236**, 3072–5.

Lang, U. (1986). Melatonin and puberty. *Pineal Research Reviews* **4**, 199–243.

Lang, U., Kornemark, M., Aubert, M. L., Paunier, L., and Sizonenko, P. C. (1981*a*). Radioimmunological determination of urinary melatonin in human: correlation with plasma levels and typical 24 hour rhythmicity. *Journal of Clinical Endocrinology and Metabolism* **53**, 645–50.

Lang, U., Aubert, M. L., and Sizonenko, P. C. (1981*b*). Location of melatonin receptors. *Paediatrics Research* **15**, 80.

Lang, U., Theintz, G., Rivest, R. W., and Sizonenko, P. C. (1986). Nocturnal urinary melatonin excretion and plasma cortisol levels in children and adolescents after a single oral dose of dexamethasone. *Clinical Endocrinology* **25**, 165–72.

Levine, J., Zumoff, B., and Fukushima, K. (1978). Extra-adrenal effects of metyrapone in man. *Journal of Clinical Endocrinology and Metabolism* **47**, 845–9.

Lewy, A. J., Wehr, F. A., Gold, P. W., and Goodwin, F. K. (1979). Plasma melatonin in manic-depressive illness. In *Catecholamines: basic and clinical frontiers* (ed. E. Usdin, I. J. Kopin, and J. Barchas), *volume 2*, pp. 1173–5. Pergamon Press, New York.

Lieberman, K. C. (1969). Effects of metyrapone on liver microsomal drug oxidations. *Molecular Pharmacology* **5**, 1–9.

Lynch, H. J., Ho, M., and Wurtman, R. J. (1977). The adrenal medulla may mediate the increase in pineal melatonin synthesis induced by stress, but not that caused by exposure to darkness. *Journal of Neural Transmission* **40**, 87–97.

Mehdi, A. Z. and Sandor, T. (1977). The effect of melatonin on the biosynthesis of corticosteroids in beef adrenal preparations in vitro. *Journal of Steroid Biochemistry* **8**, 821–3.

Motta, M., Fraschini, F., Piva, F., and Martini, L. (1968). Hypothalamic and extra-hypothalamic mechanisms controlling adrenocorticotropin secretion. In *Memoirs of the Society of Endocrinology*, Vol. 17, pp. 3–18. Cambridge University Press, Cambridge.

New, M. I. and Levine, L. S. (1984). Steroid 21-hydroxylase deficiency. In *Adrenal diseases in childhood* (ed. I. M. New and L. S. Levine), pp. 1–46. S. Karger, Basel.

Ogle, T. F. and Kitay, J. I. (1977). Effects of melatonin and an aqueous pineal extract on adrenal secretion of reduced steroid metabolites in female rats.

Neuroendocrinology 23, 113–20.

Ogle, T. F. and Kitay, J. I. (1978). In vitro effects of melatonin and serotonin on adrenal steroidogenesis. Proceedings of the Society for Experimental and Biological Medicine 157, 1418–25.

Ozaki, Y., Lynch, H. J., and Wurtman, R. J. (1976). Melatonin in rat pineal, plasma and urine: 24-hour rhythmicity and effect of chlorpromazine. Endocrinology 94, 1418–25.

Perlow, M. J., Reppert, S. M., Boyar, R. M., and Klein, D. C. (1981). Daily rhythms in cortisol and melatonin in primate cerebrospinal fluid. Neuroendocrinology 32, 193–6.

Reiter, R. J., et al. (1982). Pineal melatonin production: endocrine and age effects. In Melatonin rhythym generating system (ed. D. C. Klein), pp. 143–54. S. Karger, Basel.

Relkin, R. (1983). Pineal-hormone interactions. In The pineal gland (ed. R. Relkin), pp. 225–46. Elsevier, New York.

Sachar, E. J. (1975). Twenty-four hour cortisol secretory patterns in depressed and manic patients. Progress in Brain Research 42, 81–91.

Sizonenko, P. C. (1978). Endocrinology in preadolescents and adolescents: Hormonal changes during normal puberty. American Journal of Childhood Diseases 132, 704–12.

Sugden, D. and Klein, D. C. (1983). Regulation of rat pineal hydroxyindole-O-methyltransferase in neonatal and adult rats. Journal of Neurochemistry 40, 1647–53.

Thompson, C., et al. (1985). The effect of desipramine upon melatonin and cortisol secretion in depressed and normal subjects. British Journal of Psychiatry 147, 389–93.

Vaughan, M. K., Vaughan, G. M., Reiter, R. J., and Benson, B. (1972). Effect of melatonin and other pineal indoles on adrenal enlargement produced in male and female mice by pinealectomy, unilateral adrenalectomy, castration and cold stress. Neuroendocrinology 10, 139–54.

Vaughan, G. M., McDonald, S. D., Jordan, R. M., Allan, J. P., Bell, R., and Stevens, E. A. (1979). Melatonin, pituitary function and stress in humans. Psychoneuroendocrinology 4, 351–62.

Waldhauser, F., Lieberman, H. J., Frisch, H., Herkner, K., and Wurtman, R. J. (1984a). Increase of human serum prolactin levels after melatonin administration. EPSG-Newsletter Supplement 5, 47.

Waldhauser, F., Weiszenbacher, G., Frisch, H., Zeitlhuber, U., Waldhauser, M., and Wurtman, R. J. (1984b). Fall in nocturnal serum melatonin during prepuberty and pubescence. Lancet i, 362–5.

Waldhauser, F., Frisch, H., Krautgasser-Gasparotti, A., Schober, E., and Bieglmayer, C. (1986). Serum melatonin is not affected by glucocorticoid replacement in congenital hyperplasia. Acta Endocrinologica 111, 355–9.

Werner, S., Brismar, K., Wetterberg, L., and Eneroth, P. (1980). Circadian rhythms of melatonin, prolactin, growth hormone and cortisol in patients with pituitary adenomes, empty sella syndrome and Cushing's syndrome due to adrenal tumours. In Melatonin: current status and perspectives (ed. N. Birau and W. Schloot), pp. 357–63. Pergamon Press, Oxford.

Wetterberg, L. (1983). The relationship between the pineal gland and the pituitary-adrenal axis in health, endocrine and psychiatric conditions. *Psychoneuroendocrinology* **8**, 75–80.

Wetterberg, L., Beck-Friis, J., Aperia, B., and Petterson, U. (1979). Melatonin/cortisol ratio in depression. *Lancet* **ii**, 1361.

Wetterberg, L., *et al.* (1981). Pineal-hypothalamic-pituitary function in patients with depressive illness. In *Steroid hormone regulation of the brain* (ed. K. Fuxe, J. A. Gustafsson, and L. Wetterberg), pp. 397–403. Pergamon Press, Oxford.

Wetterberg, L., *et al.* (1982). Melatonin and cortisol levels in psychiatric illness. *Lancet* **ii**, 100.

Wetterberg, L., Beck-Friis, J., Kjellman, B. F., and Ljunggren, J. G. (1984). Circadian rhythms in melatonin and cortisol secretion in depression. In *Frontiers in Biochemical and Pharmacological Research in Depression* (ed. E. Usdin, M. Asberg, L. Bertilsson, and F. Sjöqvist), pp. 197–205. Raven Press, New York.

Wurtman, R. J., Altschule, M. D., and Holmgren, U. (1959). Effects of pinealectomy and of a bovine pineal extract in rats. *American Journal of Physiology* **197**, 108–10.

Young, I. M. (1981). The pineal indole hormones in Cushing's disease and acromegaly. In *Pineal function* (ed. C. P. Matthews and R. F. Seamark) pp. 7–12. Elsevier North-Holland Biochemical Press, Amsterdam.

5. Melatonin and thyroid function

Jerry Vriend and Meir Steiner

Introduction

Evidence for an inhibitory effect of melatonin on thyroid function of laboratory animals has accumulated over the past decade. The role of melatonin on human thyroid function, on the other hand, has not been investigated in a systematic fashion. Since the question of a possible role for melatonin on the brain–thyroid axis remains virtually unexplored, this chapter can only review the results of studies of melatonin administration in laboratory animals and raise questions relating to the action of melatonin on thyroid function in humans. Since there is little data relating melatonin and thyroid function in humans, and many experimental protocols used in animal studies cannot be used in human studies, we will discuss the question of how melatonin can be tested on the human hypothalamic–pituitary–thyroid axis (HPT axis). Basic questions on the level of action of melatonin on the HPT axis must be addressed (if indeed, melatonin does influence the human brain–thyroid system at all). These questions will lead to the broader and more important issue of the site of action of melatonin. Speculation on the clinical usefulness of melatonin hopefully will lead to systematic testing by clinical physiologists as well as by clinical psychiatrists interested in brain–thyroid interactions in clinical conditions. As is often the case with human studies, the 'experiments of nature' provided by clinical conditions may lead to insights into the interaction of melatonin with the brain–thyroid axis and thus insights into clinical conditions in which a protocol of melatonin administration would be useful.

An understanding of the physiological role of melatonin must include information on its site of production. For years after its isolation from bovine pineal glands (Lerner *et al.* 1958, 1959) the pineal gland was considered to be the only site of melatonin synthesis. Evidence has accumulated showing that in some non-mammalian vertebrates melatonin may also be produced in the retina (Wiechmann 1986). Reports of its presence in mammalian retinas (Pang and Allen 1986; Yu *et al.* 1981) require confirmation. In virtually all species studied melatonin is synthesized in response to changes in the environmental photoperiod. In mammals the daily rhythm of melatonin secretion is characterized by a rise in melatonin synthesis during the dark phase of the daily light–dark cycle (e.g. Fig. 5.1). It is released from the pineal gland into the circulation in picogram amounts, also during the dark phase of the photoperiodic cycle. Although it is widely assumed that melatonin must be released into the circulation to be physiologically active, a local site of action in pineal gland or retina cannot be ruled out at present. Removal of either the pineal gland or the retinae will influence thyroid function in laboratory animals that respond to changes in photoperiod (see below).

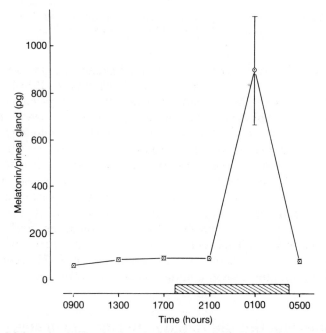

Fig. 5.1. Melatonin rhythm in the pineal gland of the Syrian hamster on a 14L/ 10D light–dark cycle. Means and standard errors are shown.

The hypothalamic–pituitary–thyroid axis

The control system which regulates mammalian thyroid function consists of several components (Fig. 5.2). The isolation and identification of thyrotrophin releasing hormone (TRH) (Folkers *et al.* 1969; Schally *et al.* 1969; Burgus *et al.* 1970) demonstrated that the thyroid control system was not a simple closed loop. TRH, a tripeptide hormone produced in the hypothalamus, stimulates TSH release from the pituitary. TRH was reported localized in and around the paraventricular nuclei of the

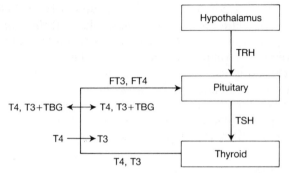

Fig. 5.2. Regulation of thyroid hormone secretion. [From Vriend, *Pineal Research Review* **1**, 183–206, 1983, with permission.]

hypothalamus (Aizawa and Greer 1981; Lechan and Jackson 1982). TRH, a product of a larger prohormone (Lechan *et al.* 1986), is transported by axons to the median eminence of the hypothalamus. The hypothalamo–hypophyseal system of portal veins is regarded as the route that TRH travels to arrive at the anterior pituitary. After the discovery of TRH, the hypothesis that thyroid hormones inhibit TSH secretion by an action at the pituitary was supported by data showing that thyroid hormones block the stimulatory effect of TRH on TSH release (Bowers *et al.* 1970; Redding *et al.* 1970) and at the same time reduce the number of TRH receptors in the pituitary gland (DeLean *et al.* 1977; Perrone and Hinkle 1978; Hinkle and Goh 1982). Although the mechanism of feedback inhibition of TSH secretion may not be completely understood at present, apparently reduction in the number of TRH receptors by thyroid hormone contributes to inhibition of TSH secretion.

Melatonin and thyroid function in subhuman species

Melatonin and the classical model of thyroid function

Experiments with melatonin administration in laboratory animals have produced results showing significant effects of melatonin on various components of the hypothalamic–pituitary–thyroid control system as shown in Fig. 5.2. Melatonin has been reported to influence TRH content of the hypothalamus, TSH content of the pituitary, iodine uptake by the thyroid gland, circulating levels of thyroid hormones, and percentage saturation of thyroid binding proteins. An assumption underlying much of this work is that melatonin has a site of action within in the brain. Miline (1963), more than 20 years ago, reported that administration of extracts of pineal glands (which contain melatonin) to rats resulted in diminished iodine uptake by the thyroid and hypertrophy of cells in the paraventricular nuclei of the hypothalamus. He concluded that the paraventricular nuclei were a link by which the pineal gland exerted an inhibitory influence on the pituitary–thyroid axis. The identification of melatonin (Lerner *et al.* 1958, 1959), and later of TRH, supported this hypothesis.

Melatonin and thyroid function in rats and mice

Early evidence for pineal gland interaction with the thyroid of mammals came from studies showing enlargement of the thyroid after pinealectomy (Demel 1927; Davis and Martin 1940). Studies carried out with rats and mice during the 1960s showed a small, but significant, enlargement of the thyroid gland after pinealectomy (Miline 1963; Scepovic 1963; Pazo *et al.* 1968; Houssay and Pazo, 1968). Losada (1977) reported that pinealectomized rats had increased height of thyroid follicular cells and decreased amount of colloid in follicles compared to controls. Houssay *et al.* (1966) found that administration of melatonin prevented the thyroid 'hypertrophy' observed in male mice after pinealectomy, but had no effect on the size of the thyroid glands of sham-operated controls. Melatonin administration was also reported to reduce thyroid weight of pinealectomized rats (DeFronzo and Roth 1972). These results suggested that melatonin administration could serve as replacement therapy in pinealectomized rodents. Pineal extracts, presumably due to their melatonin content, were reported to inhibit experimentally induced goitre in rats (Milcu *et al.* 1959; DeLuca *et al.* 1961). Baschieri and colleagues (1963) found that melatonin injection in methylthiouracil-treated rats reduced the goitrogenic effect of methyl-thiouracil. Melatonin administration reduced the increase in thyroid weight and in follicular cell height which was observed in methyl-thiouracil-treated rats.

This work lead to studies of the effects of pinealectomy and melatonin administration on tests of thyroid function. In adult rats under a conventional 12L/12D photoperiod, pinealectomy had no significant effect (Brammer *et al.* 1979) on serum TSH or on serum thyroid hormones. In immature rats, however, pinealectomy was found to increase plasma levels of TSH and T4 after 3 days, compared to sham-operated controls (Relkin 1978). Dark exposure of immature rats was found to reduce plasma TSH and T4; these effects were reversed by pinealectomy. Continuous-light treatment resulted in T4 and TSH levels similar to those observed in pinealectomized rats. Relkin (1972, 1978) concluded that dark-rearing of rats resulted in increased inhibitory activity of the pineal gland on the neuroendocrine thyroid axis. He attributed this inhibition to an action of melatonin on the brian. Niles and associates (1979) kept young rats under a 1L/23D photoperiod for 72 days. They found that pinealectomy or administration of melatonin antiserum significantly increased plasma TSH levels. The degree of difference was found to depend partly on time of day with respect to the light–dark cycle. Niles and associates (1979) also came to the conclusion that rearing rats under conditions of restricted lighting (1L/23D) resulted in activation of a pineal-dependent mechanism which was responsible for inhibition of the neuroendocrine thyroid system. He attributed the effects of pinealectomy to the loss of the source of melatonin.

Studies of exogenous melatonin administration in rats and mice provided evidence, for the most part, of an inhibitory effect of melatonin on the HPT axis. Administration of melatonin was reported to inhibit thyroid secretion rate (TSR) in young male and in young female rats (Ishibashi *et al.* 1966; Narang *et al.* 1967; Singh and Turner 1972). This inhibition of TSR was found to be reduced (Narang *et al.* 1966; Singh *et al.* 1969) after puberty.

Melatonin administered subcutaneously, intraperitoneally, or intra-ventricularly has been found to inhibit radioactive iodine uptake by the thyroid glands of rats kept in a diurnal photoperiod (Baschieri *et al.* 1963; DeProspo *et al.* 1968; DeProspo and Hurley 1971). DeProspo *et al.* (1969) found a similar inhibition of iodine uptake in rats kept under constant darkness, compared to 'normal' controls kept under a 12L/12D photoperiod. Rats kept under constant darkness were found to have iodine values less than 50 per cent of controls. DeProspo and coworkers (1969) interpreted these results as due to endogenous melatonin secretion by pineal glands activated by constant darkness. Cardinali and colleagues (Cardinali *et al.* 1982; Vacas *et al.* 1982) found that 14 hours following superior cervical ganglionectomy of rats pineal melatonin was increased several fold while circulating TSH and T4 levels were decreased. The increase in melatonin content of the pineal gland was associated with degenerating sympathetic nerve terminals in the pineal. The decrease in TSH and T4 was prevented by pinealectomy. These interesting results supported the earlier evidence that the pineal gland produced a substance (presumably melatonin) which inhibited the neuroendocrine–thryoid axis.

It is believed that the site of action of melatonin in the rat is central, probably at the level of hypothalamic secretion of TRH (Relkin 1978; Niles *et al.* 1979). The possibility of a local site of action of melatonin within the thyroid gland has been recently raised by Lewinski and Sewerynek (1986). These investigators reported inhibition of basal and TSH-stimulated mitotic activity in thyroid follicular cells by melatonin.

DeProspo *et al.* (1969) suggested that the effects of melatonin injections might depend on the time of day, relative to the photoperiod. Fiske and Huppert (1968) had reported previously that melatonin administration could shift or block a diurnal rhythm of serotonin in the pineal gland of rats. Melatonin administration just before the onset of darkness enhanced the normal day–night rhythm of serotonin secretion; melatonin administration 6 hours prior to the onset of darkness significantly decreased the amplitude of the serotonin rhythm. The concept that the sensitivity to melatonin depended on the time of administration with respect to the onset of darkness proved to be important for later research.

Melatonin and thyroid function in hamsters

The interaction of melatonin secretion and endocrine rhythms varies considerably among mammalian species (Quay 1986). The Syrian hamster has been a very useful experimental animal for endocrinologists interested in the pineal gland and the actions of melatonin. The endocrine system of the Syrian hamster is very sensitive to changes in photoperiod. Changes in melatonin secretion by the pineal gland appears to play a central role in mediating the effects of changes in photoperiod. The sensitivity to photoperiod has apparently aided the hamster in surviving seasonal changes in temperature and food supply, as well as providing a mechanism for increasing the probability of giving birth to young in the spring, at a time when food supply is adequate and temperatures not too severe for survival. In the laboratory several of the endocrine responses to changes in photoperiod can be elicited by melatonin administration. Both reproductive competence and thyroid function are influenced by melatonin administration (Reiter 1980; Vriend 1983). Ralph (1979) has suggested that melatonin may play a role in temperature regulation in mammals. The extent to which the effects of melatonin are related to temperature regulation in the hamster remains to be determined.

The data that has been generated from studies of melatonin administration in hamsters has been considerably more complicated than the data presented in reports of melatonin administration in rats. As is noted below, different doses and different protocols of administration have resulted in various effects on the neuroendocrine thyroid axis. When similar protocols of administration have been used, results have been comparable from different laboratories.

Daily variations in circulating levels of thyroid hormones must also be taken into consideration in the interpretation of experimental results of experiments with melatonin administration. Figure 5.3 shows that both T4 and T3 concentrations significantly increase during the light phase of the light–dark cycle and decrease during the dark phase of the cycle.

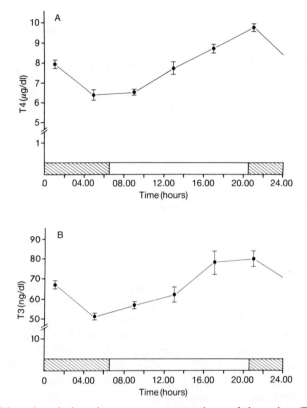

Fig. 5.3. Diurnal variations in serum concentrations of thyroxine (T4) and tri-iodothyronine (T3) in male hamsters. For T4, $F = 24.04$, df $= 5,35$, $P < 0.001$. For T3, $F = 10.13$, df $= 5,35$, $P < 0.01$. [From Vriend, *Journal of Pineal Research* **1**, 15–22, 1984, with permission.]

These data raise the issue of controls for daily rhythms in thyroid hormones, but also raise the question of whether these daily changes in thyroid hormones influence the sensitivity of other endocrine systems to melatonin.

Singh and Turner (1972) first reported that melatonin administration inhibited thyroid function of hamsters. They found that daily sub-cutaneous injection of 50 or 100 μg of melatonin for 10 days reduced the thyroid secretion rate of hamsters. They reported that hamsters were more sensitive to melatonin in this regard than rats.

As radioimmunoassay procedures for measuring thyroid hormones became readily available, reports of inhibitory effects of melatonin on circulating levels of thyroid hormones appeared. Vriend and Reiter (1977) found that administration of melatonin subcutaneously (25 μg

daily for 50 days) significantly ($P < 0.001$) depressed plasma concentrations of thyroxin compared to vehicle treated, as well as to pinealectomized controls. The percentage inhibition obtained in the experiments
reported in this study varied from 28 to 52 per cent. The free thyroxine
index (FTI), which takes into account variations in unsaturated binding
capacity of plasma binding proteins, was also reduced by melatonin (Fig.
5.4). In these experiments the hamsters were maintained on a 14L/10D
photoperiod. Plasma was sampled during the daytime (and daylight)
hours. Since these findings were reported in 1977 they have been
confirmed many times in a variety of studies. The usual daily dose of
melatonin injected in most of these studies was 25 μg, but as little as 5
μg daily was found to significantly inhibit circulating T4 levels (Vaughan
et al. 1984*a–c*).

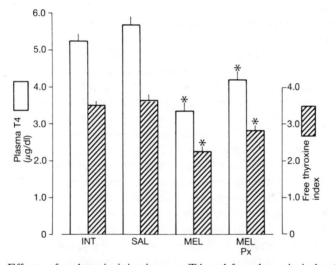

Fig. 5.4. Effects of melatonin injections on T4 and free thyroxin index in intact
and pinealectomized male hamsters. Mean and SEM are indicated. *$P < 0.01$.
[From Vriend, *Pineal Research Review* **1**, 183–206, 1983, with permission.]

Vaughan and co-workers (1983, 1984*a–c*) have tested the effects of
several compounds related to melatonin for their effectiveness in
inhibiting T4 concentrations in plasma of hamsters. Among the natural
or synthetic compounds that were tested were acetyl methoxytryptophol,
N-acetylserotonin, 6-hydroxymelatonin, hydroxytryptophol, methoxytryptophol, 5-methoxytryptamine, and the synthetic analogues, hexanoyl
methoxytryptamine, propionyl methoxytryptophol, and 6-chloromela-
tonin. None of these compounds were found to be as effective as
melatonin in reducing T4 and FTI in both male and female hamsters

when administered as daily subcutaneous injections (25 μg/day). 6-chloromelatonin was reported to be as effective as melatonin in female hamsters, but not in male hamsters. Vaughan and collaborators (1984*a,b*) regarded 6-chloromelatonin as a useful melatonin agonist. 5-methoxytryptamine was found to inhibit circulating T4 levels at doses approximately 20 times the effective melatonin dose (Vaughan *et al.* 1984*a–c*). The differences in the inhibitory effects of the two indoles was attributed to differences in the side chain attached to the 3rd carbon atom, based on the work of Heward and Hadley (1975). Thus, the inhibitory effect of melatonin appears to be relatively specific. Vaughan and colleagues (1984*a–c*) sampled the plasma of hamsters in these experiments between 08.30 and 11.30 hours, and found no inhibition of T3 concentration by melatonin. The differences between control and melatonin-injected animals would probably be greater if plasma had been sampled near the end of the daily light–dark cycle, when T4 and T3 concentrations of hamsters would normally be higher (Fig. 5.3).

A number of factors were found to influence the extent of the melatonin-induced depression of T4 in male hamsters. The inhibitory effect of melatonin, for example, was attenuated, or prevented completely, by surgical removal of the pineal gland or the superior cervical ganglia (Vaughan *et al.* 1983; Vriend and Reiter 1977; Vriend *et al.* 1979). Age was found to influence the inhibitory response as well. Hamsters that were 22 weeks of age responded to subsequent melatonin injections with a significant reduction in T4 within 3 weeks (*v.* 9 weeks for hamsters that were 6 weeks of age at the beginning of the experiment; Vriend *et al.* 1979). It appears from these data that the sensitivity to melatonin develops after puberty in the hamster. The time course of the melatonin-induced inhibition was found to be similar to the time course of the melatonin-induced inhibition of testicular weight in the hamster (Fig. 5.5) as well as the associated changes in gonadal and gonadotrophic hormones. The first evidence that pinealectomy influenced plasma T4 levels (and FTIs) came from a study showing that pinealectomy increased T4 levels in male hamsters that had been blinded by bilateral orbital enucleation. The same results were obtained by surgical removal of the superior cervical ganglia (which supply the pineal gland with its sympathetic innervation; Vriend *et al.* 1977).

Melatonin-induced reduction in plasma TSH concentrations were reported in at least one study with male hamsters (Vaughan *et al.* 1982*b,c*), using a rat TSH assay. Melatonin-induced reduction in plasma TSH was not a consistent finding, however. In hamsters made hypothyroid with thiourea in the drinking water serum TSH was increased (as a result of the induced hypothyroidism) and was easily measured by the rat TSH assay. Daily melatonin injections were found to significantly

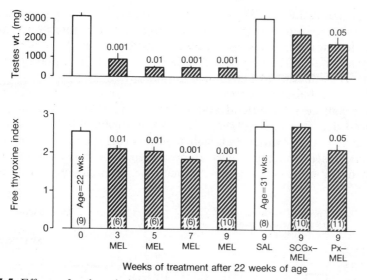

Fig. 5.5. Effects of melatonin injections on free thyroxine index and gonad weight of old male golden hamsters. Numbers above standard error bars indicate level of significant difference between melatonin-treated hamsters and control hamsters treated with saline for 9 weeks. [From Vriend, *et al. General Comparative Endocrinology* **38**, 189–95, 1979, with permission.]

reduce both serum and pituitary levels of TSH in hamsters made hypothyroid with thiourea (Vriend 1985).

T3 concentrations in plasma were reported significantly depressed by melatonin in male hamsters (Vaughan *et al.* 1982*b,c*), but this was not as consistent a finding as the melatonin-induced decrease in circulating levels of T4. Reverse T3 levels were also reported reduced by melatonin in male hamsters (Vaughan *et al.* 1982*a*). Circulating T3 levels have been reported as substantially higher in hamsters maintained under natural lighting and temperature conditions than in hamsters maintained under laboratory conditions (Vaughan *et al.* 1984*a–c*). The interaction of melatonin in temperature-related changes in thyroid hormones has not yet been adequately studied.

Melatonin-induced inhibition of circulating T4 and FTI has been reported several times as well in female hamsters (Vriend *et al.* 1982; Vaughan *et al.* 1984*a–c*). In a recent study (Vriend *et al.*, unpublished data) circulating levels of T4 were reduced by melatonin injections (25 μg daily for 10 weeks) to 47 per cent of controls ($P < 0.01$) and T3 levels to 66 per cent of controls ($P < 0.05$). In these studies melatonin was administered at the end of the daily photoperiod. Circulating TSH was reported depressed by melatonin in female hamsters (Vriend *et al.* 1982).

Similar depression of TSH, T4, and T3, obtained by chronic exposure to short photoperiod (Vaughan *et al.* 1982*b,c*) could be interpreted as due to endogenous changes in melatonin secretion by the pineal gland. Vaughan *et al.* (1982*a–c*) regarded these effects of short photoperiod as due to a central action of the pineal gland.

Melatonin-induced depression of T4 was studied in ovariectomized female hamsters by Vriend and collaborators (1982). The melatonin induced decrease in circulating T4 was not attributable to the presence or absence of the ovaries. In female hamsters cessation of oestrous cycles and decreased circulating levels of oestrogen (Vriend *et al.*, unpublished data) occur concomitantly to the decrease in circulating levels of T4 (Fig. 5.6).

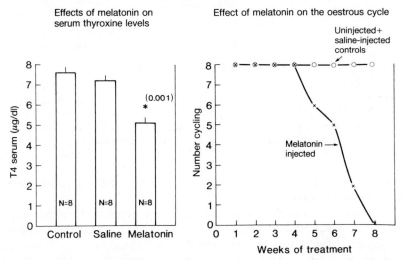

Fig. 5.6. Concomitant inhibition of thyroid and gonadal systems in female hamsters. [From Vriend, *Pineal Research Review* **1**, 183–206, 1983, with permission.]

Melatonin and TRH secretion

The decrease in circulating thyroid hormones observed in hamsters treated with melatonin was interpreted as a central effect (Vriend *et al.* 1979). Later studies which showed significantly increased hypothalamic content of TRH in melatonin-treated male hamsters (Fig. 5.7) supported this view (Vriend and Wasserman 1986). Previous work had shown an effect of pinealectomy on hypothalamic content of TRH in female hamsters (Vriend and Wilber 1983). Thus, both the presence of an active pineal gland or melatonin injections resulted in increased hypothalamic content of TRH. The increased content of TRH was interpreted as a results of decreased release. The conclusion made by Vriend and

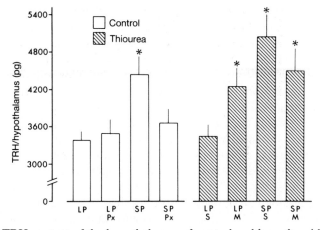

Fig. 5.7. TRH content of the hypothalamus of control and hypothyroid hamsters. LP=14L/10D; SP=2L/22D; Px=pinealectomized; S=saline injected; M= melatonin injected. *P<0.05 compared to LP controls. [From Vriend and Wasserman, *Neuroendocrinology* **42**, 498–503, 1986, with permission.]

Wasserman (1986) was that melatonin acted on one of the neural systems which regulate TRH release (as well as the release of other hypothalamic hypophysiotrophic hormones).

'Stimulatory' effects of melatonin on circulating T4

The inhibitory effects of melatonin on circulating T4 described above have been demonstrated in experiments in which single injections of melatonin (usually between 5 and 25 µg) were administered daily. With this protocol of administration circulating melatonin levels are elevated for several hours after each injection (Vaughan *et al.* 1986). To maintain circulating levels of melatonin at levels continuously higher than daytime values melatonin must be administered in a different manner. Plasma levels of melatonin have been elevated continuously using melatonin implants in beeswax pellets underneath the skin, by placing melatonin in the drinking water, or even by injecting massive doses daily. Under these conditions melatonin prevented the suppression that would otherwise occur as a result of short photoperiod, blinding, or even the suppression that would otherwise occur as a result of daily melatonin injections (Vriend *et al.* 1982; Vriend and Gibbs 1984; Vaughan *et al.* 1984*a*–*c*). At least one natural substance related to melatonin (5-methoxytryptamine, in beeswax implants) was also reported capable of preventing the suppression obtained with melatonin injections (Vaughan *et al.* 1984*a*–*c*). It is not clear whether the 'counter-inhibitory' effect of melatonin represents a pharmacological or physiological response. Reiter (1980) suggested that the counter-inhibitory effects of melatonin were due to 'down-regulation' of melatonin receptors and has presented a theoretical model relating melatonin receptor sensitivity to the mode of administration. The difficulty in testing this model is that melatonin receptors cannot be easily, or convincingly, demonstrated. This author's receptor model was based on the observation that melatonin implants could counteract pineal-mediated gonadal regression in the Syrian hamster (Reiter *et al.* 1977). This reasoning was later applied to the thyroid axis (Vaughan *et al.* 1984*a–c*; Vriend *et al.* 1982).

In blind hamsters melatonin in drinking water significantly increased circulating T4 in a dose-dependent fashion (Vriend and Gibbs 1984). Since melatonin prevented the depression that would otherwise occur as a result of blinding (presumably activating the pineal gland), this effect of melatonin was also described as 'counter-inhibitory'. At a dose of 10 μg/ml in the drinking water, plasma melatonin assayed by radioimmunoassay was 285 pg/ml. Coincidentally, with the increase in T4 which occurred in blinded hamsters receiving melatonin in the drinking water, a dramatic increase in gonadal weight occurred. A significant increase in gonadal weight of blind hamsters occurred at a dose as low as 1 μg/ml in the drinking water (and blood levels near 100 pg/ml). The blood levels, therefore, were approximately double normal night-time values and approximately 10-fold higher than normal daytime levels. It appears that the 'counter-inhibitory' effects of melatonin are the results of continuously elevated levels of circulating melatonin, a situation which physiologically does not normally occur.

Melatonin–thyroid interaction and the gonadal axes

Several similarities were noted between the effects of melatonin on circulating levels of T4 and effects of melatonin on gonads and gonadotrophic hormones (Vriend 1983). The time course of the anti-thyroid effects of melatonin and the antigonadal effects of melatonin were very similar. Both effects were attenuated by removal of the pineal gland or superior cervical ganglia surgically. Both were similarly influenced by age. Finally, 'counter-inhibitory' effects of continuously elevated serum melatonin were similar for the effects of melatonin on T4 and on pituitary–gonadal hormones (Reiter 1980; Gibbs and Vriend 1983; Vriend and Gibbs 1984).

These similarities suggested a functional relationship between melatonin–thyroid and melatonin–gonadal axes. The effects of melatonin in male hamsters receiving thiourea, with or without T4 replacement, were recently reported (Vriend 1985). Thiourea-induced hypothyroidism inhibited the action of melatonin on gonadal weight and on circulating

LH (Fig. 5.8). This effect of hypothyroidism was reversed by T4 replacement injections. These data were interpreted as evidence for a thyroid hormone requirement for melatonin-induced gonadal inhibition of gonadotrophins and melatonin-induced gonadal involution. In female hamsters, on the other hand, thiourea-induced hypothyroidism resulted in decreases in circulating oestradiol, decreases in uterine weight, decreased ovarian follicular development and ovulation, and decreases in circulating prolactin (Vriend *et al.*, unpublished data). Figure 5.9 illustrates the antagonistic effects of melatonin and T4 on circulating levels of oestradiol and on uterine weight.

The role of the thyroid in photoperiodically induced changes in gonadal hormones has been studied in birds (Dawson *et al.* 1986; Follet and Nicholls 1984*a,b*). In Japanese quail, and in starlings, development of photorefractoriness was shown to be dependent upon thyroid hormones (Follet and Nicholls 1984*a,b*). These investigators have not

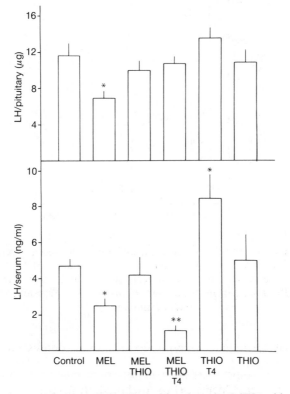

Fig. 5.8. Pituitary and plasma LH: effects of melatonin (MEL), thiourea (THIO), and T4 replacement. *$P < 0.05$; **$P < 0.01$ (compared to controls). [From Vriend, *Endocrinology* **117**, 2402–7, 1985, with permission.]

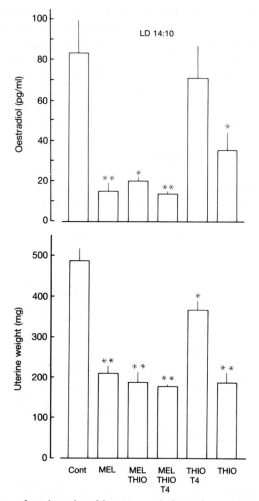

Fig. 5.9. Effects of melatonin, thiourea, and thyroxine replacement on serum oestradiol and on uterine weight of female hamsters under LD 14:10. Mean and standard error are shown. Asterisks indicate significant difference from controls: *$P < 0.05$; **$P < 0.01$. [From Vriend *et al. Biology of Reproduction* in press, with permission.]

yet shown that the gonadal response to melatonin administration requires an intact thyroid gland.

Melatonin–thyroid interaction and growth factors

Studies of seasonal growth patterns of hamsters under various photoperiods provided evidence that the pineal gland and its product melatonin influenced body weight regulation (Hoffman *et al.* 1982;

Hoffman 1983). Studies of melatonin-injected hamsters have frequently reported increases in body weight (Tamarkin *et al.* 1976; Reiter *et al.* 1977). Smythe and Lazarus (1973) reported that melatonin injections blocked increases in growth hormone (GH) secretion induced by 5-hydroxytryptophan. Vriend *et al.* (unpublished data) have studied effects of melatonin (25 μg daily s.c. for 10 weeks) on GH levels of male hamsters. Hypothyroidism induced by thiourea reduced pituitary GH, as expected. In serum, however, thiourea resulted in a 3-fold increase. This increase in serum GH was prevented by T4 replacement. The T4 replacement effect was blocked by melatonin (Fig. 5.10). The data

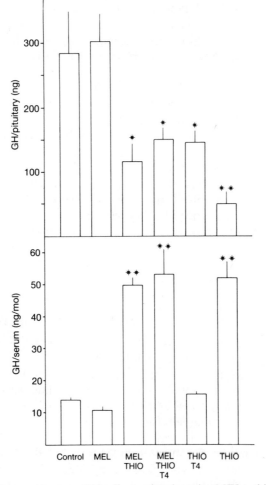

Fig. 5.10. Pituitary and serum GH: effects of melatonin (MEL), thiourea (THIO), and T4 injections. *$P < 0.05$; **$P < 0.01$; compared to controls. [From Vriend *et al. Growth*, in press, with permission.]

showing antagonistic effects of melatonin and T4 on circulating levels of GH did not provide an explanation for the site of action of melatonin on GH release. If, as generally assumed, melatonin has a CNS site of action, its effects on GH could be on one of the hypothalamic hormones influencing GH release. Thyroid hormones regulate GH gene expression in the pituitary (Eberhardt *et al.* 1980; Evans *et al.* 1982), but may influence circulating levels of GH by additional mechanisms.

Recently, T4 and melatonin have been shown to influence circulating levels of the somatomedin, IGF-1 in female hamsters (Vriend *et al.*, unpublished data). Thiourea and melatonin administration were found to inhibit circulating levels of IGF-1 in proportion to their ability to inhibit circulating levels of T4. Paradoxically, melatonin injections were found to antagonize T4 replacement effects on IGF-1 levels. Changes in somatomedin levels could explain seasonal changes in body weight and in fat deposition (Hoffman *et al.* 1982; Hoffman 1983).

Does T4 feed back on the pineal gland?

Studies of T4 uptake by pineal slices (Cady *et al.* 1971) were interpreted as evidence for a mechanism for rapid uptake of T4 by the pineal gland. Although these uptake studies were not continued, other data showed that thyroid hormones did influence pineal gland secretion. T4 and T3 were reported to stimulate melatonin production (Nir and Hirschmann 1978). Thyroidectomy has been reported to have small, but significant, inhibitory effects on pineal melatonin secretion (Reiter 1982). The hypothesis of Miline (1963) of a feedback loop between the pineal gland and the thyroid gland, however, has not been supported by strong experimental data. Champney (1986) has pointed out that compared to the hypothalamo–hypophyseal system the pineal gland is very insensitive to hormonal feedback via the circulatory system, but responds primarily to neural input.

Biogenic amines and the neuroendocrine–thyroid axes

Human clinical studies have shown that several biogenic amines influence the neuroendocrine–thyroid axis in humans. Human clinical studies have shown that several serotonin antagonists inhibit TSH secretion. These include metergoline, cyproheptadine, and methysergide (Delitala *et al.* 1978; Ferrari *et al.* 1976; Egge *et al.* 1977; Collu 1978). These studies suggest that serotonin stimulates TRH release. The action of dopamine, on the other hand, has been reported as inhibitory to TSH secretion. A number of dopamine antagonists used clinically augment TSH secretion under a variety of conditions. These drugs include metoclopramide (Healy and Burger 1977), domperidone (Pourmand *et al.* 1980; Delitala *et al.* 1980*a*,*b*), sulpiride (Massara *et al.* 1978; Zaboni *et al.* 1979), and benzerazide (Delitala *et al.* 1980*a,b*). Dopamine infusions lower serum TSH concentrations in man (Delitala 1977) and reduce TRH-induced TSH secretion (Burrow *et al.* 1977; Besses *et al.* 1975). An influence of the central noradrenergic system on TSH secretion has been inferred from studies in which the alpha-blocker phentolamine diminished TRH-induced TSH secretion (Nilsson *et al.* 1974). In addition, both alpha- and beta-adrenergic mechanisms may have effects on thyroid hormone secretion by acting directly on the thyroid gland. Ahren (1986) has concluded that noradrenalin has a dual action on thyroid hormone secretion: a stimulatory action under basal conditions and an inhibitory (cyclic AMP dependent) action during TSH stimulation.

Based on results showing an increase in serotonin content in midbrain and hypothalamus of rats administered melatonin, Anton-Tay and colleagues (1968) proposed that melatonin acted on the brainstem serotonergic system to produce its neuroendocrine effects. Other investigators have proposed the hypothesis that melatonin acts on catecholeaminergic neurons of the hypothalamus (Steger *et al.* 1984) to influence the secretion of releasing factors.

Melatonin–thyroid interaction in humans?

In contrast to the substantial body of research on the relationship of melatonin to thyroid hormone levels in laboratory animals, there is almost no information available on the effects of melatonin on thyroid function in humans. We are currently collecting serum samples of human subjects with pinealomas (and encourage other clinicians to contribute such samples) to determine whether increased circulating melatonin levels are associated with decreased levels of thyroid hormones.

Changes in both the HPT axis and the peripheral homeostasis of thyroid hormone during prolonged Antarctic residence have been described (Reed *et al.* 1986). The most significant finding in this study was the increase in the integrated TSH response to TRH over time of exposure to cold. The authors propose that their data do not seem to be related to a seasonal or daily circadian rhythm—subjects were studied during both outdoor total darkness and total sunlight, but melatonin levels were not reported.

Recently, the effects of chronic administration of exogenous melatonin in man were studied in a small group ($n = 6$) of healthy subjects (Wright *et al.* 1986). The effects of long-term (1 month), timed, low-dose melatonin treatment on endocrine function were assessed. Although there was a significantly earlier fall in the nocturnal serum prolactin peak in melatonin-treated subjects, plasma levels of growth hormone, cortisol, testosterone, luteinizing hormone, and T4 were reported as uninfluenced by the treatment. Indeed, Waldhauser and colleagues (1987) could demonstrate no alteration in TSH secretion following acute pharmacological doses of melatonin in healthy males.

Recently, the retinal–hypothalamic–pineal axis has gained attention in the study of the pathophysiology of depression. Depressed patients have been reported to be supersensitive to nocturnal suppression of melatonin by light (Lewy *et al.* 1981) and also to have low basal nocturnal circulating plasma levels of melatonin (Wetterberg *et al.* 1984; Nair *et al.* 1985; Beck-Friis *et al.* 1985; Brown *et al.* 1985; Steiner and Brown 1985; Miles *et al.* Chapter 13). It has also been postulated that melatonin plays a role in the therapeutic effect of phototherapy used for certain types of seasonal depression (Rosenthal *et al.* 1984; Sack and Lewy, Chapter 10; Thompson, Chapter 11). The release of TSH after TRH stimulation has been observed to be blunted in some cases of depression (Kirkegaard and Bjorum 1980). Recently, the response of TSH to TRH was reported to be a predictor of the outcome of treatment with anti-depressant and neuroleptic drugs (Langer *et al.* 1986). These studies have unfortunately reported neither circulating levels of thyroid hormones nor the influence of melatonin on the TRH test.

In summary, the animal data provide evidence for melatonin-induced inhibition of circulating levels of thyroid hormones, as well as evidence for interaction of thyroid hormones in several major effects of melatonin outside the hypothalamo–pituitary–thyroid axis. Whether melatonin has a specific site of action (such as the paraventricular hypothalamic nuclei) or a more general action (such as inhibition of cAMP) has not yet been determined. It appears that the fundamental question of the site and mechanism of action of melatonin must be answered before its clinical usefulness can be adequately assessed.

Acknowledgement

The authors research was supported by the Medical Research Council of Canada.

References

Ahren, B. O. (1986). Thyroid neuroendocrinology: Neural regulation of thyroid hormone secretion. *Endocrine Reviews* **7**, 149–55.

Aizawa, T. and Greer, M. A. (1981). Delineation of the hypothalamic area controlling thyrotropin secretion in the rat. *Endocrinology* **109**, 1731–8.

Anton-Tay, F., Chou, C., Anton, S., and Wurtman, R. J. (1968). Brain serotonin concentration: elevation following intraperitoneal administration of melatonin. *Science* **162**, 277–8.

Baschieri, L., DeLuca, F., Cramarosa, L., DeMartino, C., Oliverio, A., and Negri, M. (1963). Modification of thyroid activity by melatonin. *Experientia* **19**, 15–17.

Beck-Friis, J., *et al.* (1985). Serum melatonin in relation to clinical variables in patients with major depressive disorder and a hypothesis of a low melatonin syndrome. *Acta Psychiatrica Scandinavica* **71**, 319–30.

Besses, G. S., Burrow, G. N., Spaulding, S. W., and Donabedian, R. K. (1975). Dopamine infusion acutely inhibits the TSH and prolactin response to TRH. *Journal of Clinical Endocrinology and Metabolism* **41**, 985–8.

Bowers, C. Y., Schally, A. V., Enzmann, F., Boler, J., and Folkers, K. (1970). Pocine thyrotropin releasing hormone is (Pyro)glu-his-pro(NH$_2$). *Endocrinology* **86**, 1143–53.

Brammer, G. L., Morley, J. E., Geller, E., Yuwiler, A., and Hershman, J. J. M. (1979). Hypothalamus–pituitary–thyroid axis interactions with pineal gland in the rat. *American Journal of Physiology* **236**, 416–20.

Brown, R., *et al.* (1985). Differences in nocturnal melatonin secretion between melancholic depressed patients and control subjects. *American Journal of Psychiatry* **142**, 811–16.

Burgus, R., Dunn, T. F., Disiderio, D., Ward, D. N., Vale, W., and Guilleman, R. (1970). Characterization of ovine hypothalamic TSH releasing factor. *Nature* **226**, 321–5.

Burrow, G. N., May, P. B., Spaulding, S. W., and Donabedian, R. K. (1977). TRH and dopamine interactions affecting pituitary hormone secretion. *Journal of Clinical Endocrinology and Metabolism* **45**, 65–72.

Cady, P., Dillman, R. O., and Rupnik, J. K. (1971). Transport mechanisms and thyroxine uptake in pineal slices. *Neuroendocrinology* **8**, 86–93.

Cardinali, D. P., Pisarev, M. A., Barontini, M., Juvenal, G. J., Boado, R. J., and Vacas, M. I. (1982). Efferent neuroendocrine pathways of sympathetic superior cervical ganglia. *Neuroendocrinology* **35**, 248–54.

Champney, T. H. (1986). Dual neuroendocrine glands: Sensitivity of the pineal gland to hormonal feedback. *Medical Hypotheses* **20**, 109–15.

Collu, R. (1978). The effect of TRH on the release of TSH, Pr1 and GH in man under basal conditions and following methysergide. *Journal of Endocrinological Investigation* **2**, 121–4.

Davis, L. and Martin, J. (1940). Results of experimental removal of the pineal gland in young animals. *Archives of Neurological Psychiatry* **43**, 23–45.

Dawson, A., Goldsmith, A. R., Nicholls, T. J., and Follett, B. K. (1986). Endocrine changes associated with the termination of photorefractoriness by short daylengths and thyroidectomy in starlings (Sturnus vulgaris). *Journal of Endocrinology* **110**, 73–9.

DeFronzo, R. A. and Roth, W. D. (1972). Evidence for the existence of a pineal–adrenal and pineal–thyroid axis. *Acta Endocrinologica* **70**, 31–42.

DeLean, A., Ferland, L., Drouin, J., Kelly, A., and Labrie, F. (1977). Modulation of pituitary thyrotropin releasing hormone receptor levels by estrogens and thyroid hormones. *Endocrinology* **100**, 1496–504.

Delitala, G. (1977). Dopamine and TSH secretion in man. *Lancet* **ii**, 760.

Delitala, G., Devilla, L., and Lotti, G. (1980a). Domperidone, an extracerebral inhibitor of dopamine receptors, stimulates thyrotropin and prolactic release in man. *Journal of Clinical Endocrinology and Metabolism* **50**, 1127–9.

Delitala, G., Devilla, L., and Lotti, G. (1980*b*). TSH and prolactin stimulation by the decarboxylase inhibitor benserazide in primary hypothyroidism. *Clinical Endocrinology* **12**, 313–16.

Delitala, G., Rovasio, P. P., Masala, A., Alagna, S., and Devilla, L. (1978). Metergoline inhibition of thyrotropin and prolactin secretion in primary hypothyroidism. *Clinical Endocrinology* **8**, 69–73.

DeLuca, F., Cramarossa, L., Peruzy, A. D., and Oliverio, A. (1961). [Variations in thyroid function during treatment with pineal extract]. *Rassegna di Fisiopatologia Clinica e Terapeutica* **33**, 396–405.

Demel, R. (1927). Experimentelle Studie zur Funktion der Zirbeldruse. *Arbeiten aus dem neurologischen Institut an der Wiener Universitat* **30**, 13–26.

DeProspo, N. D. and Hurley, J. (1971). A comparison of intracerebral and intraperitoneal injections of melatonin and its precursors on ^{131}I uptake by the thyroid glands of rats. *Agents and Actions* **2**, 14–17.

DeProspo, N. D., DeMartino, L. J., and McGuinness, E. T. (1968). Melatonin's effect on ^{131}I uptake by the thyroid glands in normal and ovariectomized rats. *Life Sciences* **7**, 183–8.

DeProspo, N. D., Safinski, R. J., DeMartino, L. J., and McGuinness, E. T. (1969). Melatonin and its precursors' effects on ^{131}I uptake by the thyroid gland under different photo conditions. *Life Sciences* **8**, 837–42.

Eberhardt, N. L., Apriletti, J. W., and Baxter, J. D. (1980). The molecular biology of thyroid hormone action. *Biochemical Actions of Hormones* **7**, 312–94.

Egge, A. C., Rogol, A. D., Varma, M. M., and Blizzard, R. M. (1977). Effect of cyproheptadine on TRH-stimulated prolactin and TSH release in man. *Journal of Clinical Endocrinology and Metabolism* **44**, 210–13.

Evans, R. M., Birenberg, N. C., and Rosenfeld, M. G. (1982). Glucocorticoid and thyroid hormones transcriptionally regulate growth hormone gene expression. *Proceedings of the National Academy of Sciences USA* **79**, 7659–63.

Ferrari, C., Paracchi, A., Rondena, M., Beck-Peccoz, P., and Faglia, G. (1976). Effect of two serotonin antagonists on prolactin and thyrotropin in man. *Clinical Endorcrinology* **5**, 575–8.

Fiske, V. M. and Huppert, L. C. (1968). Melatonin action on pineal varies with photoperiod. *Science* **162**, 279.

Folkers, K., Enzmann, F., Boler, J., Bowers, C. Y., and Schally, A. V. (1969). Discovery of modification of the synthetic tripeptide-sequence of the thyrotropin releasing hormone having activity. *Biochemical and Biophysical Research Communications* **37**, 123–6.

Follett, B. K. and Nicholls, T. J. (1984*a*). Photorefractoriness in Japanese quail: possible involvement of the thyroid gland. *Journal of Experimental Zoology* **232**, 573–80.

Follet, B. K. and Nicholls, T. J. (1984*b*). Influences of thyroidectomy and thyroxine replacement on photoperiodically controlled reproduction in quail. *Journal of Endocrinology* **107**, 211–21.

Gibbs, F. P. and Vriend, J. (1983). Counterantigonadotropic effect of melatonin administered via the drinking water. *Endocrinology* **113**, 1447–51.

Healy, D. L. and Burger, H. G. (1977). Increased prolactin and thyrotropin secretion following oral metoclopramid: dose–response relationships. *Clinical Endocrinology* 7, 195–201.

Heward, C. B. and Hadley, M. E. (1975). Structure–activity relationships of melatonin and related indoleamines. *Life Sciences* 17, 1167–78.

Hinkle, P. M. and Goh, K. B. (1982). Regulation of thyrotropin releasing hormone receptors and responses to L-triiodothyronine in dispersed rat pituitary cell cultures. *Endocrinology* 110, 1725–31.

Hoffman, R. A. (1983). Seasonal growth and development and the influence of the eyes and pineal gland on body weight of golden hamsters (M. Auratus). *Growth* 47, 109–21.

Hoffman, R. A., Davidson, K., and Steinberg, K. (1982). Influence of photoperiod and temperature on weight gain, food consumption, fat pads, and thyroxine in male golden hamsters. *Growth* 46, 150–62.

Houssay, A. B., Pazo, J. H., and Epper, C. E. (1966). Effects of the pineal gland upon the hair cycles in mice. *Journal of Investigative Dermatology* 47, 230–4.

Ishibashi, T., Hahn, P. W., Srivastava, L., Kumaresan, P., and Turner, C. W. (1966). Effect of pinealectomy and melatonin on feed consumption and thyroid hormone secreting rate. *Proceedings of the Society for Experimental Biology and Medicine* 122, 644–7.

Kirkegaard, C. and Bjorum, N. (1980). TSH responses to TRH in endogenous depression. *Lancet* i, 52.

Langer, G., Koinig, G., Hatzinger, R., Schonbeck, G., Resch, F., Aschauer, H., Keshavan, M. S., and Sieghart, W. (1986). Response of thyrotropin to thyrotropin-releasing hormone as predictor of treatment outcome. *Archives of General Psychiatry* 43, 861–8.

Lechan, R. M. and Jackson, I. M. D. (1982). Immunohistochemical localization of thyrotropin-releasing hormone in the rat hypothalamus and pituitary. *Endocrinology* 111, 55–65.

Lechan, R. M. et al. (1986). Thyrotropin-releasing hormone precursor: characterization in rat brain. *Science* 231, 159–61.

Lerner, A. B., Case, J. D., Lee, T. H., and Mori, W. (1958). Isolation of melatonin, the pineal factor that lightens melanocytes. *Journal of the American Chemical Society* 80, 2587.

Lerner, A. B., Case, J. D., and Heinzelman, R. V. (1959). Structure of melatonin. *Journal of the American Chemical Society* 81, 6084–5.

Lewinski, A. and Sewerynek, E. (1986). Melatonin inhibits the basal and TSH-stimulated mitotic activity of thyroid follicular cells in vivo and in organ culture. *Journal of Pineal Research* 3, 291–5.

Lewy, A. J., Wehr, T. A., Goodwin, F. K., Newsome, D. A., and Rosenthal, N. E. (1981). Manic-depressive patients may be supersensitive to light. *Lancet* i, 383–4.

Losada, J. (1977). Effects of experimental pinealectomy. *Annals of Anatomy* 26, 133–53.

Massara, F., Camanni, F., Belforte, L., Virgano, V., and Molinatti, M. (1978). Increased thyrotropin secretion induced by sulpiride in man. *Clinical Endocrinology* 9, 419–28.

Milcu, S. M., Lupulescu, A., Negoescu, I., and Cocu, F., (1959). Influence of the epiphyseal hormone on experimental goitre produced by an iodine-deficient diet. *Communicatia Academia Republica Populare Romine* **9**, 515–20.

Miline, R. (1963). La part du noyau paraventriculaire dans l'histophysiologie correlative de la glande thyroide et de la gland pineale. *Annals d'Endocrinologie* **24**, 255–69.

Nair, N. P. V., Hariharasubramanian, N., and Pilapil, C. (1985). Circadian rhythm of plasma melatonin and cortisol in endogenous depression. *Advances in the Biosciences* **53**, 339–45.

Narang, G. D. and Turner, C. W. (1966). Effect of advancing age on thyroid hormone secretion rate of female rats. *Proceedings of the Society for Experimental Biology and Medicine* **121**, 203–5.

Narang, G. D., Singh, D. V., and Turner, C. W. (1967). Effect of melatonin on thyroid hormone secretion rate and feed consumption of female rats. *Proceedings of the Society for Experimental Biology and Medicine* **125**, 184–8.

Niles, L. P., Brown, G., and Grota, L. J. (1979). Role of the pineal gland in diurnal endocrine secretion and rhythm regulation. *Neuroendocrinology* **29**, 14–21.

Nilsson, K. O., Thorell, J. I., and Hofkeldt, B. (1974). The effect of thyrotropin releasing hormone on the release of thyrotropin and other pituitary hormones in man under basal conditions and following adrenergic blocking agents. *Acta Endocrinologica* **76**, 24–34.

Nir, I. and Hirschmann, N. (1978). The effect of thyroid hormones on rat pineal indoleamine metabolism *in vitro*. *Journal of Neural Transmission* **42**, 117–26.

Nordlund, J. and Lerner, A. B. (1977). The effects of oral melatonin on skin color and on the release of pituitary hormones. *Journal of Clinical Endocrinology and Metabolism* **45**, 768–74.

Pang, S. F., and Allen, A. E. (1986). Extra-pineal melatonin in the retina: its regulation and physiological function. *Pineal Research Reviews* **4**, 55–96.

Pazo, J. H., Houssay, A. B., Davison, T. A., and Chait, R. J. (1968). On the mechanism of thyroid hypertrophy in pinealectomized rats. *Acta. Physiologia Latino-Americana* **18**, 332–76.

Perrone, M. H. and Hinkle, P. M. (1978). Regulation of pituitary receptors for thyrotropin releasing hormone by thyroid hormones. *Journal of Biological Chemistry* **253**, 5168–73.

Pourmand, M., *et al.* (1980). Domperidone: A novel agent for the investigation of anterior pituitary function and control in man. *Clinical Endocrinology* **12**, 211–15.

Quay, W. B. (1986). Pineal and biorhythms. *Pineal Research Reviews* **4**, 183–97.

Ralph, C. L. (1979). Pineal complex and thermoregulation. *Biological Reviews* **54**, 41–72.

Redding, T. W., Arimura, A., and Schally, A. V. (1970). Effect of porcine thyrotropin (TSH)-releasing hormone (TRH) in release of TSH from sheep and goat anterior pituitary tissue in vitro. *General and Comparative Endocrinology* **14**, 598–601.

Reed, H. L., Burman, K. D., Shakir, K. M. M. and O'Brian, J. T. (1986). Alterations in the hypothalamic–pituitary–thyroid axis after prolonged residence in Antarctica. *Clinical Endocrinology* **25**, 55–65.

Reiter, R. J. (1980). The pineal and its hormones in the control of reproduction in mammals. *Endocrine Reviews* **1**, 109–31.

Reiter, R. J. (1982). Pineal melatonin production: Endocrine and age effects. In *Melatonin rhythm generating system* (ed. D. C. Klein), pp. 143–54. S. Karger, Basel.

Reiter, R. J., Rudeen, P. K., Sackman, J. W., Vaughan, M. K., Johnson, L. Y., and Little, J. C. (1977). Subcutaneous melatonin implants inhibit reproductive atrophy in male hamsters induced by daily melatonin injections. *Endocrine Research Communications* **4**, 35–44.

Relkin, R. (1972). Effects of pinealectomy, constant light and darkness on thyrotropin levels in the pituitary and plasma of the rat. *Neuroendocrinology* **10**, 46–52.

Relkin, R. (1978). Use of melatonin and synthetic TRH to determine site of pineal inhibition of TSH secretion. *Neuroendocrinology* **25**, 310–18.

Rosenthal, N. E., *et al.* (1984). Seasonal affective disorders. *Archives of General Psychiatry* **41**, 72–80.

Scepovic, M. (1963). Contribution a l'etude histophysiologique de la glande thyroide et de la glande pineale. *Annals d'Endocrinology* **24**, 371–6.

Schally, A. V., Redding, T. W., Bowers, C. Y., and Barrett, J. F. (1969). Isolation and purification of porcine thyrotropin-releasing hormone. *Journal of Biological Chemistry* **244**, 4077–88.

Singh, D. V., Narang, G. D., and Turner, C. W. (1969). Effect of melatonin and its withdrawal on thyroid hormone secretion rate of female rats. *Journal of Endocrinology* **43**, 489–90.

Singh, D. V. and Turner, C. W. (1972). Effect of melatonin upon thyroid hormone secretion rate in female hamsters and adult rats. *Acta Endocrinologica* **69**, 35–40.

Smythe, G. A. and Lazarus, L. (1973). Growth hormone regulation by melatonin and serotonin. *Nature* **244**, 230–1.

Steger, R. W., Reiter, R. J., and Siler-Khodr, T. M. (1984). Interactions of pinealectomy and short-photoperiod exposure on the neuroendocrine axis of the male Syrian hamster. *Neuroendocrinology* **38**, 158–63.

Steiner, M. and Brown, G. M. (1985). Melatonin/cortisol ratio and the dexamethasone suppression test in newly admitted psychiatric inpatients. *Advances in the Biosciences* **53**, 347–53.

Tamarkin, L., Westrom, A. I., Hamill, A. I., and Goldman, B. D. (1976). Effect of melatonin on the reproductive systems of male and female Syrian hamsters: a diurnal rhythm in sensitivity to melatonin. *Endocrinology* **114**, 1344–51.

Vacas, M. I., Keller-Sarmiento, M. I., Pereyra, E., and Cardinali, D. P. (1982). Early changes in cAMP and melatonin levels of rat pineal gland after superior cervical ganglionectomy. *Neuroendocrinology Letters* **4**, 267–71.

Vaughan, G. M., Vaughan, M. K., Seraile, L. G., and Reiter, R. J. (1982*a*). Thyroid hormones in male hamsters with activated pineals or melatonin treatment. In *The pineal and its hormones* (ed. R. J. Reiter), pp. 187–96. Alan R. Liss, New York.

Vaughan, G. M., Mason, A. D., and Reiter, R. J. (1986). Serum melatonin after a single aqueous subcutaneous injection in Syrian hamsters. *Neuroendocrinology*

42, 124–7.

Vaughan, M. K., Powanda, M. C., Brainard, G. C., Johnson, L. Y., and Reiter, R. J. (1982*b*). Effects of blinding or afternoon melatonin injections on plasma cholesterol, triglycerides, glucose, TSH and thyroid hormone levels in male and female Syrian hamsters. In *The pineal and its hormones* (ed. R. J. Reiter), pp. 177–86. Alan R. Liss, New York.

Vaughan, M. K., Powanda, M. C., Richardson, B. A., King, T. S., Johnson, L. Y., and Reiter, R. J. (1982*c*). Chronic exposure to short photoperiod inhibits free thyroxine index and plasma levels of TSH, T4, triiodothyronine (T3) and cholesterol in female Syrian hamsters. *Comparative Biochemistry and Physiology* **71**, 615–18.

Vaughan, M. K., *et al.* (1983). Natural and synthetic analogues of melatonin and related compounds II. Effects on plasma thyroid hormones and cholesterol levels in male Syrian hamsters. *Journal of Neural Transmission* **56**, 279–91.

Vaughan, M. K., Brainard, G. C., and Reiter, R. J. (1984*a*). The influence of natural short photoperiodic and temperature conditions on plasma thyroid hormones and cholesterol in male Syrian hamsters. *International Journal of Biometeorology* **28**, 201–10.

Vaughan, M. K. *et al.* (1984*b*). Effects of injections and/or chronic implants of melatonin and 5-methoxytryptamine on plasma thyroid hormones in male and female Syrian hamsters. *Neuroendocrinology* **39**, 361–6.,

Vaughan, M. K., Richardson, B. A., Petterborg, L. J., Vaughan, G. M., and Reiter, R. J. (1984*c*). Injections and/or implants of 6-chloromelatonin and melatonin: effects on plasma thyroid hormones in male and female Syrian hamsters. *Biomedical Research* **5**, 413–18.

Vriend, J. (1983). Evidence for pineal gland modulation of the neuroendocrine-thyroid axis. *Neuroendocrinology* **36**, 68–78.

Vriend, J. (1985). Effects of melatonin and thyroxine replacement on thyrotropin, luteinizing hormone, and prolactin in male hypothyroid hamsters. *Endocrinology* **117**, 2402–7.

Vriend, J. and Gibbs, F. P. (1984). Coincidence of counter-antigonadal and counter-antithyroid action of melatonin administration via the drinking water in male golden hamsters. *Life Sciences* **34**, 617–23.

Vriend, J. and Reiter, R. J. (1977). Free thyroxin index in normal, melatonin-treated and blind hamsters. *Hormone and Metabolic Research* **9**, 231–4.

Vriend, J. and Wasserman, R. A. (1986). Effects of afternoon injections of melatonin in hypothyroid male Syrian hamsters. *Neuroendocrinology* **42**, 498–503.

Vriend, J. and Wilber, J. F. (1983). Influence of the pineal gland on hypothalamic content of TRH in the Syrian hamster. *Hormone Research* **17**, 108–13.

Vriend, J., Sackman, J. W., and Reiter, R. J. (1977). Effects of blinding, pinealectomy and superior cervical ganglionectomy on free thyroxin index of male golden hamsters. *Acta Endocrinologica* **86**, 758–62.

Vriend, J., Reiter, R. J., and Anderson, G. R. (1979). Effects of the pineal and melatonin on thyroid activity of male golden hamsters. *General and Comparative Endocrinology* **38**, 189–95.

Vriend, J., Richardson, B. A., Vaughan, M. K., Johnson, L. Y., and Reiter, R. J.

(1982). Effects of melatonin on thyroid physiology of female hamsters. *Neuro-endocrinology* **35**, 79–85.

Waldhauser, F., *et al.* (1987). A pharmacological dose of melatonin increases PRL levels in males without altering those of GH, LH, FSH, TSH, testosterone or cortisol. *Neuroendocrinology* **46**, 125–30.

Wetterberg, L., Beck-Friis, J., Kjellman, B. F., and Ljungren, J. G. (1984). Circadian rhythms in melatonin and cortisol secretion in depression. *Advances in Biochemical Pharmacology* **39**, 197–205.

Wiechmann, A. F. (1986). Melatonin: parallels in pineal gland and retina. *Experimental Eye Research* **42**, 507–27.

Woeber, K. A. and Ingbar, S. H. (1974). Interactions of thyroid hormones with binding proteins. In *Handbook of physiology,* Vol. 3. (ed. R. O. Greep and E. B. Astwood), pp. 187–96. Williams and Wilkins, Baltimore.

Wright, J., Aldhous, M., Franey, C., English, J., and Arendt, J. (1986). The effects of exogeneous melatonin on endocrine function in man. *Clinical Endocrinology* **24**, 375–82.

Yu, H. S., Pang, S. F., and Tang, S. L. (1981). Increase in the level of retinal melatonin and persistence of its diurnal rhythm in rats after pinealectomy. *Journal of Endocrinology* **91**, 477–81.

Zaboni, A., Zaboni-Musiaccia, W., and Zanussi, C. (1979). Enhanced TSH stimulating effect of TRH by sulpiride. *Acta Endocrinologica* **91**, 257–63.

6. Melatonin: effects on sleep and behaviour in man

Harris R. Lieberman and Anne E. Lea

Introduction

Melatonin, when administered to humans or other animals in pharmacological quantities, has definite effects on behaviour, although these have
not as yet been thoroughly evaluated. Indeed, a substantial number of
animal studies have been conducted with melatonin, but these have been
limited in scope and have not fully characterized the psychopharmacological properties of this substance. Many of these studies have focused
on possible anti-anxiety and hypnotic effects of melatonin. Melatonin has
also been reported to decrease pain sensitivity in animals (Lakin *et al.*
1981) and alter circadian rhythms of activity (Armstrong *et al.* 1986).
The relatively small number of human studies with exogenously
administered melatonin have, for the most part, focused on properties
related to sleep and sleepiness. In addition, little research has been done
to determine whether administration of melatonin could have beneficial
effects when given to appropriate patient populations.

Melatonin: effects on sleep and behaviour in man

Melatonin and sleep

The ability of melatonin to facilitate sleep was first noted in animals
(Marczynski *et al.* 1964) soon after its isolation more than two decades
ago (Lerner *et al.* 1958). Early human studies also suggested a possible
role for melatonin in the regulation of the onset of sleep (Anton-Tay *et
al.* 1971; Cramer *et al.* 1974) and it is now known that melatonin affects
human mood and performance as well (Vollrath *et al.* 1981; Arendt *et al.*
1984; Lieberman *et al.* 1984; Arendt, Chapter 2, this volume).

To date, the effects of melatonin on sleepiness and on sleep have been
the most widely investigated, with the effects of acute administration the
most often reported. In one of the earliest studies Anton-Tay and
collaborators (1971) administered melatonin intravenously (1.25 mg/kg)

to five healthy adults while simultaneously monitoring the electro-encephalogram (EEG), heart rate, respiration, and skin reflex (GSR). Shortly after administration the EEG showed a slight deactiviation and within 15–20 minutes volunteers fell asleep. Cramer and colleagues (1974) also reported rapid sleep onset (15–40 minutes) after intravenous injection of 50 mg of melatonin to healthy young men in the daytime. All-night sleep recordings were also obtained from 15 male volunteers following administration of 50 mg intravenously at 21.30 hours. Melatonin significantly decreased mean latency of sleep onset from 23.3 to 10.4 minutes. No significant differences in total sleep or in the amount of time spent in sleep stages 1 through 4 or in REM sleep were found. However, individuals displayed a slight increase in the total number of sleep stages and more frequent awakenings were observed.

More recently, in a double-blind cross-over study with 10 healthy volunteers, Vollrath and associates (1981) administered 1.7 mg of melatonin in the form of a nasal spray. Of the 10 subjects, seven fell asleep within approximately 40–60 minutes following melatonin administration compared to only one following placebo administration. Those who did not fall asleep reported feeling mildly tired or sedated. The study was repeated two weeks later with eight of the original 10 volunteers and all subjects fell asleep, with the first incidence of self-reported tiredness occurring 30 minutes after administration. The apparent potency of this relatively small dose of melatonin (compared to previous studies) may have been related to the route of administration employed. Vollrath and co-workers (1981) attribute the potency of intranasal melatonin administration to the fact that the enterohepatic circulation is avoided. However, to date, the relative behavioural potency of various routes of melatonin administration have not been directly compared nor have dose–response studies been conducted with single routes of administration. The available evidence suggests that acute doses of melatonin administered through various routes in pharmaco-logical doses have hypnotic, sedative-like effects.

Melatonin has also been reported to have residual effects on mood following sleep that is induced by its administration. Anton-Tay and collaborators (1971) reported that, following 45 minutes of melatonin-induced sleep, all subjects reported a sense of well-being and moderate elation. Vollrath *et al.* (1981) also reported that all subjects who responded to melatonin reported feelings of well-being and felt emotionally well-balanced after melatonin-induced sleep, with the exception of one subject who reported mild depression.

Although exogenous melatonin clearly affects sleep and sleepiness, it is not known whether endogenous release of melatonin by the pineal gland alters human sleep. Akersted and colleagues (1982) reported a circadian covariation of fatigue and urinary melatonin, and suggested that melatonin was involved in the regulation of the human sleep–wake cycle. Waldhauser *et al.* (1986) recently demonstrated that the normal pattern of melatonin secretion is disrupted in individuals who are engaged in shift-work and their endogenous rhythm of melatonin release depends on their daily pattern of activity. Based on the temporal pattern of melatonin secretion in humans and the effects of exogenous melatonin administration, Wurtman and Lieberman (1985) hypothesized that melatonin acts as a mediator of circadian variations in sleep and sleepiness. Birkeland (1982) reported that melatonin was released in pulses during the night and that spontaneous waking episodes were significantly correlated with the occurrence of melatonin peaks. He therefore suggested that melatonin may act to restore sleep in humans. However, the results of a similar study conducted by Claustrat and associates (1986) do not support a direct relationship between melatonin secretion and the sleep–wake cycle in humans.

While it is clear that, in general, the release of melatonin in humans is correlated with the occurrence of sleep, it has yet to be established that this relationship is causal in any direct sense. Although pharmacological doses of melatonin have clear hypnotic-sedative properties, doses that produce physiological changes in melatonin levels have not been adequately investigated. The demonstration that an endogenously released substance has a specific behavioural function is a difficult undertaking. For example, even potent drugs, unless they are administered in massive doses, only act as potentiators of sleep. Also, if a substance like melatonin does have endogenous effects, they almost certainly occur within the brain and it is not currently possible to simultaneously assess CNS melatonin concentration and behaviour. Furthermore, what appear to be large pharmacological doses of melatonin, at least as far as plasma levels are concerned, may not, for a variety of reasons, produce the necessary changes in brain at the particular sites where this hormone exerts its effects. Therefore, the effects of melatonin on the CNS may not always be appropriately defined as a certain plasma concentration.

Melatonin and mood

The acute administration of melatonin has also been demonstrated to have effects on various aspects of human mood state, particularly those related to level of arousal. To examine its effects we administered 240 mg of melatonin in a divided dose over a period of 2 hours (12.00–14.00 hours) to 14 healthy male volunteers (Lieberman *et al.* 1984). Mood was assessed hourly using standardized self-report questionnaires known to be sensitive to the subtle effects of a variety of classes of drugs, including hypnotics and stimulants. Melatonin significantly reduced vigour and increased fatigue as measured by subscales of one of these questionnaires; the Profile of Mood States (POMS; Fig. 6.1). Confusion was also significantly elevated as measured by another sub-scale of this questionnaire. Sleepiness as measured by another questionnaire, the Stanford Sleepiness Scale (SSS), also increased following melatonin administration (Fig. 6.2). The greatest effect of melatonin on mood occurred at 15.00 and 16.00 hours after all three doses had been administered and well after onset of elevated plasma concentration. Plasma levels were still elevated at 17.00 hours although mood had returned to normal by this time.

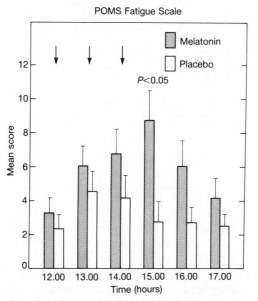

Fig. 6.1. Effect of melatonin or placebo on the Fatigue scale of the Profile of Mood State (POMS; Lieberman *et al.* 1984). The POMS is a 65-adjective, self-report mood scale that, when analysed, yields six factors: Tension–Anxiety, Depression–Dejection, Anger–Hostility, Vigour–Activity, Fatigue–Inertia, and Confusion–Bewilderment. The times of melatonin administration (80 mg) are indicated by the arrows.

In a recent study, conducted at the NIMH, Sherer and colleagues (1985) administered melatonin orally (2–2.4 mg daily) to six patients with Seasonal Affective Disorder (SAD) and assessed mood and performance. Melatonin was found to increase subjective ratings of fatigue and the need for sleep, as well as decreasing self-reported energy levels as measured by the Weekly Mood Inventory which was

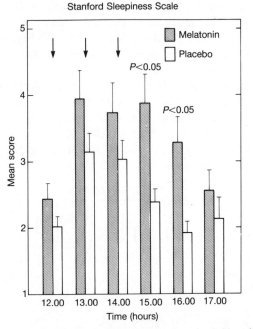

Fig. 6.2. Effect of melatonin or placebo on the Stanford Sleepiness Scale (SSS; Lieberman *et al.* 1984). The times of melatonin administration (80 mg) are indicated by the arrows.

administered on the last day of each treatment. The Hamilton Rating Scale for depression was also administered, together with an addendum sensitive to the unique symptoms of SAD (see Thompson, Chapter 11, this volume). The investigators did not find an effect of treatment condition on overall Hamilton Scores. However, after 1 week of melatonin treatment a significant increase in the severity of some of the symptoms characteristic of SAD was noted (Sherer *et al.* 1985; Rosenthal *et al.* 1985).

These findings were consistent with those of a previous clinical trial in which melatonin was reported to exacerbate the symptoms of individuals with primary depressive illness (Carman *et al.* 1976) when it was administered during the day in high doses. This may be attributable to its sedative-like effects which could be amplifying some of the symptoms usually associated with depression, such as lethargy, fatigue, and psychomotor retardation (see Miles *et al.*, Chapter 13, this volume).

The behavioural effects of chronic administration of melatonin have also been investigated. Arendt and co-workers (1984) administered small doses of melatonin (2 mg) in a 50 ml milk solution at 17.00 hours to 12 healthy volunteers for 4 weeks. Fatigue was measured using self-ratings on a visual analogue scale 4 times a day (08.00, 12.00, 19.00 22.00 hours). Seven subjects had significantly increased fatigue at 19.00 or 22.00 hours with the first incidence of increased fatigue apparently occurring on about the fifth day of the treatment condition. Five subjects reported reduced tiredness in the morning (08.00 hours). The results of this study suggest that chronic low-dose melatonin administration can produce a phase shift in the circadian rhythm of arousal. The slow onset of the effect is consistent with modulation of circadian rhythms of alertness rather than an acute sleep-inducing effect. However, it may be that sleep-induction and fatigue induced by high doses of melatonin, as well as the phase-shift induced by lower doses of melatonin given in the evening, can be attributed to the same underlying mechanism of action. Arendt and colleagues (1984) also reported that appropriately-timed doses of melatonin speed re-entrainment following a forced phase-shift, confirming results from an earlier study conducted with only a few volunteers (Armstrong *et al.* 1986). For a detailed review of the effects of melatonin on the human circadian system, the reader is referred to Chapter 2 of the present volume.

Melatonin and performance

Various aspects of human performance have also been found to be affected by the administration of melatonin. The effects of melatonin on reaction time and time estimation were apparently first studied by Anton-Tay and associates (1971) in 11 healthy adult volunteers. Reaction time (RT) was assessed by requiring subjects to respond as rapidly as possible to a visual stimulus. To assess time estimation ability, subjects were required to estimate the time intervals at which light stimuli were administered. Following melatonin administration average reaction time did not change, but the estimated duration of the time intervals between delivery of the light stimuli increased.

In a more recent double-blind, placebo-controlled study conducted at MIT and described above, 14 healthy male subjects received 240 mg of melatonin divided into three equal doses taken at 12.00, 13.00, and 14.00 hours (Lieberman *et al.* 1984). Melatonin was administered in a divided dose to ensure that pharmacological plasma levels would be maintained throughout the afternoon test session (12.00–17.00 hours). This permitted us to administer a wide variety of behavioural tests while individuals had very high levels of melatonin present in their plasma. Baseline measurements of all behavioural parameters were taken before treatment began. Testing consisted of a standardized battery of mood questionnaires (the results of these have been described above) and tests that assessed memory, mental performance, and visual sensitivity, some of which were administered repeatedly throughout the test session. Simple auditory reaction time was measured by a microcomputer-administered test that recorded subjects' responses to the onset of a 75 db (SPL) 1900 Hz tone. Five warm-up trials were presented followed by 125 test trials in rapid succession. Both commission errors (responding prior to the onset of a stimulus) and errors of omission (response latency greater than one second) were recorded. Sustained vigilance performance was assessed using the four choice visual RT test (Wilkinson and Houghton 1975). In this test one of four different spatial locations on a CRT screen is randomly illuminated and the subject is asked to correctly identify the location of each stimulus by striking one of four adjacent keys on a microcomputer keyboard. Sustained complex motor performance was measured using a modified version of the grooved pegboard test. Subjects were required to insert a series of 25 grooved pegs into a correspondingly grooved board. The test was administered eight times in rapid succession to measure sustained performance. Visual sensitivity was measured using a critical flicker fusion test that determines the threshold at which a flickering light is perceived to be steadily on. Recall and recognition memory were tested by administering word lists of 10 words pertaining to one specific category followed by a string of six digits presented via a tape recorder. Blood samples were drawn hourly from an indwelling heparinized catheter and used to assess plasma melatonin concentration.

In addition to its substantial effects on mood described above, melatonin also altered specific aspects of performance. Melatonin significantly increased response latency on the four-choice visual reaction time task. Although reaction time was slower, the number of incorrect responses decreased. This apparent disassociation between speed and accuracy is not unprecedented and has occasionally been noted in studies of other drugs with sedative properties (Lieberman *et al.* 1985). Melatonin also significantly reduced the number of errors on the simple auditory RT task, although latency was not significantly increased. Melatonin administration did not impair anterograde memory on a test that had previously been shown to be sensitive to the acute effects of the benzodiazepines on this function. One hour after administration of the first dose (13.00 hours) a substantial increase in plasma melatonin was apparent. Plasma levels remained elevated throughout the afternoon.

In another study which assessed human performance Sherer and his colleagues at the NIMH (1985) orally administered 200 μg of melatonin to subjects with SAD 30 minutes before and every hour during daily light treatment for a total of 2–2.4 mg daily. The objective of this dosing regimen was to extend the duration of nocturnal melatonin secretion and produce somewhat higher than normal plasma levels. Subjects were tested at baseline and then weekly at 15.00 hours. Continuous performance task (CPT), reaction time (RT), cognition, and memory were assessed. RT was significantly faster during melatonin treatment compared to placebo. No differences were found in CPT or on the cognition and memory tests. Although apparently inconsistent with respect to reaction time changes induced by melatonin, direct comparison of the results of the NIMH study with the MIT study (Lieberman *et al.* 1984) could be misleading. Sherer and associates (1985) tested a comparatively low dose of melatonin administered over several days to a specific patient population undergoing a treatment known to suppress the release of endogenous melatonin. The MIT study administered much higher doses of melatonin on one occasion to a group of normal volunteers. There were also substantial differences in the tests that detected the different effects. The simple visual reaction time task, a test in which only one possible stimulus is presented, was improved by melatonin administration in the NIMH study. However, it was choice visual reaction time (on each trial one of four possible stimuli are presented) that was significantly slowed in the MIT study. Simple auditory reaction time was also reported to be somewhat slower in the MIT study but that difference was not statistically significant. It should be noted that the results from the mood questionnaires and Hamilton Rating Scale that Sherer and colleagues (1985) administered were in agreement with the MIT finding. As discussed above, both studies found that melatonin administration during the daytime produces sedation. Typically, sedatives slow performance as well, although Sherer *et al.* (1985) point out that benzodiazepines may facilitate certain aspects of performance in low doses. Additional studies in normal individuals and in patients with SAD will be necessary to resolve these issues.

Conclusion

Although when melatonin is administered in pharmacological doses it has potent hypnotic-sedative properties, these have not been adequately described in normal or patient populations. In normal volunteers when it is administered in appropriate doses, it can produce sedative effects of the same magnitude as those produced by clinically effective doses of hypnotics (Lieberman *et al.* 1985). However, dose–response studies with sleep measured using electroencephalographic techniques and subjective questionnaires have not been performed nor have melatonin's sedative effects during the day been adequately examined at a wide range of doses. Additionally, studies with patients with various sleep disorders have not, as yet, been conducted. The results from the various behavioural studies reviewed above certainly suggest that melatonin should be evaluated as a hypnotic in appropriately selected patient populations. The apparently brief duration of the sedative effects of melatonin suggests that, unlike the classic hypnotics such as the longer-acting benzodiazepines, melatonin would not produce adverse residual effects the next day. Melatonin also appears to lack the potent amnesia-inducing effects of the benzodiazepines, the most widely prescribed hypnotics. Due to its short pharmacological and behavioural half-life it might be necessary to employ a timed-release preparation if melatonin was to be tested as a hypnotic in clinical trials.

References

Akerstedt, T., Gillberg, M., and Wetterberg, L. (1982). The circadian covariation of fatigue and urinary melatonin. *Biological Psychiatry* **17**, 547–54.

Anton-Tay, F., Diaz, J. L., and Fernandez-Guardiola, G. (1971). On the effects of melatonin upon human brains: its possible therapeutic implications. *Life Science* **10**, 841–50.

Arendt, J., Borbely, A. A., Franey, C., and Wright, J. (1984). The effects of chronic, small doses of melatonin given in the late afternoon on fatigue in man: a preliminary study. *Neuroscience Letters* **45**, 317–21.

Armstrong, S. M., Cassone, V. M., Chesworth, M. J., Redman, J. R., and Short, R. V. (1986). Synchronization of mammalian circadian rhythms by melatonin. *Journal of Neural Transmission* (Suppl.) **21**, 375–94.

Birkeland, A. J. (1982). Plasma melatonin levels and nocturnal transitions between sleep and wakefulness. *Neuroendocrinology* **34**, 126–31.

Carman, J. S., Post, R. M., Buswell, R., and Goodwin, F. K. (1976). Negative effects of melatonin on depression. *American Journal of Psychiatry* **133**, 1181–6.

Claustrat, B., Brun, J., Garry, P., Roussel, B., and Sassolas, G. (1986). A once-repeated study of nocturnal plasma melatonin patterns and sleep recordings in six normal young men. *Journal of Pineal Research* **3**, 301–10.

Cramer, H., Rudolph, J., Consbruch, U., and Kendel, K. (1974). On the effects of melatonin on sleep and behavior in man. *Advances in Biochemical Psychopharmacology* **11**, 187–91.

Lakin, M. L., Miller, C. H., Stott, M. L., and Winters, W. D. (1981). Involvement of the pineal gland and melatonin in murine analgesia. *Life Sciences* **29**, 2543–51.

Lerner, A. B., Case, J. D., Takahashi, Y., Lee, T. H., and Mori, W. (1958). Isolation of melatonin, the pineal gland factor that lightens melanocytes. *Journal of the American Chemical Society* **80**, 2587.

Lieberman, H. R., Waldhauser, F., Garfield, G., Lynch, H., and Wurtman, R. J. (1984). Effects of melatonin on human mood and performance. *Brain Research* **323**, 201–7.

Lieberman, H. R., Garfield, G., Waldhauser, F., Lynch, H. J., and Wurtman, R. J. (1985). Possible behavioral consequences of light-induced changes in melatonin availability. In *The medical and biological effects of light* (ed. R. J. Wurtman, M. J. Baum, and J. T. Potts, Jr), pp. 242–52. The New York Academy of Sciences, New York.

Marczynski, T. J., Yamaguchi, N., Ling, G. M., and Grodzinska, L. (1964). Sleep induced by the administration of melatonin (5-methoxy-N-acetyltryptamine) to the hypothalamus in unrestrained cats. *Experientia* **20**, 435–7.

Rosenthal, N. E., *et al.* (1985). Seasonal affective disorder and phototherapy. In *The medical and biological effects of light* (ed. R. J. Wurtman, M. J. Baum, and J. T. Potts, Jr), pp. 260–9. The New York Academy of Sciences, New York.

Rosenthal, N. E., *et al.* (1986). Melatonin in seasonal affective disorder and phototherapy. *Journal of Neural Transmission* (Suppl.) **21**, 232–40.

Sherer, M. A., Weingartner, H., James, S. P., and Rosenthal, N. E. (1985). Effects of melatonin on performance testing in patients with seasonal affective disorder. *Neuroscience Letters* **58**, 277–82.

Vollrath, L., Semm, P., and Gammel, G. (1981). Sleep induction by intranasal application of melatonin. In *Melatonin: current studies and perspectives* (ed. N. Birau and W. Schloot), pp. 327–9. Pergamon Press, London.

Waldhauser, F., Vierhapper, H., and Oirich, K. (1986). Abnormal circadian melatonin secretion in night-shift workers. *The New England Journal of Medicine* **315**, 1614.

Wilkinson, R. T. and Houghton, D. (1975). Portable four-choice reaction time test with magnetic tape memory. *Behaviour Research and Methods and Instrumentation* **7**, 441–6.

Wurtman, R. J. and Lieberman, H. R. (1985). Melatonin secretion as a mediator of circadian variations in sleep and sleepiness. *Journal of Pineal Research* **2**, 301–3.

7. Melatonin and cancer: basic and clinical aspects

David E. Blask and Steven M. Hill

Introduction

Tremendous advances have been made within the last two decades with respect to the elucidation of the functions of the pineal gland and its major hormone melatonin, particularly those relating to photoperiodic time measurement and the regulation of seasonal reproductive phenomena in mammals. In contrast to this rapid progress, advances in our understanding of the pineal's impact on human health and disease have been understandably slower. However, our current ability to measure melatonin in virtually any human body fluid (see Miles *et al.*, Chapter 13) has ushered in a new era of rapidly expanding knowledge concerning pineal functions in healthy individuals as well as in those afflicted with various diseases.

One of the most exciting aspects of pineal research, which actually had its origin nearly 60 years ago (Georgiou 1929), concerns the role of the pineal gland and more particularly melatonin in neoplastic disease. Progress in this field was slow until the 1950s and 1960s when in Europe, a veritable flurry of activity occurred in the use of pineal extracts as well as xenogeneic grafts of pineal tissue from cows and pigs to treat a variety of human malignancies (Lapin 1976). Whether melatonin was the substance actually responsible for the oncostatic effects of the pineal extracts or xenografts is unknown.

Since that time, research has intensified, particularly within the past 10 years, in the areas of the pineal-melatonin influence on hormone-responsive cancers in addition to the changes in melatonin secretory patterns that occur in malignancy (Tapp 1982; Blask 1984). Contrary to Georgiou's (1929) original proposal that the pineal stimulates neoplastic growth, the bulk of subsequent experimental and clinical evidence has given rise to the hypothesis that the pineal is primarily an 'oncostatic gland' with melatonin perhaps serving as its major oncostatic envoy (Blask 1984).

This chapter will focus on current knowledge and concepts of the role

128

of the pineal gland and its hormone melatonin in the regulation of a variety of neoplasms in both *in vivo* and *in vitro* animal and human models of cancer. Prominent attention will be given to the basic cell biology of melatonin in both normal and tumour cells. In addition to the basic science aspects of melatonin and cancer, a detailed account will be given with regard to melatonin and the clinical aspects of malignant disease including a review of studies on melatonin as an anti-cancer therapy as well as a discussion of alterations in melatonin secretory patterns in cancer patients. The concluding part of this chapter will be devoted to an integration of the basic and clinical information on melatonin and cancer with the hope of drawing some preliminary conclusions as to possible mechanisms by which melatonin exerts its oncostatic influence as well as its potential clinical usefulness both as a marker for cancer and a therapeutic agent in human malignancy.

Basic aspects

Normal cell biology of melatonin

The ability of melatonin to directly inhibit cell division was originally observed in the protozoan *Stentor coeruleus* and in onion root tips (Banerjee *et al.* 1972; Banerjee and Margulis 1973). Additionally, the fact that melatonin could block the inhibitory effects of colchicine and vinblastine on pigment granule migration in amphibian melanocytes (Malawista 1973) supported the hypothesis proposed earlier by Quay (1968) that one of melatonin's cellular mechanisms of action is to interact with microtubles. Since the microtubular components of the cytoskeletal architecture of eukaryotic cells play a critical role in cell shape, motility and proliferation, a melatonin-induced inhibition of either the formation or function of this intracellular organelle could conceivably inhibit normal cell division. However, arguing against this hypothesis is evidence that melatonin does not affect the proliferation of normal Chinese hamster ovarian cells which is readily suppressed by both colchicine and vinblastine (Poffenbarger and Fuller 1976). Furthermore, the results of studies with mammalian brain preparations have been equivocal with respect to the ability of melatonin to inhibit radio-labelled colchicine binding to microtubular protein. Additionally, melatonin has no effect on microtubule polymerization (Winston *et al.* 1974; Poffenbarger and Fuller 1976).

Our own studies (Hill 1986) with several normal animal (mouse 3T3 and 3T6, rat 1) and human (human foreskin fibroblasts) cell lines revealed that neither 10^{-7} M or 10^{-9} M melatonin, N-acetylserotonin nor serotonin inhibits cell proliferation. Interestingly, Boucek and Alvarez

(1970) reported that similar concentrations of serotonin stimulated 3T6 and human embryonic lung fibroblastic growth while N-acetylserotonin inhibited cell proliferation by about 30 per cent (melatonin was not tested); however, no effect of these compounds was observed on the growth of five other cell lines. On the other hand, Lewinski (1986) recently reported that melatonin not only completely inhibits the proliferation of thyroid follicular cells in organ-cultured rat thyroid explants, but it blocks the mitogenic effect of thyroid-stimulating hormone on the proliferation of these cells as well.

While there is a lack of consistent data *in vitro* to support the hypothesis that melatonin has a direct anti-proliferative effect on normal mammalian cell division (Blask 1984; Pawlikowski 1986) there is, however, *in vivo* evidence to suggest that melatonin may be a naturally occurring anti-proliferative hormone. For example, pinealectomy has been shown to stimulate the proliferative capacity of a variety of both endocrine and non-endocrine tissues in the rat (Bindoni and Raffaele 1968; Bindoni 1971; Pawlikowski 1986). Furthermore, Leadem and Blask (1984) have shown that 'activation' of the pineal gland by light deprivation inhibits the proliferation of anterior pituitary cells in the prepubertal rat while pinealectomy prevents this effect. Whether the pinealectomy-induced stimulation of mitotic activity in normal tissues is due to the elimination of melatonin from the circulation or to some other effect of pinealectomy is unknown. Similarly, it is unclear whether the inhibition of normal cell proliferation in light-deprived animals is due to melatonin or to some other pineal constituent. Nevertheless, these data do suggest that one of the functions of the pineal gland in normal rats and even in human beings may be to act as a natural 'brake' mechanism to restrain the proliferation of normally differentiated tissues perhaps via the melatonin signal.

Tumour biology of melatonin—in vitro studies

A little over 12 years ago, Fitzgerald and Veal (1976) first addressed the hypothesis that melatonin has a direct anti-proliferative effect on human cancer cells *in vitro*. They incubated millimolar concentrations of melatonin with either HeLa cells (derived from a human cervical carcinoma) or KB cells (derived from a human epidermoid carcinoma) and found that they failed to inhibit cell division. However, they did find that the preincubation of HeLa cells with high concentrations of melatonin completely inhibited the ability of colchicine to subsequently induce mitotic arrest in this cell line thus offering some support of earlier work indicating that melatonin may interact with the microtubular components of eukaryotic cells.

A series of studies have been performed using clonogenic melanoma cell lines from both animal and human tumours. In the first such study, millimolar concentrations of melatonin were shown to cause a modest 25 per cent inhibition in the growth of a cloned melanoma cell line (B7) derived from a spontaneously arising tumour in a male golden hamster. Interestingly, micromolar amounts of this indole caused a 37 per cent stimulation of cell growth suggesting that melatonin may have a biphasic effect (Walker *et al.* 1978). In a similar investigation using human melanoma cells, Bartsch and Bartsch (1984) found that micromolar concentrations of melatonin produced just the opposite results by causing a 60 per cent inhibition of cell proliferation while a 10-fold higher dose produced a moderate stimulation of cell growth. The discrepancies between these two studies could be ascribed to the use of tumour cells from different species as well as the possibility of different culture conditions.

Using clonogenic melanoma cells obtained from patient biopsies of malignant-melanoma metastatic nodules, Meyskens and Salmon (1981) observed that in 55 per cent of the patient biopsies, total cloning efficiency was decreased as a function of increasing concentrations of melatonin (10^{-15} M to 10^{-5} M). No change in either colony type or mean colony diameter was observed. However, at physiological concentrations of melatonin (10^{-9} M) cloning efficiency was increased 4-fold over controls in one case. Study of another patient biopsy revealed that 10^{-11} M melatonin caused a 9-fold increase in the number of dark, small-cell colonies over the total colonies seen in the control plates (predominantly light, large cell variant); however, higher concentrations caused a decrease in the number of colonies. Interestingly, three out of the 11 biopsy specimens exhibited no response to melatonin when tested in the clonogenic assay.

The differences among the studies cited above with respect to the effects of melatonin on melanoma growth notwithstanding, all three studies demonstrated either a stimulatory, inhibitory, or no effect of melatonin on cell proliferation depending upon the dose employed. Exactly how these results could be applied to the design of therapeutic trials to test the ability of melatonin to inhibit tumour growth in melanoma patients is difficult to ascertain at this point.

In our own studies (Hill and Blask 1985*a,b*; Blask and Hill 1986*a,b*) with the human breast cancer cell line MCF-7, we have found that concentrations of melatonin corresponding to the physiological range (10^{-9} M and 10^{-11} M) found in human blood cause a 75–80 per cent inhibition of cell proliferation in culture; higher or lower concentrations of the indole were without effect. We have also tested the ability of several precursors, metabolites, or analogues of melatonin (i.e. serotonin, N-acetylserotonin, 6-hydroxymelatonin, 5-methoxytryptophol, 5-methoxytryptamine, and 6-chloromelatonin) to inhibit MCF-7 cell growth. All compounds tested with the exception of the melatonin analogue 6-chloromelatonin had no effect on cell proliferation; 6-chloromelatonin (10^{-9} M) was equally as effective as melatonin in inhibiting breast cancer cell growth (Hill and Blask 1985*a,b*; Blask and Hill 1986*a,b*). These results indicate that the direct anti-proliferative effect of melatonin on human breast cancer cells is limited to its physiological range and may be a relatively specific feature of this indoleamine. Furthermore, the fact that neither of the major metabolites of melatonin, namely 6-hydroxy-melatonin and N-acetylserotonin (Young *et al.* 1985), inhibited cell growth suggests that melatonin itself rather than one of its metabolic products is actually endowed with oncostatic activity *in vitro*. This is further supported by the report that MCF-7 cells do not metabolize melatonin *in vitro* (Danforth *et al.* 1983).

We were aware of the fact that serum contains a variety of growth factors, the presence of which might be required in order for the anti-proliferative effect of melatonin to be manifested. As it turns out, melatonin's anti-proliferative effect does indeed depend on the presence of serum since cell growth is not inhibited by melatonin in medium which has less than 3 per cent fetal calf serum or in medium that is serum-free (Hill and Blask 1986*a,b*). However, the inhibitory effect of melatonin on MCF-7 cell growth in chemically-defined, serum-free medium can be partially reconstituted by the addition of either oestradiol or prolactin to the culture plates. Therefore, oestradiol and prolactin may be at least two constituents of fetal calf serum that exert a 'permissive' effect on the expression of melatonin's anti-proliferative capacity. It is important to note in this regard that MCF-7 cells possess both oestrogen and prolactin receptors, and prolactin itself can stimulate an increase in oestrogen receptors in these cells (Shafie and Brooks 1977).

In addition to containing oestrogen receptors, MCF-7 cells are able to mount a significant mitogenic response to oestradiol (Lippman *et al.* 1976). Danforth *et al.* (1984) reported that they could inhibit oestradiol-stimulated MCF-7 cell growth in charcoal-stripped medium; however, cell growth was not suppressed below that in the control plates. Surprisingly, this inhibitory effect was associated with an induction of nuclear and cytoplasmic oestrogen receptor hormone binding (Danforth *et al.* 1983). Using a chemically-defined, serum-free medium, not only were we able to inhibit oestradiol-stimulated cell growth with melatonin, but we demonstrated a further suppression of cell proliferation below that observed in the control plates lacking hormones.

Therefore, we further tested the hypothesis that one of the mechanisms by which melatonin inhibits cell proliferation is via an interaction with oestradiol. For example, a physiological concentration of melatonin completely inhibited the oestradiol-stimulated incorporation of radio-labelled-thymidine into DNA of MCF-7 cells (Fig. 7.1). Furthermore, melatonin, even in the presence of 10 per cent fetal calf serum, was completely ineffective as a growth inhibitor of the human breast cancer cell line (BT20) which is both oestrogen-unresponsive and devoid of oestrogen receptors.

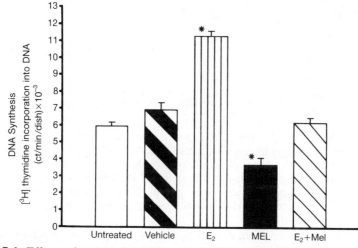

Fig. 7.1. Effects of melatonin (MEL) and 17-β-oestradiol (E$_2$) either alone or in combination on ^{3}H-thymidine incorporation into DNA of MCF-7 human breast cancer cells. Following replicate plating and incubation of cells in DMEM with 10 per cent fetal calf serum, cells were re-fed with serum-free medium for 24 hours prior to the addition of the hormones. Following 48 hours of incubation with the hormones, cells were washed in serum-free medium and then incubated in the same medium for 2 hours more. The cells were then pulsed with the labelled nucleoside (0.01 μCi/ml) for 2 hours, harvested, and the radioactivity measured. *$P < 0.01$ as compared with controls. (Reproduced with permission from Blask and Hill 1986a.)

In related experiments, we have observed that nanomolar concentrations of melatonin are more effective in inhibiting MCF-7 cell growth in serum-containing medium than micromolar concentrations of tamoxifen (Fig. 7.2) a non-steroidal anti-oestrogen used to treat oestrogen-receptor-positive breast cancer (Osborne 1985). Interestingly, the combination of melatonin and tamoxifen produces a degree of growth suppression equivalent to that produced by tamoxifen alone indicating that tamoxifen blocks the anti-proliferative action of melatonin. Since tamoxifen itself binds to the oestrogen receptor (Osborne 1985), these results together with those of Danforth and colleagues (1983) suggest

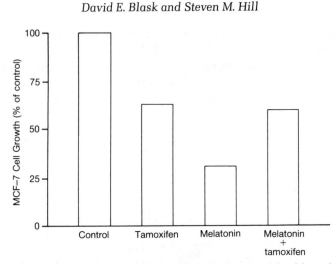

Fig. 7.2. Effects of melatonin (1 nM) and tamoxifen (1 μM) either alone or in combination on MCF-7 human breast cancer cell growth in DMEM with 10 per cent fetal serum. The data represent cell growth in each treatment group (six plates per group) as a percentage of controls by the last day of a 7-day growth curve. The melatonin and tamoxifen groups differed from their respective controls, $P < 0.001$. [From Hill and Blask, unpublished.]

that under certain experimental conditions, part of the mechanism by which the anti-proliferative effect of melatonin on oestrogen-responsive human breast cancer cells is expressed is via an interaction with oestradiol at the level of its receptor.

In preliminary experiments (S. M. Hill and D. E. Blask, unpublished) we observed that melatonin, incubated with MCF-7 cells in serum-free, chemically-defined medium, markedly inhibited the release (but not the production) of several proteins including an oestradiol-inducible 52K glycoprotein that acts in an autocrine fashion to stimulate MCF-7 cell growth (Vignon and Rochefort 1985; Fig. 7.3).

Apparently, physiological levels of melatonin lack the ability to inhibit the growth of other epithelially-derived, oestrogen-receptor-positive human cancer cell lines *in vitro*. For example, the logarithmic growth of an oestrogen-receptor-positive human endometrial cancer cell line (RL95) in serum-containing medium is unaffected by 10^{-9} M melatonin. However, a 1000-fold greater concentration of melatonin causes nearly a 40 per cent reduction in cell growth (Blask and Hill 1986*b*).

Tumour biology of melatonin—in vivo studies

In addition to reviewing the effects of melatonin *per se* on tumour growth in this section, studies on the effects of pinealectomy and photoperiodic changes on tumour growth will be included, since these manipulations

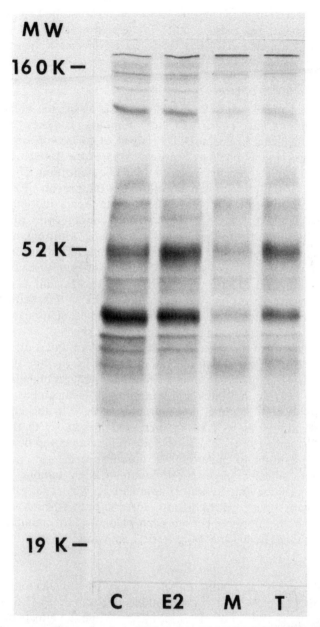

Fig. 7.3. An autoradiograph of an SDS polyacrylamide slab gel depicting the effects of melatonin (M) (1 nM), 17-β-oestradiol (E$_2$) (10^{-8} M) and tamoxifen (T) (1 μM) on the secretion of newly-synthesized ^{35}S-methionine-labelled proteins by MCF-7 human breast cancer cells grown in DMEM with 2.5 per cent fetal calf serum (charcoal-stripped). The migration of standard protein molecular weight markers is indicated on the left of the figure.

affect the secretion of melatonin. However, it should be kept in mind that these treatments may also modify the secretion of as yet unidentified pineal constituents that may also have oncostatic (or oncogenic) properties (Blask 1984).

Undifferentiated neoplasms

As reviewed previously (Lapin 1976; Blask 1984) the ablation of the pineal glands of mice bearing Erhlich's tumours promotes cell proliferation. Whether this is due to the removal of melatonin or some other pineal compound is unclear; however, studies by Bartsch and Bartsch (1981) seem to suggest that melatonin may be important. They examined the effects of daily melatonin injections on the growth of Erlich's solid tumours in mice using different injection paradigms and under different photoperiodic regimens. They found that under long photoperiodic conditions, tumour growth in mice receiving morning injections of melatonin was slightly stimulated while late afternoon injections significantly suppressed tumour growth. Exposure of tumour-bearing mice to short photoperiod slightly decreased tumour size and eliminated the diurnal rhythm of responsiveness to melatonin so that neither morning nor afternoon injections of the indole had a significant effect on neoplastic growth. To our knowledge, this is the first and thus far the only published study to address the important question of whether neoplastic growth exhibits a diurnal or nocturnal rhythm of responsiveness to either inhibitory or possibly stimulatory effects of melatonin.

In rats bearing Yoshida solid tumours, pinealectomy (as well as thymectomy) resulted in a considerable increase in the growth of the primary tumours as well as an increased metastasis to the liver and pancreas (Lapin 1974). In a later study, this same group discovered that the administration of melatonin to tumour-bearing rats not only prevented the pinealectomy-induced stimulation of tumour growth, but increased survival time as well (Lapin and Frowein 1981). Interestingly, animals with intact pineal glands and bearing Yoshida tumours are refractory to melatonin injections even when they are administered in the late afternoon (Huxley and Tapp 1972; Lapin and Frowein 1981).

Sarcomas

As is the case for undifferentiated tumours, pinealectomy markedly stimulates the growth of transplantable fibrosarcomas (Nakatani *et al.* 1940; Katugiri 1943; Barone *et al.* 1972). While the effects of melatonin on this tumour type have not been examined in pinealectomized animals, melatonin has been shown to slow fibrosarcoma growth in pineal-intact mice (Lapin and Ebels 1976; Bartsch and Bartsch 1981). In the latter study, melatonin injected during the late afternoon hours of a 13L:11D photoperiod inhibited tumour growth while morning injections actually augmented the neoplastic process; however, neither morning nor afternoon melatonin injections had a significant effect on tumourigenesis under 12L:12D photoperiodic conditions (Bartsch and Bartsch 1981).

Apparently, the combination of neonatal pinealectomy and thymectomy promotes the development of leukaemia in rats subsequently treated with a catecholamine depletor (reserpine) (Lapin 1978). Furthermore, Buswell (1975) showed that melatonin caused an 80 per cent inhibition of the growth of a transplantable leukaemia cell line (LSTRA) in BALB/c mice.

Carcinomas

A wide variety of carcinomas have been studied with respect to the effects of pinealectomy and melatonin. In the case of transplantable ovarian carcinomas (Das Gupta 1968) or the carcinomatous variant of the Walker 256 carcinosarcoma (Rodin 1963; Barone and Das Gupta 1970), pinealectomy has been shown to markedly promote tumour growth; in the latter case pinealectomy enhanced the metastatic spread of this neoplasm particularly to lung as well as to axillary and mediastinal lymph nodes. In experiments in which rats with Walker carcinomas were subjected to either constant light (melatonin synthesis and secretion inhibited) or constant darkness (melatonin rhythm free-running), it was observed that neoplastic growth was increased while host survival time was decreased in constant light as compared with constant darkness; however, no diurnal lighting controls were included (Tessman *et al.* 1972). With respect to melatonin, Kallenbach and Malz (1957) reported that this indole had no effect on the growth of this tumour type in rats.

In a recent investigation by Toma and associates (1988), daily afternoon melatonin injections were effective in slowing the growth of a transplantable, androgen-sensitive rat prostatic carcinoma. However, opposite results have been obtained with melatonin on transplantable Lewis lung carcinoma growth (Lapin and Ebels 1976). Also, in contrast to the studies cited above, pinealectomy has been reported to actually inhibit carcinogen-induced hepatocarcinoma growth in the rat (Lacassagne *et al.* 1969) calling to mind the results of Georgiou's (1929) original study.

In their work on pituitary tumours, Leadem and Burns (1987) have shown that light deprivation in female Fischer 344 rats causes a 60 per cent inhibition of diethylstilboestrol (DES)-induced pituitary prolactinoma growth, an effect partially prevented by pinealectomy. The administration of daily late afternoon injections of melatonin mimicked the effects of total light deprivation on prolactinoma growth in pineal intact animals. Similarly, constant-release melatonin implants also inhibited prolactinoma growth in this model. Since light deprivation and melatonin injections inhibit prolactin cell function in rats (Blask *et al.* 1980; Leadem and Blask 1981) it is conceivable that melatonin may inhibit prolactinoma growth by affecting the secretion of hypothalamic prolactin-regulatory factors (Blask and Reiter 1978; Blask *et al.* 1980).

Pinealectomy greatly increases the growth and metastatic spread of a transplantable and hormone-responsive hamster melanotic melanoma No. 1 (MM1) in animals maintained on long photoperiods (Das Gupta and Terz 1967; Stanberry *et al.* 1983). Similar results were obtained with pinealectomy in hamsters with carcinogen-induced melanomas (Aubert *et al.* 1970). Surprisingly, pinealectomy accentuated the slowing of MM1 growth in hamsters kept on short photoperiods (Stanberry *et al.* 1983) prompting the conclusion that the photoperiod on which hamsters are maintained dictates the growth response of melanomas to pinealectomy.

It was originally shown by Das Gupta's group (El-Domeiri and Das Gupta 1973, 1976; Ghosh *et al.* 1973) that, while daily injections of melatonin had no effect on MM1 tumour growth in pineal-intact hamsters, melatonin administration did inhibit the augmented growth and metastasis of tumours in pinealectomized animals. However, in a more recent investigation by this laboratory (Stanberry *et al.* 1983) it was reported that daily morning injections (a few hours after lights on) of anywhere from 50 μg to 5 mg of melatonin caused a stimulation of tumour growth in pineal-intact hamsters kept on long photoperiods; similar results were obtained with late afternoon injections of lower doses of melatonin. In a related study, they found that the treatment of long day-exposed hamsters with constant-release melatonin implants (Silastic capsules) potentiated tumour growth, while in short day-exposed animals melatonin implants retarded tumour growth. Although some-what confusing, these results suggest that the photoperiod as well as the method of melatonin administration are important factor(s) in deter-mining the responsiveness of the MM1 tumour line to the anti-tumouri-genic effects of melatonin.

Japanese investigators (Narita and Kudo 1985) found that the oral administration of melatonin via the drinking water had a profound effect in inhibiting the proliferation of B16 mouse melanoma cells in male BALB/c athymic mice (Fig. 7.4). A similar effect of melatonin was obtained on tumour growth in female mice; however, the magnitude of the response was less. Narita and Kudo suggested that this sexually dimorphic response to melatonin may be related to gonadal hormones since ovarian steroids exert an inhibitory effect on tumour growth. Furthermore, since both adrenal and gonadal weights were also reduced in these animals, it was hypothesized that melatonin might have retarded the growth of these steroid-responsive tumours indirectly by influencing

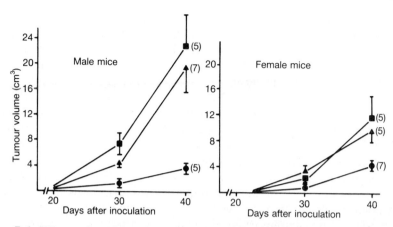

Fig. 7.4. Effects of orally administered melatonin on the growth of B16 mouse melanoma in male and female BALB/c athymic nude mice. Control group 1 (squares) was given drinking water, control group 2 (triangles) was given drinking water containing 0.5 per cent ethanol and group 3 (circles) was given drinking water containing melatonin (5 µg/g body weight/day) and 0.5 per cent ethanol.
[Reproduced with permission from Narita and Kudo 1985.]

gonadal and/or adrenal function; however, a direct antineoplastic effect could not be ruled out.

Of all the tumour types discussed thus far, experimental breast carcinomas have been the most extensively studied with respect to the influences of pinealectomy, photoperiod, and/or melatonin treatment on tumour growth (Blask 1984). For reasons that are not completely clear, the reported effects of pinealectomy on 7,12-dimethylbenzanthracene (DMBA)-induced mammary tumour growth have been highly variable not only from laboratory to laboratory, but even within the same laboratory. Indeed, some workers (Lapin 1978; Aubert *et al.* 1980; Tamarkin *et al.* 1981; Kothari *et al.* 1984) have reported no effect of pinealectomy on tumour growth while some of these same investigators have also observed either inhibitory (Kothari *et al.* 1984) or stimulatory (Lapin 1978; Tamarkin *et al.* 1981; Shah *et al.* 1984) effects of pineal removal. Unexpectedly, Chang and co-workers (1985) discovered that the dennervation of the pineal gland of the rat via bilateral superior cervical ganglionectomy actually restricted DMBA-induced tumour growth. The explanation for this wide inter- and intralaboratory variation in responsiveness of DMBA mammary tumour growth to pineal removal may relate in part to differences in the dose and route of DMBA administration and the nature of the photoperiod as well as to the timing of pinealectomy in relation to tumour initiation and promotion (Blask *et al.* 1986).

We have been investigating the effects of pinealectomy in another animal model of human breast cancer in which the carcinogen N-methylnitrosourea (NMU) is used to induce breast tumours in female rats. These tumours are believed to be more akin to human breast cancer than DMBA tumours in terms of growth characteristics and hormone-responsiveness (Welsch 1985). Pinealectomy of rats several days prior to NMU administration results in an enhancement of tumourigenesis over intact controls in terms of the incidence of rats developing tumours (Fig. 7.5), the cumulative number of tumours and the number of new tumours appearing each week over the course of the experiment (D. E. Blask, unpublished results). The exact mechanism by which pinealectomy promotes carcinogen-induced mammary tumourigenesis is currently unknown; however, it may involve an enhancement of polyamine bio-synthesis which is an important biochemical component of cell prolifera-tion in both normal and neoplastic tissues (Ferioli *et al.* 1983).

The effects of varying photoperiods and total light deprivation have been assessed on DMBA-induced mammary carcinogenesis. While most workers have documented a stimulatory effect of constant light exposure on breast cancer growth (Hamilton 1969; Kothari *et al.* 1982; Shah *et al.*

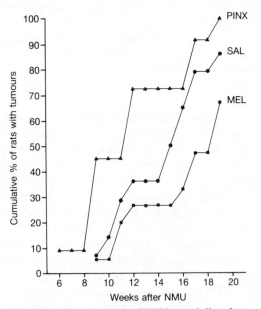

Fig. 7.5. Effects of either pinealectomy (PINX) or daily afternoon injections of melatonin (MEL) (500 µg/day) for 20 weeks on the cumulative percentage of female Sprague–Dawley rats with palpable N-methylnitrosourea (NMU)-induced mammary cancers. [From Blask, Pelletier, and Hill, unpublished.]

1984), there is one report that constant light has no effect on DMBA-induced tumour growth (Aubert *et al.* 1980). One explanation for the stimulatory effect of constant light on mammary tumourigenesis may relate to the fact that constant light exposure suppresses melatonin production throughout each 24-hour period and hence is tantamount to a 'physiological pinealectomy' (Reiter 1980). Interestingly, constant light-stimulated prolactin secretion (Vaticon *et al.* 1980; Shah *et al.* 1984) may be part of the mechanism by which mammary tumourigenesis is promoted inasmuch as DMBA mammary cancer growth is prolactin-dependent (Welsch and Nagasawa 1977). Since pinealectomy alone generally does not stimulate the synthesis or secretion of prolactin (Leadem and Blask 1981; Shah *et al.* 1984), the mechanisms by which surgical versus physiological pinealectomy enhance tumour growth may differ.

The laboratory albino rat is a relatively less photoperiodically sensitive species than the hamster. Thus, before these animals will respond to light deprivation, in terms of an inhibition of the neuroendocrine-reproductive axis, they must be either rendered anosmic by olfactory bulbectomy or underfed (Reiter 1980). We and others have shown that in fact both of these 'potentiating factors' make the rat more sensitive to the anti-gonadotrophic effects of daily afternoon melatonin injections (Blask and Nodelman 1979; Blask *et al.* 1981; Reiter *et al.* 1980). Therefore, in light-deprived (blind) and anosmic female rats, mammary tumour regression is increased (Blask 1984; Chang *et al.* 1985) while tumour incidence, size, and number is decreased (Chang *et al.* 1985). Pinealectomy of blind-anosmic rats prevents these effects and causes a further increase in tumour number (Blask 1984). In rats that are only light-deprived, there is a significant decrease in tumour incidence while no change occurs in tumour number. Furthermore, the number of tumours in blind-pinealectomized rats is greater throughout the experimental period than in both the intact and blind sham-operated animals (D. E. Blask, unpublished results; Fig. 7.6). Thus, in addition to preventing the anti-neoplastic effects of light deprivation, pinealectomy in and of itself may actually promote mammary cancer growth even in light-deprived animals.

The effects of light deprivation and underfeeding are even more dramatic than blinding and anosmia since no tumours develop in blind, underfed animals. Moreover, pinealectomy completely prevents tumour-suppressive effects of blinding coupled with underfeeding (Blask 1984).

In light-deprived rats, particularly anosmic animals, pineal levels (and presumably serum levels) of melatonin are increased perhaps because of the free-running nature of the melatonin rhythm in these animals (Reiter *et al.* 1980); serum prolactin levels are depressed in sensory-deprived rats as well (Leadem and Blask 1981; Blask 1984). Therefore, it is

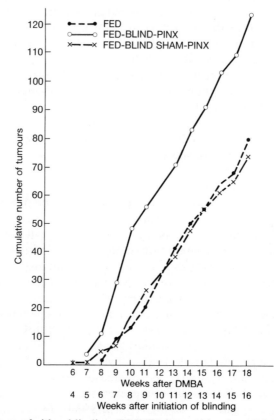

Fig. 7.6. Effects of either blinding (BLIND) in combination with either pinealec-tomy (PINX) or SHAM-PINX on the cumulative number of 7,12-dimethyl-benzanthracene (DMBA)-induced mammary cancers in *ad libitum* fed (FED) female Sprague–Dawley rats. Blinding and pinealectomy were performed 3 weeks after the administration of DMBA. [From Blask, Hill, and Massa, unpublished.]

possible that a pineal-melatonin mediated suppression of prolactin secretion may be responsible for the inhibition of growth of these prolactin-dependent mammary tumours. A similar mechanism may be operating in blind, underfed rats since nocturnal circulating prolactin levels are suppressed (Blask 1984).

The first study to examine the effects of melatonin on DMBA mammary tumourigenesis revealed that while melatonin had no effect on tumour incidence it actually 'enhanced' the genesis of adenocarcinomas (Hamilton 1969). However, in virtually every study performed in this tumour model since that time, it has been shown that daily afternoon injections of melatonin started prior to or at the time of carcinogen administration have a marked inhibitory effect on mammary tumour incidence regardless of whether these animals are maintained in diurnal lighting (Aubert *et al.* 1980; Tamarkin *et al.* 1981) or constant light (Shah *et al.* 1984). Interestingly, in one study (Aubert *et al.* 1980) melatonin was effective as an oncostatic neurohormone in pinealectomized rats (in which the endogenous night-time melatonin signal is absent), whereas in the other investigations it was ineffective (Tamarkin *et al.* 1981; Shah *et al.* 1984). These latter results prompt the suggestion that an intact pineal is required in order for single daily injections of melatonin to be effective. However, that an intact endogenous melatonin signal *per se* may not be required is suggested by the effectiveness of melatonin as an oncostatic agent in constant light-exposed rats in which the melatonin signal is abolished as well (Shah *et al.* 1984).

In another model of human mammary cancer, we have shown that daily afternoon melatonin injections commencing on the first day of NMU administration and continued throughout the experiment, not only decrease mammary cancer incidence (Fig. 7.5), but tumour number as well (D. E. Blask, unpublished results). Since melatonin administration in all of the studies mentioned above, including our own, encompassed both the initiation and promotional phases of tumourigenesis, it is difficult to ascertain whether melatonin is affecting tumour initiation, promotion, or both.

In an attempt to address this issue, we administered daily injections of melatonin to female rats 3 weeks following the injection of DMBA and prior to the appearance of palpable tumours; this injection scheme therefore restricted melatonin injections to the promotional phase of tumourigenesis (Blask *et al.* 1896). Although melatonin inhibited tumourigenesis, the effect was modest by comparison to previously published studies unless the animals were also underfed; however, the combination of underfeeding and melatonin injections caused a dramatic suppression of oncogenesis (Fig. 7.7). These results suggest that melatonin is effective as an anti-tumourigenic agent even when restricted to the promotional phase. This is further supported by the studies of Aubert and associates (1980) who showed that melatonin could actually cause the partial and in some cases the complete regression of established DMBA tumours. Along similar lines, it has also been demonstrated that melatonin injections can inhibit the growth of transplantable breast tumours in both mice. (Anisimov *et al.* 1973) and rats (Karmali *et al.* 1978).

As regards the potential mechanisms by which melatonin inhibits mammary tumourigenesis, it has been postulated that a melatonin-induced suppression of circulating prolactin (Tamarkin *et al.* 1981; Shah *et al.* 1984) and/or oestradiol levels (Shah *et al.* 1984) may be responsible for the inhibition of the growth of these hormone-responsive

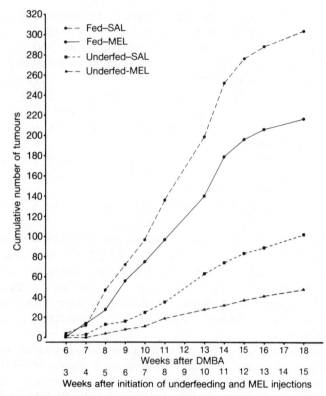

Fig. 7.7. Cumulative number of 7,12-dimethylbenzanthracene (DMBA)-induced mammary cancers in fed or underfed female rats treated with either daily afternoon injections of melatonin (MEL) (250 μg/day) or saline (SAL). Injections were begun 3 weeks after DMBA administration and continued for the duration of the experiment. [Reproduced with permission from Blask *et al.* 1986.]

cancers. However, we have not observed a clear-cut melatonin-induced suppression in either prolactin or oestradiol levels in either the DMBA or NMU tumour models (Blask *et al.* 1986; D. E. Blask, unpublished observations). Clearly, the levels of these important hormones should be monitored around the clock at all stages of the tumourigenic process particularly during the early phases.

Clinical aspects

Melatonin therapy in human malignancy

As alluded to in the introduction, pineal gland extracts and xenografts were used to treat a number of human diseases including cancer. What is fascinating about this development is that it occurred before very much was actually known about pineal function in general, and melatonin in particular. The potential therapeutic value of pineal extracts in treating a variety of human malignancies was vigorously explored during the 1950s particularly as an adjunct to surgery, irradiation, or hormonal therapy (Lapin 1976). In some cases, pineal extract therapy was successful (Svet-Moldavsky *et al.* 1976; Tapp 1980); however, since a variety of extraction procedures were used in these studies, it is virtually impossible to ascertain whether or not melatonin was the active oncostatic constituent of these extracts. To our knowledge, there have been only two published reports on the efficacy of melatonin in the treatment of human cancers.

The first report on the clinical effectiveness and usefulness of melatonin in the treatment of human cancers appeared over 15 years ago (Starr 1970). Unfortunately, no well-controlled clinical trials with large numbers of patients were conducted, but rather a series of anecdotal accounts were given of the responses of individual patients with particular types of cancer to melatonin infusion. In spite of its lack of scientific rigor, this report nevertheless provided the first glimpse of the potential usefulness of melatonin in the treatment of cancer and thus from an historic perspective is an important piece of work.

Starr's rationale for treating cancer patients with melatonin was based on evidence suggesting that the pineal gland in humans may be inhibitory to cancer growth. This hypothesis was derived from the observation of Rodin and Overall (1967) that the pineal glands of patients dying from malignant tumours were larger than those of patients dying from other causes. These workers came to the interesting conclusion that the presence of a tumour may stimulate the pineal gland to exert an inhibitory effect on the hormonal output of endocrine glands that promote tumour growth.

Starr typically treated his patients with a constant infusion of melatonin at a rate of 1 mg/kg/24-hours usually over a 1–2-week period, but sometimes for shorter periods of a day or two. It apparently took 2–3 days before melatonin infusion began to cause cancer regression. The best response to melatonin infusion was observed in patients with malignancies that he classified as somatotrophic cancers based on the finding that serum growth hormone levels were higher than normal in these individuals. Tumours that fit into this classification scheme included osteogenic sarcoma, rhabdomyosarcoma, carcinoma of the colon, glioma, lymphadenoma, and leukaemia. Interestingly, he correlated melatonin's ability to cause a drop in serum growth hormone levels with its effectiveness in causing tumour regression.

In one case, a patient, who had suffered for 3 years with a rhabdomyo-sarcoma of the leg recurrent for the third time after excision, was treated with an infusion of melatonin not only for the primary tumour but for metastatic disease in the groin and neck. Apparently, all tumour masses disappeared within 10 days of the beginning of the treatment, and growth hormone levels decreased during the period of treatment. A similar melatonin-induced tumour regression was observed in a patient with a mixed giant-cell tumour on the thumb.

A 16-year-old male patient with an osteogenic sarcoma of the left upper tibia was given a melatonin infusion while awaiting radiation therapy. The cancer began to regress and growth hormone levels fell dramatically within 2 weeks of treatment. Melatonin treatment was discontinued and then followed by 4 months of radiation therapy resulting in a disappearance of the primary disease and suspected lung metastases. A similar success was achieved in a 25-year-old male with a synovial sarcoma of the knee who was treated with melatonin infusion followed by radiation therapy. According to Starr, if melatonin infusion is begun early enough in the disease process, particularly in patients with sarcomas, metastatic lesions of the lungs could be avoided.

The significance of the drop in growth hormone levels in response to melatonin infusion in patients with so-called somatotrophic cancers and its correlation with the clinical regression of the tumours is unclear. The implication from Starr's monograph was that the melatonin-induced decrease in the high levels of growth hormone in these patients was not only a neuroendocrine marker of the clinical response of the tumour to treatment but perhaps the actual cause of the regression itself. This is inferred from the following statement he made regarding melatonin therapy:

In the inhibitory phase, the object of treatment is the reduction of the serum growth hormone level and an assessment of the clinical result of such inhibition. For this purpose, the secretions of the postpubertal pineal and the prepubertal basophils of the adenohypophysis appear to be most suited.

A more recent, but totally anecdotal report by DiBella and colleagues (1979) revealed that his group has also had success in treating patients with osteogenic sarcomas as well as those with lung, stomach and breast cancer, lymphomas, and acute and chronic lymphoblastic leukaemia. In his melatonin-treated patients survival time was apparently extended and symptoms became less severe. Furthermore, in leukaemic patients melatonin suppressed the abnormal growth of blast or immature myeloid or lymphoid cells as manifested by lower cell numbers in both the circulation and bone marrow. They also implied that some leukaemic patients treated with melatonin went into complete remission. Unfortunately, no details about their clinical oncology studies with melatonin were provided.

Apparently, the oncostatic effect of melatonin in DiBella's study was markedly enhanced by pharmacologically lowering serum levels of both growth hormone and prolactin at the same time that melatonin was administered. These results are interesting in light of Starr's data on the effectiveness of melatonin in causing the regression of tumours in patients with elevated growth hormone levels in addition to the fact that it caused an actual reduction in growth hormone levels. The approach of DiBella and associates (1979) ostensibly produced good results in causing tumour regression unless the tumour mass was very extensive and/or poorly vascularized. These workers concluded that:

(1) melatonin is cytostatic rather than cytotoxic;
(2) cancer growth may result from an imbalance in circulating levels of melatonin, growth hormone, and prolactin.

In an abstract, Burns (1973) reported that melatonin administered for 2 months at a dosage of 1 mg/kg/day to patients with advanced breast carcinoma caused a significant reduction in their levels of urinary oestrogen presumably as compared to pretreatment levels. Unfortunately, no patient controls were included in this study nor were the effects (if any) of melatonin on breast cancer growth reported.

Melatonin levels in human malignancy

Cancers in general

In a preliminary study by Pico and co-workers (1979) plasma levels of both melatonin and 5-methoxytryptophol were compared between carcinoma patients (types of carcinomas not specified) and normal volunteers. The levels of these indoles were measured by gas chromatography–mass spectrometry (GCMS) presumably during the daytime hours since melatonin levels were reportedly low (30 pg/ml) in the normal controls. Although these workers reported that in 60 samples from cancer patients the mean levels of melatonin and 5-methoxytryptophol were reduced (below detectable limits in three patients) as compared with controls, they did not report the degree of this difference in the remaining 57 patients. Similarly, Schloot and colleagues (1981) reported, albeit anecdotally, that daytime immunoassay levels of melatonin were undetectable, with or without therapy, in patients with either bladder cancer, lymphocytic leukaemia, or neurofibromatosis. Birau (1981) found that, in patients with uterine cancer, daytime levels of melatonin were either high (50–100 pg/ml) or undetectable.

Also by using a melatonin radioimmunoassay, Fraschini and collaborators (1985) evaluated pineal-melatonin function in 29 patients of both sexes (mean age 39 years) with various types and clinical stages of cancer as compared with 26 age-matched healthy subjects. The assay of both day and night serum levels of melatonin revealed that the cancer group had daytime levels which were 2–4-fold higher than the healthy subjects. Furthermore, the increase in serum melatonin levels from daytime to nocturnal values in the controls was 2300 per cent while in the oncological group it was less than 100 per cent which could be ascribed to the already elevated daytime hormone titres.

In a corroborative study (Touitou *et al.* 1985), plasma melatonin assays were performed on plasma samples obtained over a 7-month period in several hundred unselected elderly patients (mean age 81 years), 80 per cent of whom were women. All subjects were on the same lighting (lights on 07.00–21.00 hours) and meal schedules. Of the total number of patients studied, 12.5 per cent were found to have a variety of cancers either diagnosed clinically or at autopsy. Blood samples were obtained from cancer patients during the morning hours about 1 hour after lights on (08.30 and 09.00 hours).

When all patients (both male and female) with clinically diagnosed cancer were considered, their mean daytime melatonin levels (32 pg/ml) were significantly higher than those free of cancer, but having other diseases (23 pg/ml). Further analysis of these data reveals that in men with cancer discovered at autopsy, melatonin levels (13 pg/ml) were significantly lower than levels (21 pg/ml) in men without cancer. On the other hand, melatonin levels in women with incidental cancer at autopsy (23 pg/ml) did not differ from those without cancer (24 pg/ml); however, women clinically diagnosed as having various types of cancer had significantly elevated mean plasma levels of melatonin (34 pg/ml). When melatonin levels were analysed with respect to the type of cancer patients possessed, it was found that levels were high but quite variable in patients with cancer of the uterus, colon, bladder, ovary, and in patients with metastatic disease of unknown origin. Conversely, melatonin levels were low in those patients with cancers of the liver, kidney, skin, and upper respiratory tract.

In contrast to the studies cited above, Tapp and colleagues (1980) published preliminary data indicating that daytime levels of immuno-reactive melatonin in cancer patients were no different from those in tumour-free control subjects.

Breast cancer

The evidence is now very strong that the amount of biologically active oestrogen available is probably one of the most important factors in the aetiology of breast cancer (Cuzick *et al.* 1986). Cohen and associates (1978) hypothesized that since the pineal gland and its hormone melatonin inhibit ovarian function in experimental animals, decreased pineal function may promote the development of breast cancer in humans; that is, diminished pineal activity and hence melatonin production may lead to breast cancer by inducing a state of hyper-oestrogenism. They cite epidemiological evidence linking a higher occurrence of pineal calcification, and presumably decreased pineal function, in countries with high rates of breast cancer. Interestingly, in both men and women, the amplitude of the nocturnal rise in melatonin decreases with advancing age (Arendt 1985*a*,*b*) including the age range during which breast cancer is most likely to develop. Therefore, a diminution in at least one aspect of pineal-melatonin function in an age group of women at greater risk for developing breast cancer may contribute in some way to the aetiology of this disease.

Wetterberg and coworkers (1979*a*) evaluated the question of whether the urinary excretion of melatonin varies among populations with different risks for breast cancer. They compared urinary melatonin excretion in divided 24-hour samples in adult American women from Minnesota at high risk for developing breast cancer with those in Japanese women from Kyushu at low risk for developing breast cancer; each population included both premenopausal normally cycling women as well as post-menopausal subjects. Urine samples were collected approximately every 2–3 hours while the subject was awake and whenever a subject awoke from sleep. Both groups elicited a nocturnal surge of melatonin excretion; however, the amplitude of this surge was in general lower in the Japanese group than in the American population. As pointed out by Vaughan (1984) the interpretation of this study is confounded by the fact that these workers did not evaluate breast cancer risk itself as a potential source of variation in their analyses.

Bartsch and colleagues (1981) studied the 24-hour urinary excretion of immunoreactive melatonin in 10 post-menopausal Indian women most of whom had clinical stage III breast cancer vesus nine age-matched post-menopausal control women without breast cancer (seven out of nine suffered from uterovaginal prolapse). The patients were maintained on a controlled light–dark cycle of 16:8 hours with lights off between 22.00 and 06.00 hours. Urine samples were collected at 4-hourly intervals during lights on and while one sample was obtained during the mid-point of the dark phase. The collections were made over a period of a month from mid-December to mid-January to avoid potential variability due to seasonal changes in melatonin excretion (Arendt 1985).

As compared with controls, melatonin excretion was consistently lower in cancer patients between 14.00 and 18.00 hours as well as at night, while it was higher between 06.00 and 10.00 hours. This translates to a 31 per cent lower, but statistically insignificant, 24-hour mean melatonin excretion in cancer patients versus controls (Fig. 7.8). While

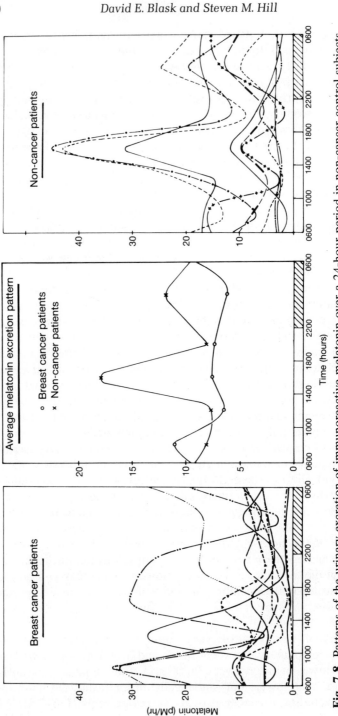

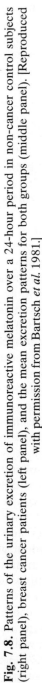

Fig. 7.8. Patterns of the urinary excretion of immunoreactive melatonin over a 24-hour period in non-cancer control subjects (right panel), breast cancer patients (left panel), and the mean excretion patterns for both groups (middle panel). [Reproduced with permission from Bartsch *et al.* 1981.]

melatonin excretion was lower in the cancer patients, serum cortisol levels were 33 per cent higher in this group than in the controls. The authors speculated that the adrenal cortex may be involved in hormonal dysfunction in cancer patients and that increased cortisol levels may be associated with decreased melatonin output as it is in certain psychiatric illnesses (Wetterberg *et al.* 1979*a*).

An interesting feature of this study is that the control subjects elicited regular late afternoon peaks of melatonin excretion of varying amplitudes in addition to the nocturnal surge. However, such a synchronous pattern of melatonin excretion was not evident in the cancer group in which surges of melatonin excretion appeared to be occurring at random throughout the 24-hour period suggesting free-running or at least asynchronous melatonin rhythms. Although five of the cancer patients apparently had oestrogen-receptor-positive (ER+) tumours, these investigators made no statistical correlation between urinary melatonin levels and ER positivity which is considered to be indicative of hormonal responsiveness. However, it is interesting to note that melatonin excretion was conspicuously the highest in the two patients with ER− cancer.

In a related investigation, Tamarkin and colleagues (1982) assayed plasma immunoreactive melatonin levels every 3 hours over a 24-hour period in 20 women (both pre- and post-menopausal) with clinical stage I and II breast cancer as well as in eight age-matched normal control women. Some patients were sampled 1 day prior to surgery while others were sampled 5 days following tumourectomy. A normal melatonin rhythm was observed in both cancer patients and normal subjects with low levels being present during the day (lighting conditions were unspecified) while peak plasma concentrations were reached during the night at 02.00 hours. However, when the cancer patients were divided into two groups, one with ER+ disease and the other with ER− disease, some distinct differences emerged. While the amplitude of the 02.00 hours nocturnal surge of melatonin in ER− patients was virtually identical to that observed in the controls, it was significantly reduced by 41 per cent in the ER+ patients (Fig. 7.9). These results contrast to those in two anecdotal reports (Birau 1981; Schloot *et al.* 1981) which stated that daytime levels of melatonin were either undetectable, normal, or very high (greater than 100 pg/ml) in breast cancer patients.

Upon further examination of the data from these patients, Tamarkin and associates (1982) found that tumour ER concentrations were significantly, but inversely correlated with peak night-time melatonin levels indicating that women with lower peak melatonin titres have higher ER levels in their tumours. No significant correlation was found between ER concentration and low daytime levels of melatonin.

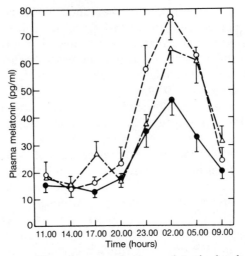

Fig. 7.9. 24-hour profiles of mean plasma melatonin levels in women with oestrogen-receptor-positive (solid circles) or oestrogen-receptor-negative (open circles) breast cancer and in normal age-matched women. [Reproduced with permission from Tamarkin *et al.* 1982.]

In a subsequent study, Danforth *et al.* (1985) confirmed these results and in addition found the same relationship between nocturnal peak melatonin levels and progesterone receptor positive (PR+) and negative (PR−) tumours. The relationship between the melatonin rhythm and steroid receptors appears to be specific for ER and PR since no correlation between melatonin and glucocorticoid receptors was found.

In this same study, a statistical comparison of the day–night differences in melatonin levels was performed among normal controls, breast cancer patients, and women at high risk for breast cancer. This latter group of women was recruited from 10 kindreds with familial breast cancer. The relative risk of breast cancer for each subject was determined with methods employing both additive and multiplicative models of combining separate risk estimates derived from age, pedigree type (mother, sister, or second-degree relative), age of menarche, first full-term pregnancy, parity, and history of fibrocystic disease. Not only was the range of day–night differences in melatonin the same among these three groups, but the mean day–night melatonin differences for each group was quite similar and not statistically different (Fig. 7.10). These results suggested to the authors that the melatonin profile would not be appropriate as a screening procedure for women with breast cancer. Additionally, there was no correlation between plasma melatonin concentrations and age, weight, parity, menopausal status, staging of the

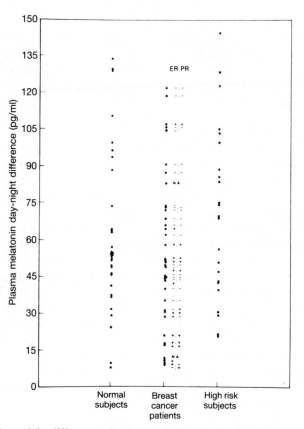

Fig. 7.10. Day–night differences in plasma melatonin levels in women with breast cancer, women at high risk for developing breast cancer, and in normal women. Each point represents the day–night melatonin difference (peak night-time levels minus the mean daytime level) for a single individual. Oestrogen receptor (ER) and progesterone receptor (PR) levels in each cancer patient's tumour are indicated as positive (+), negative (−), or not available (NA). [Reproduced with permission from Danforth *et al.* 1985.]

disease, or plasma levels of either oestradiol, oestrone, progesterone, follicle stimulating hormone, or luteinizing hormone among all subjects studied. Additionally, among high-risk patients, plasma melatonin was independent of the degree of risk for breast cancer. Nevertheless, the authors speculated that the diminution or absence of a nocturnal peak of melatonin might be a useful biochemical marker for an increased risk for the development of hormone-responsive breast cancer which could have a potentially important impact on the treatment and prognosis of this disease.

An intriguing study by Lehrer and associates (1985) evaluated olfactory function in 46 women with breast cancer as compared with normal control subjects matched for age, sex, race, and smoking history; olfactory function was assessed with the University of Pennsylvania smell identification test (UPSIT). Roughly half of the breast cancer patients had ER+ tumours while the other half had ER− disease. Interestingly, compared with matched controls, the ER+ patients had a significantly lower mean UPSIT score indicating diminished olfactory function in these individuals; ER− patients had normal olfactory function. On the basis of the aforementioned studies with regard to ER+/PR+ breast cancer and melatonin levels, the patients with a decreased ability to smell and hormone-responsive tumours presumably have altered pineal function as well. This is interesting from the standpoint that anosmia in rats (induced by olfactory bulbectomy) makes them more sensitive to the anti-gonadotrophic effects of the pineal and melatonin (Blask and Nodelman 1979; Reiter 1980; Reiter *et al.* 1980). Moreover, as alluded to previously, it has been shown that the combination of light deprivation and anosmia inhibits DMBA-induced breast cancer growth in rats, an effect that is in part prevented by pinealectomy (Blask 1984; Chang *et al.* 1985). Thus, as in the animal models of human breast cancer, these 'hyposmic' patients may have an intrinsic olfactory-associated increased sensitivity to melatonin which might make them particularly susceptible to the potential oncostatic effects of exogenously administered melatonin. Miles and colleagues (Chapter 13) have discussed the clinical and diagnostic importance of these findings.

Prostate cancer

Circulating melatonin levels have also been assessed by radio-immunoassay in male patients with carcinoma of the prostate gland. In one study, melatonin levels (time of day unspecified) were reported to be high (50–100 pg/ml) in untreated patients with prostate cancer; however, in patients treated with methyloestradiol, melatonin was apparently undetectable (Schloot *et al.* 1981). Birau (1981) anecdotally reported that daytime levels of melatonin in patients with prostatic carcinoma were either undetectable, normal, or very high (greater than 100 pg/ml). Touitou and colleagues (1985) also found normal daytime levels of melatonin in the plasma of two elderly men hospitalized for prostate cancer.

In a more extensive study, Bartsch and co-workers (1985) measured serum levels of immunoreactive melatonin over a 24-hour period in elderly men (mean age 70 years) with either benign prostatic hyperplasia (BPH), incidental prostate carcinoma (IPC), or clinical prostate carcinoma (CPC). Since most elderly men have evidence of prostate enlargement, they used normal young men (mean age 33 years) without prostatic enlargement as a control group. All tumour patients had not been treated and were awaiting surgery at the time of the blood collections. All subjects were exposed to both natural and fluorescent light with lights on from 05.00 to 10.00 hours. Blood samples were obtained via venous catheters at fixed 4-hour intervals over the 24-hour period; in all groups, specimens were collected throughout the year depending on the availability of suitable patients.

The control group evinced a nocturnal surge of melatonin (acrophase at 02.30 hours) from low daytime levels (Fig. 7.11). In the BPH patients, the melatonin rhythm was similar except the amplitude of the surge was

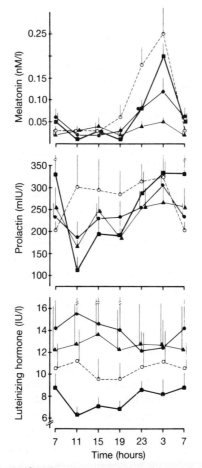

Fig. 7.11. Mean serum levels of immunoreactive melatonin, prolactin, and luteinizing hormone over a 24-hour period in young men (squares), men with benign prostatic hyperplasia (solid circles), men with clinical prostate carcinoma (triangles), and men with incidental carcinoma of the prostate (open circles). [Reproduced with permission from Bartsch and Bartsch 1985.]

lower, but not statistically so. In IPC patients there was a significant melatonin rhythm comparable to that observed in the controls. However, in CPC patients, melatonin rhythms were completely absent and the mean 24-hour levels were very low as compared with the other groups. While the controls and BPH patients exhibited circadian fluctuations in circulating prolactin, in both the IPC and CPC patients the prolactin rhythms appeared to be ultradian rather than circadian in nature. The general conclusion reached by these workers was that the pineal gland and more specifically melatonin, influences the development of prostate cancer in humans. They further speculated that the somewhat higher amplitude of the nocturnal melatonin peak in IPC patients reflects a hyperactivity of the pineal gland in its attempt to inhibit the further growth of some malignant cells that remained dormant in these patients. On the other hand, in CPC patients with progressive prostate carcinoma, tumour growth may have exceeded the pineal's ability to inhibit oncogenesis thus perhaps resulting in an exhaustion of pineal function and melatonin secretion in these patients (see Miles *et al.*, Chapter 13, for a review of the diagnostic importance of these observations).

Pineal tumours

It is well known that tumours of the pineal gland are quite rare, accounting for only about 1 per cent of all intracranial neoplasms. In this regard, the pineal gland itself is highly resistant to the establishment of growth of primary and metastatic tumours (Tapp 1982). It would be logical for one to expect abnormalities in melatonin secretion in patients with pineal tumours; however, the results obtained in various studies of melatonin secretion in patients with pineal tumours have been inconsistent as they probably reflect the histopathological heterogeneity of these neoplasms (Kerenyi 1979; Tapp 1982).

In one study, immunoreactive melatonin levels were measured every 4 hours in a 37-year-old man who had a histopathologically diagnosed cystic pineocytoma 2 months before death (Barber *et al.* 1978). Compared with 24-hour melatonin secretion in five normal controls there was a more prominent nocturnal rise in serum melatonin levels such that the peak level measured at 22.00 hours was 238 pg/ml versus mean levels of about 75 pg/ml in the controls at about 02.30 hours. What was more striking about these results is that the daytime levels in the tumour patient were about 5-fold greater than those in the controls at the same time-points. These workers were optimistic that elevated daytime levels of circulating melatonin as determined by midday serum sampling might be useful as a marker for pineal tumours.

Other investigators have experienced variable results in their analyses of melatonin levels in patients with either suspected pinealomas or histopathologically diagnosed tumours. Arendt (1978) reported anecdotally that, in two young men with suspected pinealoma, plasma melatonin levels were extremely high during both the day and night in one case and low in the other patient. Kennaway and associates (1979) found immunoreactive melatonin to be absent from the plasma at several different time-points during the day and night in a 15-year-old prepubertal boy with a histologically unconfirmed pinealoma as well as in a 43-year-old caucasian male with a confirmed pinealoblastoma. On the basis of the heterogeneity of pineal tumours, these investigators questioned the clinical utility of elevated melatonin levels as a marker for pineal tumours in all cases.

A number of workers have examined the effects of the surgical removal of suspected pineal tumours on melatonin levels as compared with preoperative levels. For example, a 30-year-old man with a post-operatively confirmed pineal germinoma had a preoperative 10.00 hour melatonin level of 185 pg/ml which dropped to 19 pg/ml by 4 weeks after treatment with surgery and radiation therapy and was undetectable by 3 months after treatment (Tapp 1978). These results indicate that daytime levels of melatonin can be considerably elevated even in patients with non-parenchymal tumours of the pineal region. According to Tapp (1978) it is unlikely that there was pineal tissue in the germinoma and therefore it might be assumed that the post-therapeutic decrease in melatonin levels was due to the complete removal of the calcified pineal gland. This suggests that the germinoma may have been affecting the function of the patient's intact pineal gland in some way.

Neuwelt and Lewy (1983) evaluated a young man with a pineal region tumour pre- and post-operatively with respect to his 24-hour plasma melatonin profiles using GCMS; blood samples were withdrawn every 2 hours. The melatonin secretory profile prior to surgery displayed a typical normal pattern with low daytime levels (1.0–10 pg/ml) and an 02.00 hour nocturnal peak of a little over 70 pg/ml (Fig. 7.12). However, melatonin was undetectable in the plasma at any time over a 24-hour period 6 weeks following the complete surgical removal of the tumour which was well encapsulated and contained elements of both low grade astrocytoma and pinealoblastoma. These authors concluded that since the pineal gland appears to be the sole source of plasma melatonin in man, melatonin measurements may provide a reliable means of determining the completeness of pinealectomy.

In a similar situation, radiotherapy was used to treat a 17-year-old female patient for a pinealoma; 2 years later she was readmitted for evaluation and treatment of an intracranial growth in the left cerebral hemisphere (Miles *et al.* 1985). Prior to therapy she had a late morning (11.30 hours) plasma melatonin level of 120 pg/ml while in a normal

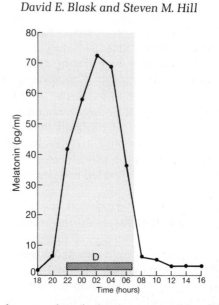

Fig. 7.12. Human plasma melatonin levels measured by gas chromatography–mass spectrometry prior to and following the surgical removal of a tumour-infiltrated pineal gland. Plasma samples were obtained ever 2-hours over a 24-hour period preceding (solid line) and following (dashed line) surgery. The shaded area represents the night-time. [Reproduced with permission from Neuwelt and Lewy 1983.]

age-matched woman it was 14 pg/ml at this time. Following therapy and the disappearance of the tumour, her 11.30 hours plasma melatonin level was below the level of detectability. Miles and colleagues (1985) suggested that the high pretreatment level of melatonin in this patient was due to its secretion from the tumour which was probably a metastatic lesion from the original primary neoplasm. Irradiation of the tumour produced a 'functional pinealectomy' and resulted in the disappearance of a melatonin from the circulation. These results emphasize the potential usefulness of melatonin as a marker for metastatic melatonin-secreting tumours of the pineal gland as well as primary pinealomas of this type. Miles and colleagues (Chapter 13) have discussed in detail the potential diagnostic significance of melatonin in patients with pinealomas.

Seasonal variations in cancer and melatonin

There have been a number of reports on seasonal variations in the incidence of various cancers including osteogeneic sarcoma, lymphadenoma, and breast cancer. In all three categories, the highest incidence of tumours occurred during the summer time (Starr 1970).

Recently, several investigators have studied more extensively the relationship between seasonality and the occurrence of breast cancer in particular in various locations around the world. For example, in the Middle East, more cases of breast cancer were diagnosed during June through August and less during September through December (Cohen *et al.* 1983). Similar findings were obtained in the United Kingdom by Kirkham *et al.* (1985) particularly in premenopausal women with breast cancer. In extremes of latitude such as in New Zealand, breast cancer is diagnosed more often in late spring and early summer and least often in late autumn and early winter (Mason *et al.* 1985). Interestingly, this group found a higher frequency of ER+/PR+ breast tumours in the summer in young women while another group of workers (Hartveit *et al.* 1983) found more ER+ tumours during the winter months. Others have found that more cases of primary breast cancers greater than 2 cm in diameter occur during the spring while during the winter months more cases were diagnosed with small tumours less than 2 cm in diameter (Ownby *et al.* 1986).

Whether the pineal gland and thus melatonin secretion undergo predictable seasonal variations in humans is still controversial. Arendt and co-workers (1977) described a bimodal circannual rhythm of melatonin in humans with lower levels of plasma melatonin in spring and autumn than in summer and winter in a temperate region; however, these workers only sampled their subjects once or twice daily at monthly intervals. In a later study, Arendt (1985) reported that no major change in the 24-hour rhythm of melatonin secretion occurred in humans in either the spring or fall. In a similar study Touitou and associates (1984) measured plasma melatonin every 4 hours over a 24-hour period in both young and elderly subjects in January, March, June, and October. Plasma melatonin levels were significantly lower in January than in June in young, men whereas in elderly individuals October melatonin levels were lower than in January and March.

Illnerova and associates (1985) found that the only seasonal difference in human nyctohemeral melatonin rhythms was that the nocturnal peak of melatonin was phase-delayed by 1.5 hours in the winter versus the summer. Apparently, in individuals stationed in Antarctica, the only seasonal change in melatonin secretory rhythms was a winter delay in the morning decline from peak night-time levels, indicating that the duration of melatonin secretion was increased during this season (Arendt 1985).

Although there is as yet no general agreement with respect to the precise nature of seasonal variations in melatonin in humans, it is probably safe to conclude that such variations do exist. How seasonal variations in pineal function and melatonin secretion relate to the increased incidence of cancers, particularly those of the breast, during the spring and early summer is uncertain. Nevertheless, it might be hypothesized that in certain populations living in particular geographic locations, a springtime depression in circulating melatonin levels may be permissive to the development and growth of certain types of cancer. Thus, certain groups of individuals might be at more or less risk for developing various neoplasms depending on the seasonal status of their melatonin secretory patterns.

Conclusions

Mechanisms of melatonin-induced oncostasis

With few exceptions, the bulk of the experimental evidence to date not only indicates an important influence of the pineal gland, and its major hormone melatonin on cancer formation and growth, but it very strongly supports the hypothesis that this influence is inhibitory in nature. Inroads are just beginning to be made with respect to uncovering some of the numerous and complex mechanisms by which melatonin inhibits tumourigenesis.

Although there are several lines of evidence that melatonin can retard the growth of hormone-dependent tumours, the most cogent argument for melatonin as an oncostatic hormone can be made by considering its effects on hormonally responsive neoplasms, particularly experimental breast cancer. Using this type of cancer as a model, both indirect neuro-endocrine as well as direct peripheral endocrine mechanisms at the cellular level can be easily envisaged in order to explain the oncostatic action of melatonin. For example, since carcinogen-induced breast cancers *in vivo* are responsive to both prolactin and oestradiol, a melatonin-induced inhibition of the neuroendocrine mechanisms controlling prolactin and gonadotrophin output could lower the circulating levels of prolactin and oestradiol below a critical level required to promote tumourigenesis. Since oestradiol normally supports pituitary prolactin production and secretion to a large degree, a neuro-endocrine or peripheral endocrine supression of ovarian steroid output (Vaughan *et al.* 1978) in and of itself could also lower the levels of these tumour-promoters which are necessary to sustain continued cancer growth (Fig. 7.13).

In addition to a neuroendocrine site of action, the action of melatonin directly at the cancer cell level itself would provide this molecule with a variety of options for inhibiting tumour cell proliferation. One particu-larly attractive mechanism would involve the antagonism by melatonin of the action of other hormones that promote tumour growth. Again using breast cancer as an example, melatonin can inhibit the mitogenic

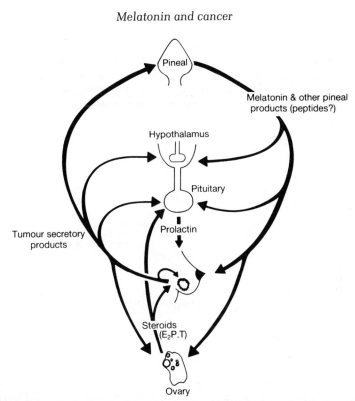

Fig. 7.13. Schematic representation of potential neuroendocrine and peripheral endocrine mechanisms by which melatonin may affect breast cancer growth. This scheme also depicts potential mechanisms by which tumour secretory proteins might feedback to affect pineal-neuroendocrine function. [Reproduced with permission from Blask *et al.* 1986.]

action of oestradiol directly on breast cancer growth. In this regard, melatonin appears to be similar to the non-steroidal anti-oestrogen tamoxifen which also inhibits oestradiol-stimulated human breast cancer cell (MCF-7) proliferation. Like tamoxifen, melatonin exerts a cytostatic effect perhaps by interacting with the ER in some way to inhibit the cell cycle. Exactly at what stage of the cell cycle melatonin inhibits proliferation is unknown; however, like tamoxifen it may cause a transition delay from G_1 to S phase of the cycle (Osborne 1985). Further studies might possibly reveal that melatonin may be the pineal gland's and thus the body's own natural version of tamoxifen.

The cellular-molecular mechanisms by which melatonin may inhibit hormone-responsive breast cancer growth are also unknown. However, it is tempting to speculate that such mechanisms may include either an inhibition of endogenous growth factor release by the tumour cells, the stimulation of endogenous growth inhibitory factor production, and/or release by tumour cells or a combination of these mechanisms (Fig. 7.14). Other possible mechanisms might also include the inhibition of oncogene expression, the action of growth factors, and/or polyamine biosynthesis, to name a few.

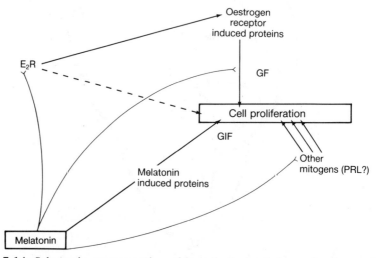

Fig. 7.14. Schematic representation of hypothetical cellular-molecular mechanisms by which melatonin might inhibit human breast cancer cell proliferation in culture. Oestrogen receptor=E$_2$R; GF=growth factors elaborated by breast cancer cells; GIF=hypothetical growth inhibitory factors elaborated by breast cancer cells; PRL=prolactin.

The ability of an individual to mount an appropriate immune response to the presence of a malignant neoplasm is another mechanism which might be targeted by melatonin in order to inhibit tumour growth. Support for such a hypothesis comes from studies in which the elimination of the melatonin signal by pinealectomy causes an incomplete and temporary impairment of immunocompetence in rats (Jankovic *et al.* 1970). More recently, it has been demonstrated that the pharmacological inhibition of melatonin synthesis dramatically inhibits humoral immunity in rodents (Maestroni *et al.* 1986). Additionally, evening injections of melatonin are able to completely restore the humoral immune response in 'pharmacologically pinealectomized' mice as well as in mice in which antibody production was suppressed by corticosterone (Maestroni *et al.* 1986).

Still another potential mechanism by which melatonin may inhibit tumour growth is via an inhibition of metabolism and glucose homeostasis (Quay and Gorray 1980). Support for this postulate is derived from experiments in which pinealectomy not only caused a marked hyperinsulinaemia (Gorray and Quay 1978), but an inhibition of the nyctohemeral increase in the adrenomedullary production of noradrenalin known to be important in the control of gluconeogenesis (Banerji and Quay 1976). However, it is not clear whether these effects were due to a lack of melatonin since melatonin replacement experiments were not performed.

Melatonin's potential as a diagnostic and evaluative tool in oncology

There has been an increasing amount of interest in the use of melatonin levels as a biological marker in the evaluation of the clinical and epidemiological aspects of cancer including its diagnosis, prognosis, post-therapeutic evaluation and follow-up, and risk. At first glance, the use of melatonin as a marker for cancer in terms of its diagnosis, evaluation, and treatment seems very attractive. However, to date, no consistent alteration(s) in the melatonin secretory profile has emerged in patients with cancer as compared with healthy individuals or those with other diseases. In fact, there is evidence to suggest that a particular profile of melatonin secretion may be specific for different types of cancer. Other factors that may contribute to the variability in melatonin levels and potentially confound the analysis and interpretation of melatonin secretory rhythms in cancer patients include age, sex, season, lighting cycles, and other environmental factors, presence of other diseases, drug therapy, and the frequency and timing of sampling.

It must be kept in mind that, in virtually all of the investigations of melatonin levels in cancer patients, hormone measurements were made during the advanced stages of the disease. Therefore, it is not exactly clear what alterations in melatonin secretion actually mean in terms of the pathobiology of oncogenesis. Do they reflect primary changes in pineal function that are causally related to the oncogenesis? Or, are they instead a secondary result of abnormal hormonal feedback relationships due to the presence of a cancerous growth (Blask and Leadem 1987). This latter question may be particularly applicable to hormone-responsive tumours such as breast cancers which are known to elaborate their own growth factors that might feedback on pineal melatonin synthesis and secretion via the circulation (Fig. 7.13).

Even beyond these considerations, precisely which aspect(s) of the daily melatonin secretory profile is the most important parameter to evaluate in cancer patients is open for discussion. For example, in certain kinds of malignant disease the phasing of the nyctohemeral rise in melatonin may be important while in other cancers the amplitude of the night-time rise might be the more important factor. Still, in other cases the presence of elevated daytime levels of melatonin may be of critical value (Bartsch and Bartsch 1984).

The measurement of melatonin levels in breast cancer patients in order to determine the presence or absence of hormone-responsive disease as well as melatonin measurement, pre- and post-operatively, in patients with pineal tumours currently holds the most promise. However, in order for melatonin analysis to become a clinically acceptable and useful marker for various aspects of oncogenesis, 24-hour melatonin profiles must first be studied longitudinally, under standardized conditions throughout the continuum of the oncogenic process in patients of both sexes and different ages encompassing the gamut of cancer types. This would include melatonin measurements in:

(1) individuals at high risk for the development of a particular cancer prior to the appearance of disease;

(2) patients with clinical disease, particularly during the earliest as well as the later stages of tumourigenesis including the post-therapeutic phase.

Only under these circumstances can appropriate clinical and statistical criteria be established and standardized with respect to the assessment of melatonin profiles for each type and stage of cancer and its treatment.

Melatonin's potential as a therapeutic tool in oncology

In our opinion, melatonin has great potential as a therapeutic tool in the treatment of cancer, particularly hormonally responsive tumours. The attractiveness of melatonin as a potential oncostatic therapy resides in the fact that it is a naturally occurring hormonal molecule that is normally released into the bloodstream of every individual in the form of a nightly neuroendocrine signal. Through indirect neuroendocrine processes and via more direct peripheral endocrine mechanisms, this normal homeostatic hormonal signal in a healthy individual may become an oncostatic message in a cancer patient.

If one considers the rhythmic nature of cell division in both normal and neoplastic tissues, it becomes evident that cancer may be, in part, a disease of desynchronization in the rhythmicity of both hormonal and non-hormonal factors involved in the control of normal cell division in various tissues. A chronobiological shift in the normal rhythmic and homeostatic balance of a variety of both stimulatory and inhibitory growth control mechanisms may result in the abnormal phasing of these hypothetical rhythms such that the scales are tipped in favour of malignant neoplasia. Thus, an approach similar to the one being used to treat patients with seasonal affective disorder (Lewy *et al.* 1985) might be applied to the treatment of some cancer patients. That is, the non-invasive manipulation of a controlled schedule of high-intensity light might be used to alter the phasing of the melatonin rhythm in certain cancer patients. In doing so, a new phase relationship could be established between one of the body's own oncostatic messengers, namely melatonin, and a biological rhythm of maximal tumour sensitivity to endogenously occurring oncostatic signals.

Another approach that has been employed to a limited extent and with some degree of success is the exogenous administration of melatonin to selected cancer patients to prevent the further growth and/or cause the regression of malignant tumours (DiBella *et al.* 1979; Starr 1970). In view of the circadian rhythms of sensitivity to inhibition by melatonin in normal rodent species (Reiter 1980) as well as in tumour-bearing animals (Bartsch and Bartsch 1981), there is a critical need for long-term studies on the efficacy of melatonin administered at certain times of the day in the treatment of cancer patients. This chrono-oncological approach to cancer therapy with melatonin could be explored as a primary therapeutic modality or as a hormonal scheme to 'synchronize' cancer cells at a particular stage of the cell cycle that is more easily blocked by still another anti-cancer agent(s). In some cancer patients with either faulty or completely defective pineal function, properly-timed melatonin administrations might be used to either modify, supplement, or replace an aberrant or absent endogenous melatonin oncostatic signal. Since melatonin is an effective anti-cancer substance when administered to experimental animals *per os* (Narita and Kudo 1985), it might eventually be employed as an 'anti-cancer pill' to high-risk populations or perhaps even to healthy individuals.

In any event, we may soon be able to add melatonin to the growing list of 'biologicals' that are currently generating so much excitement in cancer research and therapy. As one of the body's own naturally produced molecules, melatonin may emerge as an important 'chrono-oncostatic' hormone in the treatment of various forms of cancer.

Acknowledgements

We would like to thank Dr James A. Clemens for the supply of 6-chloro-melatonin used in some of the studies cited in this chapter. The work presented by the authors was supported in part by PHS Grant No. 5R23 CA27653 and IROICA 42424 from the National Cancer Institute.

References

Anisimov, V. N., Morozov, V. G., Kbavinson, V. K., and Dilman, V. M. (1973). Correlations of anti-tumour activity of pineal and hypothalamic extract, melatonin and sygethin in mouse transplantable mammary tumours. *Voprosky Onkologii* **19**, 99–101.

Arendt, J. (1978). Melatonin as a tumour marker in a patient with pineal tumour. *British Medical Journal* **2**, 635–6.

Arendt, J. (1985). Assay of melatonin and its metabolites: results in normal and unusual environments. In *Melatonin in Humans* (ed. R. J. Wurtman and F. Waldhauser), pp. 11–13. Centre for Brain Sciences and Metabolism Charitable Trust, Cambridge.

Arendt, J. (1985). Mammalian pineal rhythms. In *Pineal research reviews* (ed. R. J. Reiter), pp. 161–213. Alan R. Liss Inc., New York.

Arendt, J., Wirz-Justice, A., and Bradtke, J. (1977). Annual rhythm of serum melatonin in man. *Neuroscience Letters* **7**, 327–30.

Aubert, C., Janiaud, P., and Lecalvez, J. (1980). Effect of pinealectomy and melatonin on mammary tumour growth in Sprague–Dawley rats under conditions of lighting. *Journal of Neural Transmission* **47**, 121–30.

Aubert, C., Prade, M., and Bohuon, C. (1970). Effet de la pinealecotmie sur les tumeurs melaniques du hamster dore induites par l'administration (*per os*) d'une seule dose de 9-10 dimethyl-1-2 benzanthracene. *Comptes Rendus L'Academie Des Sciences* **271**, 2465–8.

Banerjee, S. and Margulis, L. (1973). Mitotic arrest by melatonin. *Experimental Cell Research* **78**, 314–18.

Banerjee, S., Kerr, V., Winston, M., Kelleher, J. K., and Margulis, L. (1972). Melatonin: inhibition of microtubule-based oral morphogenesis in *Stentor coerulues. Journal of Protozoology* **19**, 108–12.

Banerji, T. K. and Quay, W. B. (1976). Adrenal dopamine-beta-hydroxylase activity: 24-hour rhythmicity and evidence for pineal control. *Experientia* **32**, 253–5.

Barber, S. G., Smith, J. A., and Hughes, R. C. (1978). Melatonin as a tumour marker in a patient with pineal tumour. *British Medical Journal* **2**, 328.

Barone, R. M. and Das Gupta, T. K. (1970). Role of pinealectomy on Walker 256 carcinoma in rats. *Journal of Surgical Oncology* **2**, 313–22.

Barone R. M., Abe, R., and Das Gupta, T. K. (1972). Pineal ablation in methylcholanthrene-induced fibrosarcoma. *Surgical Forum* **23**, 115–16.

Bartsch, C. and Bartsch, H. (1984). The link between the pineal gland and cancer: an interaction involving chronobiological mechanisms. In *Chronobiological approach to social medicine*, pp. 105–26. Istituto Italiano Di Medicina Sociale, Roma.

Bartsch, C., Bartsch, H., Jain, A. K., Laumas, K. R., and Wetterberg, L. (1981). Urinary melatonin levels in human breast cancer patients. *Journal of Neural Transmission* **52**, 281–94.

Bartsch, C., Bartsch, H., Fluchter, S.-H., Attanasio, A., and Gupta, D. (1985). Evidence for modulation of melatonin secretion in men with benign and malignant tumors of the prostate: relationship with the pituitary hormones.

Journal of Pineal Research **2**, 121–32.

Bartsch, H. and Bartsch, C. (1981). Effect of melatonin on experimental tumours under different photoperiods and times of administration. *Journal of Neural Transmission* **52**, 269–79.

Bindoni, M. (1971). Relationship between the pineal gland and the mitotic activity of some tissues. *Archives of Science and Biology* **55**, 3–21.

Bindoni, M. and Raffaele, R. (1968). Mitotic activity in the adenohypophysis of rats after pinealectomy. *Journal of Endocrinology* **41**, 451–2.

Birau, N. (1981). Melatonin in human serum: progress in screening investigation and clinic. In *Melatonin—current status and perspectives, advances in the biosciences,* Vol. 29 (ed. N. Birau and W. Schloot), pp. 297–326. Pergamon Press, Elmsford.

Blask, D. E. (1984). The pineal: an oncostatic gland? In *The pineal gland* (ed. R. J. Reiter), pp. 253–84. Raven Press, New York.

Blask, D. E. and Hill, S. M. (1986*a*). Effects of melatonin on cancer: studies on MCF-7 human breast cancer cells in culture. *Journal of Neural Transmission* **21** (Suppl.) 433–49.

Blask, D. E. and Hill, S. M. (1986*b*). Inhibition of human breast cancer and endometrial cancer cell growth in culture by the pineal hormone melatonin and its analogue 6-chloromelatonin. *Biology of Reproduction* **34** (Suppl. 1), 176, Abstract 254.

Blask, D. E. and Leadem, C. A. (1987). Neuroendocrine aspects of neoplasia: a review. *Neuroendocrinology Letters* **9**, 1–11.

Blask, D. E. and Nodelman, J. L. (1979). Antigonadotrophic and prolactin-inhibitory effects of melatonin in anosmic male rats. *Neuroendocrinology* **29**, 406–12.

Blask, D. E. and Reiter, R. J. (1978). Pineal removal or denervation: effects on hypothalamic PRF activity in the rat. *Molecular and Cellular Endocrinology* **11**, 243–8.

Blask, D. E., Nodelman, J. L., and Leadem, C. A. (1980). Preliminary evidence that a dopamine receptor antagonist blocks the prolactin-inhibiting effects of melatonin in anosmic rats. *Experientia* **36**, 1008–9.

Blask, D. E., Leadem, C. A., and Richardson, B. A. (1981). Nutritional status, time of day and pinealectomy: factors influencing the sensitivity of the neuroendocrine-reproductive axis of the rat to melatonin. *Hormone Research* **14**, 104–12.

Blask, D. E., Hill, S. M., Orstead, K. M., and Massa, J. S. (1986). Inhibitory effects of the pineal hormone melatonin and underfeeding during the promotional phase of 7,12-dimethylbenzanthracene (DMBA)-induced mammary tumourigenesis. *Journal of Neural Transmission* **67**, 12–138.

Boucek, R. J. and Alvarez, T. R. (1970). 5-Hydroxytryptamine: a cytospecific growth stimulator of cultured fibroblasts. *Science* **167**, 898–9.

Burns, J. K. (1973). Administration of melatonin to non-human primates and to women with breast carcinoma. *Journal of Physiology (London)* **229**, 38P–9P.

Buswell, R. S. (1975). The pineal and neoplasia. *Lancet* **ii**, 34–5.

Chang, N., Spaulding, T. S., and Tseng, M. T. (1985). Inhibitory effects of superior cervical ganglionectomy on dimethylbenz(a)anthracene-induced mammary tumours in the rat. *Journal of Pineal Research* **2**, 331–40.

Cohen, M., Lippman, M., and Chabner, B. (1978). Role of pineal gland in aetiology and treatment of breast cancer. *Lancet* **ii**, 814.

Cohen, P., Wax, Y., and Modan, B. (1983). Seasonality in the occurrence of breast cancer. *Cancer Research* **43**, 892–6.

Cuzick, J., Wang, D. Y., and Bulbrook, R. D. (1986). The prevention of breast cancer. *Lancet* **ii**, 83–6.

Danforth, D. N., Tamarkin, L., and Lippman, M. (1983). Melatonin increases oestrogen receptor hormone binding activity of human breast cancer cells. *Nature (London)* **305**, 323–5.

Danforth, D. N., Tamarkin, L., and Lippman, M. (1984). Melatonin-induction of oestrogen receptor hormone binding activity is associated with inhibition of E_2-stimulated growth of MCF-7 human breast cancer cells. *Proceedings of the 7th International Congress of Endocrinology*, p. 507, Abstract 494.

Danforth, D. N., Tamarkin, L., Mulvihill, J. J., Bagley, C. S., and Lippman, M. (1985). Plasma melatonin and the hormone-dependency of human breast cancer. *Journal of Clinical Oncology* **3**, 941–8.

Das Gupta, T. K. (1968). Influence of the pineal gland on the growth and spread of malignant tumours. *Surgical Forum* **19**, 83–4.

Das Gupta, T. K. and Terz, J. (1967). Influence of the pineal gland on the growth and spread of melanoma in the hamster. *Cancer Research* **27**, 1306–11.

DiBella, L., Scalera, G., and Rossi, M. T. (1979). Perspectives in pineal function. In *The pineal gland of vertebrates including man, progress in brain research*, Vol. 52 (ed. J. A. Kappers and P. Pevet), pp. 475–8. Elsevier, Amsterdam.

El-Domieri, A. A. H. and Das Gupta, T. K. (1973). Reversal by melatonin of the effect of pinealectomy on tumour growth. *Cancer Research* **33**, 2830–3.

El-Domieri, A. A. H. and Das Gupta, T. K. (1976). The influence of pineal ablation and administration of melatonin on growth and spread of hamster melanoma. *Journal of Surgical Oncology* **8**, 197–205.

Ferioli, M. E., Scalabrino, G., and Fraschini, F. (1983). Influence of the pineal gland on tumour growth in mammalians: a reappraisal from a biochemical point of view. In *The pineal gland and its endocrine role* (ed. J. Axelrod, F. Fraschini, and G. P. Velo), pp. 467–76. Plenum Publishing Corp., New York.

Fitzgerald, T. J. and Veal, A. (1976). Melatonin antagonizes colchicine-induced mitotic arrest. *Experientia* **32**, 372–3.

Fraschini, F., Lissoni, P., Mauri, R., Esposti, G., and Esposti, D. (1985). Light/dark rhythm of serum melatonin levels in oncological patients. *Proceedings of the 67th Annual Meeting of the Endocrine Society*, p. 253, Abstract 1009.

Georgiou, E. (1929). Über die Natur und kie Pathogenese der Krebstumoren, Radiale Heilung des Krebses bei weißen Mausen. *Zeitschrift für Krebsforshung* **38**, 562–72.

Ghosh, B. C., El-Domeiri, A. A. H., and Das Gupta, T. K. (1973). Effect of melatonin on hamster melanoma. *Surgical Forum* **24**, 121–2.

Gorray, K. C. and Quay, W. B. (1978). Effects of pinealectomy and of sham-pinealectomy on blood glucose levels in the alloxan-diabetic rat. *Hormone and Metabolic Research* **10**, 380–92.

Hamilton, T. (1969). Influence of environmental light and melatonin upon mammary tumour induction. *British Journal of Surgery* **56**, 764–6.

Hartveit, F., Thoresen, S., Tangen, M., and Halvorsen, J. (1983). Variation in histology and estrogen receptor content in breast carcinoma related to tumor size and time of presentation. *Clinical Oncology* **9**, 233–8.

Hill, S. M. (1986). Antiproliferative effect of the pineal hormone melatonin on human breast cancer cells *in vitro*. PhD Dissertation, University of Arizona, Tucson.

Hill, S. M. and Blask, D. E. (1985a). Physiological concentrations of melatonin inhibit the proliferation of MCF-7 human breast cancer cells *in vitro*. *Proceedings of the 67th Annual Meeting of the Endocrine Society*, p. 253, Abstract 1010.

Hill, S. M. and Blask, D. E. (1985b). Melatonin inhibition of MCF-7 breast cancer proliferation: influence of serum factors, prolactin and estadiol. *Proceedings of the 68th Annual Meeting of the Endocrine Society*, p. 246, Abstract 863.

Huxely, M. and Tapp, E. (1972). Effect of biogenic amines on the growth of rat tumours. *Life Sciences* **11**, 19–23.

Illnerova, H., Zvolsky, P., and Vanecek, J. (1985). The circadian rhythm in plasma melatonin concentration of the urbanized man: the effect of summer and winter time. *Brain Research* **328**, 186–9.

Jankovic, B. D., Isakovic, K., and Petrovic, S. (1970). Effect of pinealectomy on immune reactions in the rat. *Immunology* **18**, 1–16.

Kallenbach, H. V. and Malz, W. (1957). Uber den Einflußder Epiphyse auf das Wachstum transplantabler Impftumoren. *Endokrinologie* **34**, 293–5.

Karmali, R. A., Horrobin, D. F., and Ghayur, T. (1978). Role of the pineal gland in the aetiology and treatment of breast cancer. *Lancet* **ii**, 1002.

Katugiri, E. (1943). Studies on the pineal gland. Tumor proliferation and the pineal gland. *Osaka igakkai zasshi* **42**, 935–8.

Kennaway, D. J., McCullouch, G., Matthews, C. D., and Seamark, R. F. (1979). Plasma melatonin, luteinizing hormone, follicle-stimulating hormone, prolactin, and corticoids in two patients with pinealoma. *Journal of Clinical Endocrinology and Metabolism* **49**, 144–5.

Kerenyi, N. A. (1979). Tumors and the pineal gland. In *Influences of hormones on tumour development*, Vol. 1 (ec. J. A. Kellen and R. Hilf), pp. 155–65. CRC Press, Boca Raton, Florida.

Kirkham, N., Machin, D., Cotton, D., and Pike, J. (1985). Seasonality and breast cancer. *European Journal of Surgical Oncology* **11**, 143–6.

Kothari, L. S., Shah, P. N., and Mhatre, M. C. (1982). Effect of continuous light on the incidence of 9,10-dimethyl-1,2-benzanthracene induced mammary tumors in female Holtzman rats. *Cancer Letters* **16**, 313–17.

Kothari, L. S., Shah, P. N., and Mhatre, M. C. (1984). Pineal ablation in varying photoperiods and the incidence of 9,10-dimethyl-1,2-benzanthracence induced mammary cancer in rats. *Cancer Letters* **22**, 99–102.

Lacassagne, A., Chamorro, A., Hurst, L., and Ba Giao, N. (1969). Effect de l'epiphysectomie sur l'hepatocancerogenese chimiques chez le rat. *Comptes Rendus L'Academie Des Sciences (Paris)* **269**, 1043–6.

Lapin, V. (1974). Influence of simultaneous pinealectomy and thymectomy on the growth and formation of metastases of the Yoshida sarcoma in rats. *Experimental Pathology* **9**, 108–12.

Lapin, V. (1976). Pineal gland and malignancy. *Osterreichische Zeitschrift für Onkologie* **3**, 51–60.

Lapin, V. (1978). Effects of reserpine on the incidence of 9,10-dimethyl-1,2-benzanthracene induced tumors in pinealectomized and thymectomized rats. *Oncology* **35**, 132–5.

Lapin, V. and Ebels, I. (1976). Effects of some low molecular weight sheep pineal fractions and melatonin on different tumors in rats and mice. *Oncology* **33**, 110–13.

Lapin, V. and Frowein, A. (1981). Effects of growing tumors on pineal melatonin levels in male rats. *Journal of Neural Transmission* **52**, 123–36.

Leadem, C. A. and Blask, D. E. (1981). Evidence for an inhibitory influence of the pineal on prolactin in the female rat. *Neuroendocrinology* **33**, 268–75.

Leadem, C. A. and Blask, D. E. (1984). Characterization of a pineal-mediated inhibition of pubertal prolactin cell development in blind-anosmic female rats. *Journal of Pineal Research* **1**, 263–72.

Leadem, C. A. and Burns, D. M. (1987). Pineal-induced inhibition of pro-lactinoma growth in F344 rats: effects of binding and melatonin treatment. *Proceedings of the 69th Annual Meeting of the Endocrine Society*. Abstract 344.

Lehrer, D., Levine, E., and Bloomer, W. D. (1985). Abnormally diminished sense of smell in women with oestrogen receptor positive breast cancer. *Lancet* **ii**, 333.

Lewinski, A. (1986). Evidence for pineal gland inhibition of thyroid growth: contribution to the hypothesis of a negative feedback between the thyroid and pineal. In *Advances in pineal research* (ed. R. J. Reiter and M. Karasek), pp. 167–76. John Libbey & Co, London.

Lewy, A. J., *et al.* (1985). The use of plasma melatonin levels and light in the assessment and treatment of chronobiologic sleep and mood disorder. In *Melatonin in humans* (ed. R. J. Wurtman and F. Waldhauser), pp. 279–89. Centre for Brain Sciences and Metabolism Charitable Trust, Cambridge.

Lippman, M. E., Bolan, G., and Huff, K. (1976). The effects of estrogens and antiestrogens on hormone-responsive human breast cancer in long-term tissue culture. *Cancer Research* **36**, 4595–601.

Maestroni, G. J. M., Conti, A., and Pierpaoli, W. (1986). Role of the pineal gland in immunity. Circadian synthesis and release of melatonin modulates the antibody response and antagonizes the immunosuppressive effect of cortico-sterone. *Journal of Neuroimmunology* **13**, 19–30.

Malawista, S. E. (1973). The effects of colchicine and of cytochalasin B on the hormone-induced movement on melanin granules in frog dermal melanocytes. In *Endocrinology* (ed. R. O. Scow, F. J. G. Ebling, and I. W. Henderson), pp. 288–93. Excerpta Medica, Amsterdam.

Mason, B., Holdaway, I., Mullins, P., Kay, R., and Skinner, S. (1985). Seasonal variation in breast cancer detection: correlation with tumour progesterone receptor status. *Breast Cancer Research and Treatment* **5**, 171–6.

Meyskens, F. L. and Salmon, S. F. (1981). Modulation of clonogenic human melanoma cells by follicle-stimulating hormone, melatonin and nerve growth factor. *British Journal of Cancer* **43**, 111–15.

Miles, A., Tidmarsh, S. F., Philbrick, D., and Shaw, D. (1985). Diagnostic potential of melatonin analysis in pineal tumours. *The New England Journal of Medicine* **313**, 329–30.

Nakatani, M., Ohara, Y., Katagiri, E., and Nakona, K. (1940). Studies on the pinealectomized white rat. *Nippon Bioriagakkai Kaishi* **30**, 232–6.

Narita, T. and Kudo, H. (1985). Effect of melatonin on B16 melanoma growth in athymic mice. *Cancer Research* **45**, 4175–7.

Neuwelt, E. A. and Lewy, A. (1983). Disappearance of plasma melatonin after removal of a neoplastic pineal gland. *The New England Journal of Medicine* **308**, 1132–5.

Osborne, C. K. (1985). Effects of oestrogens and antioestrogens on human breast cancer cell proliferation: *in vitro* studies in tissue culture and *in vivo* studies in athymic mice. In *Hormonally responsive tumors* (ed. V. P. Hollander), pp. 93–113. Academic Press, Orlando, Florida.

Ownby, H. E., Frederick, J., Mortensen, R. F., Ownby, D. R., Russo, J., and the Breast Cancer Prognostic Study Associates (1986). Seasonal variation in tumor size at diagnosis and immunologic responses in human breast cancer. *Invasion Metastasis* **6**, 246–56.

Pawlikowski, M. (1986). The pineal gland and cell proliferation. In *Advances in pineal research* (ed. R. J. Reiter and M. Karasek), pp. 27–30. John Libbey and Co, London.

Pico, J. L., Mathe, G., Young, I. M., Leone, R. M., Hooper, J., and Silman, R. E. (1979). Role of hormones in the etiology of human cancer. Pineal indole hormones and cancer. *Cancer Treatment Reports* **63**, 1204.

Poffenbarger, M. and Fuller, G. M. (1976). Is melatonin a microtubule inhibitor? *Experimental Cell Research* **103**, 135–41.

Quay, W. B. (1968). Comparative physiology of serotonin and melatonin. *Advances in Pharmacology* **6A**, 283–97.

Quay, W. B. (1980). Pineal effects on metabolism and glucose homeostasis: evidence for lines of humoral mediation of pineal influences on tumor growth. *Journal of Neural Transmission* **47**, 107–20.

Quay, W. B. and Gorray, K. C. (1980). Pineal effects on metabolism and glucose homeostasis: evidence for lines of humoral mediation on pineal influences on tumour growth. *Journal of Neural Transmission* **47**, 107–20.

Reiter, R. J. (1980). The pineal and its hormones in the control of reproduction in mammals. *Endocrine Reviews* **1**, 109–31.

Reiter, R. J., Petterborg, L. J., Trakulrungsi, C., and Trakulrungsi, W. K. (1980). Surgical removal of the olfactory bulbs increases sensitivity of the reproductive system of female rats to the inhibitory effects of late afternoon melatonin injections. *Journal of Experimental Zoology* **212**, 47–52.

Rodin, A. E. (1963). The growth and spread of Walker 256 carcinoma in pinealectomized rats. *Cancer Research* **23**, 1545–50.

Rodin, A. E. and Overall, J. (1967). Statistical relationships of weight of the human pineal to age and malignancy. *Cancer* **20**, 1203–14.

Schloot, W., Dubbels, R., and Birau, N. (1981). Genetics of melatonin. In *Melatonin—current status and perspectives* (ed. N. Birau and W. Schloot), pp.

269–84. Pergamon Press, Elmsford.

Shafie, S. and Brooks, S. C. (1977). Effect of prolactin on growth and the estrogen receptor level of human breast cancer cells (MCF-7). *Cancer Research* **37**, 792–9.

Shah, P. N., Mhatre, M. C., and Kothari, L. S. (1984). Effect of melatonin on mammary carcinogenesis in intact and pinealectomized rats in varying photoperiods. *Cancer Research* **44**, 3403–7.

Stanberry, L. R., Das Gupta, T. K., and Beattie, C. W. (1983). Photoperiodic control of melanoma growth in hamster: influence of pinealectomy and melatonin. *Endocrinology* **113**, 469–75.

Starr, K. W. (1970). Growth and new growth: environmental carcinogens in the process of human ontogeny. *Progress in Clinical Cancer* **4**, 1–29.

Svet-Moldavsky, G. J., *et al.* (1976). An attempt of treatment of malignant melanoma with bovine pineal tissues. *Biomedicine* **25**, 7–10.

Tamarkin, L., Cohen, M., Roselle, D., Reichert, C., Lippman, M., and Chabner, B. (1981). Melatonin inhibition and pinealectomy enhancement of 7,12-dimethyl-benz(a)anthracene-induced mammary tumors in the rat. *Cancer Research* **41**, 4432–6.

Tamarkin, L., *et al.* (1982). Decreased nocturnal plasma melatonin peak in patients with estrogen receptor positive breast cancer. *Science* **216**, 1003–5.

Tapp, E. (1978). Melatonin as a marker in a patient with pineal tumour. *British Medical Journal* **2**, 636.

Tapp, E. (1980). The human pineal gland in malignancy. *Journal of Neural Transmission* **48**, 131–5.

Tapp, E. (1982). The pineal gland in malignancy. In *The pineal gland*, Vol. III (ed. R. J. Reiter), pp. 171–88. CRC Press, Boca Raton, Florida.

Tapp, E., Skinner, R. G., and Phillips, V. (1980). Radioimmunoassay for melatonin. *Journal of Neural Transmission* **48**, 137–41.

Tessman, D., Gocke, H., and Salditt, G. (1972). The influence of light and darkness on male and female Walker carcinosarcoma. *Archiv für Geschwidforsh* **39**, 300–3.

Toma, J. G., Amerongen, H. M., Hennes, S. C., O'Brien, M. G., McBlain, W. A., and Buzzell, G. R. (1988). The effects of olfactory bulbectomy, melatonin and/or pinealectomy on three sublines of the Dunning R3327 rat prostatic adenocarcinoma. *Journal of Pineal Research* (in press).

Touitou, Y., Fevre, M., Bogdan, A., Reinberg, A., De Prins, J., Beck, H., and Touitou, C. (1984). Patterns of plasma melatonin with ageing and mental condition: stability of nyctohemeral rhythms and differences in seasonal variations. *Acta Endocrinologica* **106**, 145–51.

Touitou, Y., Fevre-Montange, M., Proust, J., Klinger, E., and Nakache, J. P. (1985). Age- and sex-associated modification of plasma melatonin concentrations in man. Relationship to pathology, malignant or not, and autopsy findings. *Acta Endocrinologica* **108**, 135–44.

Vaticon, M. D., Fernandez-Galez, C., Esquifino, A., Tejero, A., and Aguilar, E. (1980). Effects of constant light on prolactin secretion in adult female rats. *Hormone Research* **12**, 277–88.

Vaughan, G. M. (1984). Melatonin in humans. In *Pineal research reviews* (ed. R. J. Reiter), pp. 141–201. Alan R. Liss Inc, New York.

Vaughan, G. M., *et al.* (1978). Influence of pinealectomy on serum oestrogen and progesterone levels in blind-anosmic female rats. *Experientia* **34**, 1378–9.

Vignon, F. and Rochefort, H. (1985). Estrogen-specific proteins released by breast cancer cells in culture and control of cell proliferation. In *Hormonally responsive tumors* (ed. V. P. Hollander), pp. 135–53. Academic Press, Orlando, Florida.

Walker, M. J., Chauduri, P. K., Beattie, C. W., Tito, W. A., and Das Gupta, T. K. (1978). Neuroendocrine and endocrine correlates to hamster melanoma growth *in vitro. Surgical Forum* **29**, 151–2.

Welsch, C. W. (1985). Host factors affecting the growth of carcinogen-induced rat mammary carcinomas: a review and tribute to Charles Brenton Huggins. *Cancer Research* **45**, 3415–43.

Welsch, C. W. and Nagasawa, H. (1977). Prolactin and murine mammary tumourigenesis: a review. *Cancer Research* **37**, 951–63.

Wetterberg, L., Beck-Friis, J., Aperia, B., and Petterson, U. (1979*a*). Melatonin/ cortisol ratio in depression. *Lancet* **ii**, 1361.

Wetterberg, L., *et al.* (1979*b*). Circadian variation in urinary melatonin in clinically healthy women in Japan and the United States of America. *Experientia* **35**, 416–18.

Winston, M., Johnson, E., Kelleher, J. K., Banerjee, S., and Margulis, L. (1974). Melatonin: cellular effects on live stentors correlated with the inhibition of colchicine-binding to microtubule protein. *Cytobios* **9**, 237–43.

Young, I. M., Leone, R. M., Stovell, F. P., and Silman, R. E. (1985). Melatonin is metabolized to N-acetylserotonin and 6-hydroxymelatonin in man. *Journal of Clinical Endocrinology and Metabolism* **60**, 114–19.

8. Melatonin and ageing

Franz Waldhauser and Maria Waldhauser

Introduction

Since the first reports (Tamarkin *et al.* 1980; Reiter *et al.* 1980) on age-related alterations of pineal melatonin, it became established that the pineal content of melatonin (Tamarkin *et al.* 1980; Reiter *et al.* 1980, 1981; Pang and Tang 1983; Pang *et al.* 1984; Yellon *et al.* 1985), as well as the rate-limiting enzymes for its biosynthesis (Klein and Lines 1969; Ellison *et al.* 1972), are subject to changes during development and ageing. The melatonin concentration in serum is subject to even more dramatic alterations during ageing than the concentration in the pineal gland (Pang and Tang 1983; Pang *et al.* 1984; Tang *et al.* 1985; Reuss *et al.* 1986). A close relationship between serum and pineal hormone concentrations is, however, detectable if levels in the pineal are related somehow to the body size of the animal (Fig. 8.1; Klein and Lines 1969; Pang *et al.* 1984). In the following, age-related melatonin alterations in rats will be discussed in detail prior to description of human and clinical studies, since these species have been examined most thoroughly.

Age-related changes of melatonin in the rat

Serum melatonin

There are no data available on rat serum melatonin in the first 2 weeks of life, probably for technical reasons, since young pups have a low volume of blood, which complicates estimation of melatonin concentrations. The earliest and also highest nocturnal blood levels of this hormone have been reported at 3 weeks of life (prepubertal stage; Pang *et al.* 1984; Tang *et al.* 1985). Subsequently, nocturnal blood levels seem to decrease according to a biphasic pattern with a rapid drop of some 40–60 per cent until 8 weeks, and later on with a more moderate decline of approximately 10 per cent until old age (Fig. 8.1; Pang *et al.* 1984; Tang *et al.* 1985). Data on a possible age-related alteration of daytime serum melatonin levels are inconsistent (Pang *et al.* 1984; Tang *et al.* 1985;

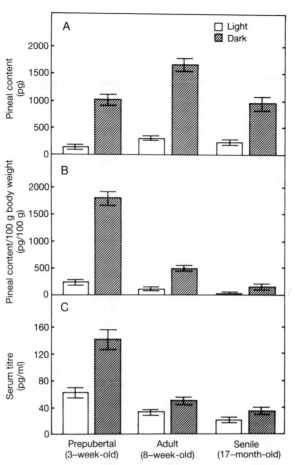

Fig. 8.1. Pineal (A, B) and blood (C) melatonin content in prepubertal, adult and senile rats. Melatonin concentrations in the pineal are expressed either in (A) absolute terms (pg/pineal gland) or (B) related to body size (pg/100 g body weight). (C) The blood melatonin content is related to body size (pg/ml). Each bar represents the mean ± SEM of six to nine samples. [By courtesy of Pang *et al.* 1984.]

Reuss *et al.* 1986). Since daytime levels are often at the limit of the sensitivity of current assay systems (Miles *et al.*, Chapter 13), it may be difficult to detect and assess possible age-associated alterations in such specimens.

Pineal melatonin, HIOMT, and NAT

Within the first week of life in the rat pineal gland, the enzymes N-Acetyltransferase (NAT) and Hydroxyindole-O-methyltransferase

(HIOMT), as well as melatonin are low, with a lack of circadian rhythmicity. Then, in the second week of life, nocturnal concentrations of the compounds begin to rise (Klein and Lines 1969; Ellison *et al.* 1972; Tamarkin *et al.* 1980) and by 3 weeks of age, the highest melatonin and enzyme levels in the pineal per unit body weight have been reported (Klein and Lines 1969; Pang *et al.* 1984). Subsequently, the concentrations of melatonin and HIOMT in the pineal drop fast until age 8 weeks. From then on, melatonin concentrations decline rather moderately until old age (Pang *et al.* 1984); as already noted, to identify this characteristic biphasic decline in concentrations of pineal melatonin and HIOMT after 3 weeks of age, it is most important to relate the levels of the compounds to the body size of the animal (Fig. 8.1; Klein and Lines 1969; Pang *et al.* 1984).

Urinary melatonin and 6-hydroxymelatonin

To our knowledge, urinary excretion of melatonin or of 6-hydroxymelatonin, its major metabolite, has not been studied in animals for age-related alterations.

The rationale for the biphasic decline of serum melatonin with ageing may be due to the fact that the pineal gland of rats grows only by about 150 per cent from age 3 weeks to senium (Legait and Oboussier 1977). During this period the melatonin and enzyme content per gland first increases moderately and then drops moderately (Fig. 8.1; Klein and Lines 1969; Ellison *et al.* 1972; Tamarkin *et al.* 1980; Reiter *et al.* 1981; Pang *et al.* 1984). This may be evidence for only minor changes in melatonin production after 3 weeks of age. The body size of the animal, however, increases by about 600 per cent from age 3–8 weeks and then by a further 100 per cent until old age (Legait and Oboussier 1977; Pang *et al.* 1984). Hence, after 3 weeks of age, a rather stable melatonin production by the pineal gland contrasts with an initially fast and then slowly growing rat. Therefore, the described biphasic alterations of serum melatonin during ageing may be mainly the consequence of the characteristic growth pattern of the animals, or, in other terms, the result of the biphasic expansion of the volume of distribution for melatonin. Several other mechanisms may possibly contribute, however, to the decrease of melatonin levels, particularly in aged animals, i.e. reduction of the number of β-adrenergic receptors in the pineal (Greenberg and Weiss 1978) or reduced rate of glucose and amino-acid metabolism in the pineal (Quay 1974). The low pineal concentrations of enzymes and melatonin during the first week of life are probably due to insufficient differentiation of the pinealocytes at this age (for review see Vollrath 1981).

The melatonin data of species other than rats, i.e. hamster (Tamarkin *et al.* 1980; Reiter *et al.* 1980, 1981; Pang and Tang 1983; Yellon *et al.* 1985) sheep (Forster *et al.* 1986), and gerbil (King *et al.* 1981), are more fragmentary, but in general, they tend to support the findings in rats.

Age-related changes of melatonin in the human

Daytime serum melatonin

A progressive decline in daytime serum melatonin levels with ageing is described by Iguchi *et al.* (1982) in 81 subjects aged 1–92 years. This is in agreement with another report on age-related reduction of melatonin in human cerebrospinal fluid (Brown *et al.* 1979). Silman *et al.* (1979) found an abrupt fall in circulating melatonin in young boys, but not in girls with advancing sexual maturation. In some additional reports, daytime levels tended to be higher in prepubertal children than in adults, but no statistics are available on these data (Touitou *et al.* 1981; Gupta *et al.* 1983; Pang 1985; Attanasio *et al.* 1985). Other laboratories, however, including our own, were not able to observe any alterations in daytime melatonin with advancing age (Lenko *et al.* 1982; Waldhauser *et al.* 1984). Finally, one group reported increasing daytime melatonin concentrations with progressing puberty (Penny 1985). Melatonin levels during daytime are known to be low, however, mostly below the limit of detection. The situation in humans may, therefore, be similar to that outlined above for rats; by examining daytime specimens, age-dependent melatonin alterations, if they exist, may be difficult to detect (see Miles *et al.*, Chapter 13).

Nocturnal serum melatonin

There are only preliminary data available concerning neonatal melatonin secretion. Melatonin levels in the umbilical artery and the umbilical vein of 39 neonates were not different from the serum melatonin concentrations of their mothers, irrespective of the time of sample collection or the kind of delivery (normal delivery versus Caesarian section). These results indicate a similar circadian melatonin variation in fetuses as in adults; it may be due to free transport of hormone between the maternal and fetal compartment of the placenta (Lang *et al.* 1986). However, as delivery usually occurs under extraordinary conditions, i.e. stress, medication, bright light, etc., it was not clear as to whether these factors had influenced the outcome in the investigation cited.

In studying 26 male infants aged 1 day to 1 year, Hartmann and colleagues (1982) found low or undetectable nocturnal serum melatonin levels up to 3 months of age. Nocturnal serum melatonin in the following nine months averaged 588 pg/ml. In a recent publication, Attanasio and colleagues (1986) confirmed this observation. We examined nocturnal serum melatonin levels of 89 children and adults aged 1–35 years. Highest values were observed in children under 5 years; from this age, serum melatonin decreased progressively until early adulthood (Fig. 8.2). Corresponding data were later reported by Attanasio and colleagues (1985) and by our group (Waldhauser *et al.* 1986). Since an inter-relationship between pineal products and human puberty has been suggested for nearly a century (Kitay and Altschule 1954), we related melatonin levels in our study to the individuals' stage of sexual maturation and found a progressive fall of nocturnal melatonin secretion during the course of sexual maturation (Waldhauser *et al.* 1984). Similarly, the net diurnal increments of plasma hormone (dark-phase value minus daytime value) showed a significant decrease during sexual maturation (Gupta *et al.* 1983). Lissoni and associates (1983) reported significantly higher values in prepubertal males than in adults. This applied also to prepubertal females compared to women during the periovulatory phase. Other laboratories, however, could not find any changes in the nocturnal melatonin values during puberty, though values tended to be higher in prepubertal subjects than in adolescents, but the results were not significantly different (Tamarkin *et al.* 1982; Ehrenkranz *et al.* 1982). Cavallo and associates (1985) examined peak and mean 24-hour melatonin concentrations in 10 prepubertal and pubertal boys (4–15 years). Both variables were higher before puberty but levels of significance were again not achieved. In a recent study, nocturnal serum

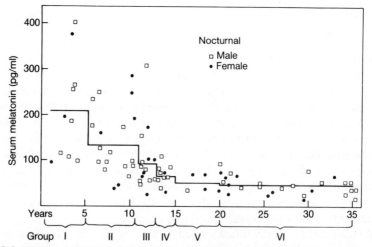

Fig. 8.2. Nocturnal serum melatonin levels of human subjects aged 1 to 35 years. The black line represents mean values for age groups.

melatonin levels of four prepubertal boys aged 11–12 years were compared to those of four post-pubertal boys aged 17.5–18.5 years. No difference was noticed between the two groups (Sizonenko *et al.* 1985). Since human sexual maturation occurs over the course of 3–5 years, a progressive decrease of nocturnal serum melatonin during that period is likely, from some of the outlined studies. Whether this decline in melatonin secretion has a causal connection to puberty or whether it occurs simply coincidentally with sexual maturation cannot be determined from these experiments.

In several other publications, possible secretory changes during adulthood, i.e. age 20–60 years, were examined. Arendt and co-workers (1982), who investigated peak and 24-hour melatonin secretion in 12 volunteers aged 20–39 years, found no correlation between age and melatonin levels. In addition, Claustrat *et al.* (1984) also found no age-associated alterations in nocturnal melatonin in another experiment involving 24 subjects aged 25–66 years. Only one laboratory reported a nearly linear negative correlation between age and peak nocturnal values (Nair *et al.* 1986).

Another significant decline in serum melatonin concentration has been reported unanimously in the elderly. Touitou and colleagues (1981, 1984) reported 24-hour mean plasma levels in elderly humans (62–91 years) which were about half of those in young men (24 ± 3.9 years). Similarly, the nocturnal rise of melatonin was significantly reduced in elderly subjects by comparing young (26.4 ± 2.3 years) and old (84.8 ± 1.8 years) men (Iguchi *et al.* 1982). This correlates well with recent results of Pang (1985), and most recently with Thomas and Miles (1988). The latter investigators observed a highly significant reduction of peak melatonin levels in elderly individuals compared to young.

To determine the precise pattern of age-associated alterations in melatonin secretion over the entire lifespan, we are performing a cross-sectional study on human subjects of all ages (3 days to 90 years). Until now, nocturnal melatonin levels of 280 individuals have been analysed (Fig. 8.3). From these data melatonin levels appear to be low during the first months of life and to increase to peak values at the age 1–3 years. Thereafter, it appears to drop progressively until the end of adolescence. In adults, melatonin secretion is rather stable with an additional drop at old age. In this ongoing study, melatonin displays a biphasic decrease after infancy with a rapid drop of approximately 75 per cent from early childhood to the end of adolescence, with an additional decline of some 10 per cent from there to senium (Waldhauser and Steger 1986). These results are in accordance with much of the relevant literature, and in our opinion there is no question that nocturnal melatonin secretion is subject to major changes during the human lifetime.

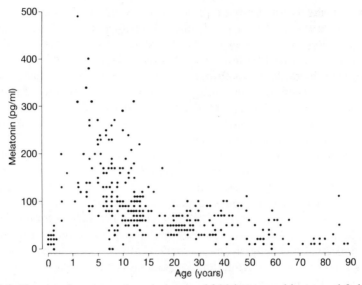

Fig. 8.3. Nocturnal serum melatonin levels of 280 human subjects aged 3 days to 90 years.

The discrepancies on this subject in the literature may derive from several factors. Serum melatonin is known to exhibit large interindividual differences, and the low number of subjects involved in some of these studies may have prevented the detection of significant differences among groups. In addition, the study of subjects of a narrow age range may also result in failure to recognize an age-associated alteration in serum melatonin. Finally, there are still major problems with the specificity of the antiserum used in some melatonin radioimmunoassay (RIA) systems. The great difference in mean melatonin values of normal subjects reported by different groups may illustrate this shortcoming (Waldhauser and Wurtman 1983; Waldhauser and Steger 1986). Unspecific melatonin measurements, may thus have led to incorrect conclusions.

Urinary melatonin and 6-hydroxymelatonin

Approximately 70 per cent of the total amount of melatonin secreted by the pineal is metabolized by the liver to 6-hydroxymelatonin which becomes conjugated to sulphate or glucuronic acid for urinary excretion (Kopin *et al.* 1961). One per cent or less of circulating melatonin is excreted unchanged into the urine (Reppert and Klein 1978; Cardinali 1981; Miles *et al.*, Chapter 13). The excretion rate of both compounds has been suggested as a good indicator for endogenous melatonin production, and in adults it accurately reflects plasma levels (Lynch *et al.* 1978; Lang *et al.* 1981; Markey *et al.* 1985; Miles *et al.*, Chapter 13). However, in using the excretion rate of these indoles as an indicator of blood melatonin concentrations in children, consideration of the body size of the individuals is again stressed. Clearly, if serum levels of a substance in children and adults are of similar altitude and the clearance for this substance does not alter during development, children would excrete less of the substance than adults. Indeed, in using urinary melatonin and 6-hydroxymelatonin excretion as an index of peripheral melatonin levels, consideration of the body size is mandatory.

When examining 26 male infants, Hartmann and associates (1982) found no alteration in urinary melatonin excretion within the first year of life following correction of excretion rate for body weight. Neonates and infants up to 3 months, however, had considerably less melatonin in their serum than older subjects. The apparent discrepancy between serum and urinary melatonin was explained by an age-dependent alteration in the clearance for this hormone which was calculated to be five times higher in the younger infants than in the older. The same group of scientists (Lemaitre *et al.* 1981) measured melatonin excretion corrected for body weight in 58 male subjects aged 1–30 years. In this study the children excreted significantly more melatonin than the adults. A similar trend was reported by Wetterberg (1979), although the excretion data were not related to body weight and no statistics were performed. Conflicting data were recorded by Sizonenko and co-workers (1985), who studied 43 normal children during sexual development. There was no alteration in urinary melatonin excretion, corrected for body size, during the period of sexual maturation. In another experiment an increase of urinary melatonin with ageing and sexual maturation was reported, when 170 children aged 9–16 years were investigated. Interestingly, adults ($n = 16$) excreted less melatonin than children, and, in this study, too, melatonin excretion data were corrected for body size (Penny 1982).

Even when the urinary melatonin excretion rate is corrected for the body size of examined individuals, such data may suffer from two short-comings. Since only 1 per cent or less of circulating melatonin is excreted, small alterations in the metabolic clearance of this compound could produce major changes in melatonin excretion; this may occur in pathological conditions or during human development and ageing. In addition, specific measurements of melatonin in urine may provide even more problems than in blood (Lynch 1983). In general, urinary 6-hydroxymelatonin excretion must be considered a more reliable indicator of endogenous melatonin production and serum melatonin concentration than estimation of urinary melatonin for the reasons outlined. In addition, 6-hydroxymelatonin concentrations in urine are 3 powers of 10 higher than those of unmetabolized compound which permits a more precise quantitation of the metabolite.

The total amount of conjugated 6-hydroxymelatonin that was excreted during 24 hours did not correlate with the age of subject, when 101 children or adolescents (age 3–16 years) and 20 adults of unspecified age were examined. Only at the time of first breast development was a significant increase of 6-hydroxymelatonin excretion noted (Tetsuo et al. 1982). This, however, proved to be the result of group selection, since the same group of individuals continued to excrete increased amounts of the indole at later stages of development (Tamarkin 1985), and these excretion data were unfortunately not related to the body size of the subjects. However, if small children and fully developed adults produce the same amount of 6-hydroxymelatonin per time unit, one would predict considerably higher blood melatonin concentrations in children than in adults. In a recent study, Sack and colleagues (1986) also found a significant decline of 6-hydroxymelatonin excretion with ageing, when adults of 22–94 years were investigated.

In spite of some inconsistency in the excretion data, there is some support for the assumption of rather constant melatonin production from early childhood until the onset of senium, and for the notion of a major change in serum melatonin concentrations during a lifetime, particularly at childhood.

As in rats, the mechanism responsible for the decline of serum melatonin levels in humans has not been identified, but several possibilities have been proposed. In fact, some of them are likely to contribute to the melatonin decrease. Rodin and Overall (1967) examined the absolute pineal weight in humans from early childhood to senium and found only slight alterations over that period, and Wurtman and associates (1964), had been unable to detect any age-related changes in HIOMT activity per gland when pineals of humans aged 3–70 years were investigated. As mentioned above, Tetsuo et al. (1982) described a similar excretion rate of 6-hydroxymelatonin in children (3–16 years) and in adults. This may be evidence that the pineal gland is fully developed in small children and that its endocrine activity is similar in very young individuals and adults. The human body size, however, increases by 500–800 per cent from early childhood to adulthood. Thus, if melatonin production is similar in small children and in adults, but the volume of distribution for the pineal hormone increases considerably during maturation, its concentration in body fluids should drop. Hence, the growth of the human body may explain the rapid fall of serum melatonin during childhood. Other factors, such as reduction of the number of β-adrenergic receptors (Greenberg et al. 1978) or reduced fuel metabolism in the pineal (Quay 1974), originally observed in aged rats, may also contribute to the decline of human serum melatonin and apply particularly to the elderly.

The possible significance, clinical or otherwise, of the age-related decrease in human melatonin secretion is one of the burning and still unsolved problems. Very few suggestions have been put forward on this topic. An old hypothesis, originally postulated by Marburg (1909) and later modified several times (Kitay 1954; Kitay and Altschule 1954), proposed that the pineal produces a substance that exerts an inhibitory effect on sexual maturation. In fact, melatonin could be such a substance with an antigonadotropic effect in humans and thus could influence sexual maturation. Since melatonin levels decrease during childhood, this hormone might lose its ability to suppress gonadotropins, and the onset of puberty might occur after melatonin levels passed below a certain threshold (Waldhauser and Gisinger 1986). This hypothesis has been challenged by studying patients with disturbances of sexual maturation; the results obtained are unfortunately inconsistent.

No difference in mean 24-hour plasma melatonin concentrations or in the 24-hour profile of plasma hormone was demonstrable when children suffering from precocious puberty ($n=5$) and normal prepubertal children ($n=8$) were studied (Ehrenkranz *et al.* 1982). On the other hand, lower day–night increments were reported in subjects with idiopathic precocious puberty ($n=5$) than in age-matched controls (Attanasio *et al.* 1983).

Cohen and colleagues (1982) observed significantly higher daytime melatonin concentrations in 33 boys with delayed puberty than in normals of similar age. However, in the patients, the decrease of melatonin levels did not occur at the onset of puberty, but rather in the mid- and late-phase of pubertal development. Attanasio and associates (1983) found day–night increments in subjects with delayed puberty ($n=5$) comparable to those of pre-school children. This is compatible with another study (Lissoni *et al.* 1983) where patients with delayed puberty had significantly higher nocturnal melatonin values than pre-pubertal controls. On the contrary, melatonin excretion per square metre body surface estimated in the morning urine did not differ significantly between boys with delayed puberty and normal prepubertal children (Sizonenko *et al.* 1985).

The assumption of an antigonadotropic action of melatonin in humans is not supported by a recent publication on primates, which shows no alteration in the physiological gonadotropin suppression during pre-puberty among normal rhesus monkeys and monkeys pinealectomized during infancy (Plant and Zarub 1986). In addition, from studies in Djungarian hamsters (Carter and Goldman 1983) and in sheep (Bittman and Karsch 1984) there is evidence that it is neither the amplitude of the circadian melatonin secretion, nor the absolute blood melatonin levels or the total amount of melatonin produced, but the duration of the nocturnal melatonin elevation which is the critical factor governing gonadal function in these species. However, available studies in normal children and patients with disorders of sexual maturation do not provide sufficient data for the reliable evaluation of the duration of the nocturnal melatonin peak. There are pros and cons for the outlined hypothesis, but an adequate evaluation is not possible to date mainly because of a lack of sufficient data in humans.

In a recent study, it was suggested that the age-related decline of blood melatonin might be useful as an indicator for human brain ageing. This notion is supported by preliminary data on patients with presenile dementia (Alzheimer's disease; $n=4$), who secreted significantly less melatonin within 24 hours than age-matched, healthy controls (Nair *et al.* 1986). In another study, however, no difference was observed in mesor, amplitude or acrophase of the circadian and seasonal rhythms when plasma melatonin from patients with senile dementia ($n=6$) and normal elderly subjects were examined (Touitou *et al.* 1984).

Conclusion

Many of the present reports and our own data support the notion that in human nocturnal blood, melatonin reaches peak levels in early childhood after a period of relative melatonin deficiency during early infancy. From there these levels decrease, apparently according to a biphasic pattern, with a rapid fall until adolescence and a moderate decline later on, perhaps mainly during senium. There is some evidence from pineal and urinary melatonin data that indicates only moderate alteration in total melatonin production after infancy. The huge drop in serum melatonin during childhood is probably the result of the increase in size of the human body during that period. The additional decline of melatonin with higher age may be due to an ageing effect of the pineal itself. These assumptions are supported by the fact that pineal content of melatonin and its rate-limiting enzymes follow a similar pattern as blood levels, if they are related to body size. Some of the preliminary data acquired on melatonin and 6-hydroxymelatonin excretion point in the same direction. The hypotheses proposed so far on the potential action of melatonin in human sexual maturation or in brain ageing are only partly consistent with collected data. Further studies are clearly necessary before a definitive statement on the clinical significance of the age-related decline in melatonin secretion can be made. Indeed, very recently, two papers were published providing additional and compelling evidence for a biphasic decline in human nocturnal serum melatonin levels with ageing

(Waldhauser *et al.* 1988; Young *et al.* 1988). In both reports, the close connection between decreasing melatonin levels and increasing body weight during growth is stressed. In addition, Young and co-workers (1988) confirm the assumption that melatonin secretion is rather constant from early childhood to adulthood.

Acknowledgement

The study was supported by a grant # P6193 from Fonds zur Foederung der Wissenschaftlichen Forschung.

References

Arendt, J., Hampton, S., English, J., Kwasowski, P., and Marks, V. (1982). 24-Hour profiles of melatonin, cortisol, insulin, c-peptide, and GIP following a meal and subsequent fasting. *Clinical Endocrinology* **16**, 89–95.

Attanasio, A., Borrelli, P., Marini, R., Cambiaso, P., Cappa, M., and Gupta, D. (1983). Serum melatonin in children with early and delayed puberty. *Neuroendocrinology Letters* **5**, 387–92.

Attanasio, A., Borrelli, P., and Gupta, D. (1985). Circadian rhythms in serum melatonin from infancy to adolescence. *Journal of Clinical Endocrinology and Metabolism* **61**, 388–90.

Attanasio, A., Rager, K., and Gupta, D. (1986). Ontogeny of circadian rhythmicity for melatonin, serotonin and N-acetyl-serotonin in humans. *Journal of Pineal Research*, **3**, 251–6.

Bittman, E. L. and Karsch, F. J. (1984). Nightly duration of pineal melatonin secretion determines the reproductive response to inhibitory day length in the ewe. *Biology of Reproduction* **30**, 585–93.

Brown, G. M., Young, S. N., Gauthier, S., Tsui, H., and Grota, L. J. (1979). Melatonin in human cerebrospinal fluid in daytime; its origin and variation with age. *Life Science* **25**, 929–36.

Cardinali, D. P. (1981). Melatonin: a mammalian pineal hormone. *Endocrine Reviews* **2**, 327–46.

Carter, D. S. and Goldman, B. C. (1983). Antigonadal effects of timed melatonin infusion in pinealectomized male Djungarian hamsters (*Phodopus sungorus sungorus*): duration is the critical parameter. *Endocrinology* **113**, 1261–7.

Cavallo, A., Richards, G. E., and Smith, E. (1985). Relationship between melatonin and luteinizing hormone in human puberty. Poster Presentation at the International Congress on Melatonin in Humans, Vienna, November 7–9, 1985.

Claustrat, B., Chazot, G., Brun, J., Jordan, D., and Sassolas, G. (1984). A chrono-biological study of melatonin and cortisol secretion in depressed subjects: plasma melatonin, a biochemical marker in major depression. *Biological Psychiatry* **19**, 1215–28.

Cohen, H. N. *et al.* (1982). Serum immunoreactive melatonin in boys with delayed puberty. *Clinical Endocrinology* **17**, 517–21.

Ehrenkranz, J. R., *et al.* (1982). Daily rhythm of plasma melatonin in normal and precocious puberty. *Journal of Clinical Endocrinology and Metabolism* **55**, 307–10.

Ellison, N., Weller, J. L., and Klein, D. C. (1972). Development of a circadian rhythm in the activity of pineal serotonin N-acetyltransferase. *Journal of Neurochemistry* **19**, 1335–41.

Foster, D. L., Karsch, F. J., Olster, D. H., Ryan, K. D., and Yellon, S. M. (1986). Determinants of puberty in a seasonal breeder. *Recent Progress in Hormone Research* **42**, 331–84.

Greenberg, L. H. and Weiss, B. (1978). β-adrenergic receptors in aged rat brain: reduced number and capacity of pineal gland to develop supersensitivity. *Science* **201**, 61–3.

Gupta, D., Riedel, L., Frick, H. J., Attanasio, A., and Ranke, M. B. (1983). Circulating melatonin in children: in relation to puberty, endocrine disorders, functional tests and racial origin. *Neuroendocrinology Letters* **5**, 63–78.

Hartmann, L., Roger, M., Lemaitre, B. J., Massias, J. F., and Chaussain, J. L. (1982). Plasma and urinary melatonin in male infants during the first 12 months of life. *Clinica Chimica Acta* **121**, 37–42.

Iguchi, H., Kato, K., and Ibayashi, H. (1982). Age-dependent reduction in serum melatonin concentrations in healthy human subjects. *Journal of Clinical Endocrinology and Metabolism* **55**, 27–9.

King, T. S., Richardson, B. A., and Reiter, R. J. (1981). Age-associated changes in pineal serotonin N-acetyltransferase activity and melatonin content in the male gerbil. *Endocrinological Research Communications* **8**, 253–62.

Kitay, J. I. (1954). Pineal lesions and precocious puberty: a review. *Journal of Clinical Endocrinology and Metabolism* **14**, 622–5.

Kitay, J. I. and Altschule, M. D. (1954). *The pineal gland: a review of the physiologic literature.* Harvard University Press, Cambridge, Massachusetts.

Klein, D. C. and Lines, S. V. (1969). Pineal hydroxyindole-O-methyltransferase activity in the growing rat. *Endocrinology* **84**, 1523–5.

Kopin, I. J., Pare, C. M. B., Axelrod, J., and Weissbach, M. (1961). The fall of melatonin in animals. *Journal of Biological Chemistry* **236**, 3072–5.

Lang, U., Kornemark, M., Aubert, M. L., Paunier, L., and Sizonenko, P. C. (1981). Radioimmunological determination of urinary melatonin in humans: correlation with plasma levels and typical 24-hour rhythmicity. *Journal of Clinical Endocrinology and Metabolism* **53**, 645–50.

Lang, U., Begnin, P. C., and Sizonenko, P. C. (1986). Fetal and maternal melatonin concentrations at birth in humans. *Journal of Neural Transmission* (Suppl.) **21**, 479–80.

Legait, H. and Oboussier, H. (1977). Etude interspecifique des correlations statistiques existant entre le volume ou le poids de la glande pineale et les poids encephalique et somatique dans deux groupes de mammiferes etude intraspecifique des correlations entre glande pineale et encephale chez le rat et chez l'homme. *Bulletin de l'association des Anatomistes* **61**, 123–32.

Lemaitre, B. J., Bouillie, J., and Hartmann, L. (1981). Variations of urinary melatonin excretion in humans during the first 30 years of life. *Clinica Chimica Acta* **110**, 77–82.

Lenko, H. L., Lang, U., Aubert, M. L. Paunier, L., and Sizonenko, P. C. (1982). Hormonal changes in puberty. VII. Lack of variation of daytime plasma melatonin. *Journal of Clinical Endocrinology and Metabolism* **54**, 1056–8.

Lissoni, P., Resentini, M., Mauri, R., Morabito, F., Djemal, S., and Fraschini, F. (1983). Studio del ritmo circardiano della melatonina in sogetti normale nelle puberta ritardate. *Journal of Endocrinological Investigation* **6**, (Suppl. I), 25.

Lynch, H. J. (1983). Assay methodology. In *The pineal gland* (ed. R. Relkin), pp. 129–50. Elsevier, New York.

Lynch, H. J., Jimerson, D. C., Ozaki, Y., Post, R. M., Bunney, W. E., and Wurtman, R. J. (1978). Entrainment of rhythmic melatonin secretion in man to a 12-hour phase shift in the light/dark cycle. *Life Sciences* **23**, 1557–64.

Marburg, O. (1909). Zur Kenntnis der normalen und pathologischen Histologie der Zirbeldrüse. *Arbeiten aus dem Neurologischen Institut an der Wiener Universität* **17**, 217–79.

Markey, S. P., Higa, S., Shih, M., Danforth, D. N. Jr, and Tamarkin, L. (1985). The correlation between human plasma melatonin levels and urinary 6-hydroxy-melatonin excretion. *Clinica Chimica Acta* **150**, 221–5.

Nair, N. P. V., Hariharasubramanian, M., Pilapil, C., Isaac, I., and Thavundayil, J. X. (1986). Plasma melatonin—an index of brain ageing in humans? *Biological Psychiatry* **21**, 141–50.

Pang, S. F. (1985). Melatonin concentrations in blood and pineal gland. *Pineal Research Review* **3**, 115–60.

Pang, S. F. and Tang, P. L. (1983). Decreased serum and pineal concentrations of melatonin and N-acetylserotonin in aged male hamsters. *Hormone Research* **17**, 228–34.

Pang, S. F., Tang, F., and Tang, P. L. (1984). Negative correlation of age and the levels of pineal melatonin, pineal N-acetylserotonin, and serum melatonin in male rats. *Journal of Experimental Zoology* **229**, 41–7.

Penny, R. (1982). Melatonin excretion in normal males and females: increase during puberty. *Metabolism* **31**, 816–23.

Penny, R. (1985). Episodic secretion of melatonin in pre- and postpubertal girls and boys. *Journal of Clinical Endocrinology and Metabolism* **60**, 751–6.

Plant, T. M. and Zorub, D. S. (1986). Pinealectomy in agonadal infantile male rhesus monkeys (*Macaca mulatta*) does not interrupt initiation of the prepubertal hiatus in gonadotropin secretion. *Endocrinology* **118**, 227–32.

Quay, W. B. (1974). *Pineal chemistry*. C. C. Thomas, Springfield.

Reiter, R. J., Richardson, B. A., Johnson, L. Y., Ferguson, B. N., and Dinn, D. T. (1980). Pineal melatonin rhythm: reduction in ageing Syrian hamsters. *Science* **210**, 1372–3.

Reiter, R. J., *et al.* (1981). Age-associated reduction in nocturnal pineal melatonin levels in female rats. *Endocrinology* **109**, 1295–7.

Reiter, R. J., Vriend, J., Brainard, G. C., Matthews, S. A., and Craft, C. M. (1982). Reduced pineal and plasma melatonin levels and gonadal atrophy in old hamsters kept under winter photoperiods. *Experimental Aging Research* **8**, 27–30.

Reppert, S. M. and Klein, D. C. (1978). Transport of maternal (^{3}H) melatonin to suckling rats and the fate of (^{3}H) melatonin in neonatal rats. *Endocrinology* **102**, 582–8.

Reuss, S., Olcese, J., and Vollrath, L. (1986). Electrophysiological and endocrino-logical aspects of aging in the rat pineal gland. *Neuroendocrinology* **43**, 466–70.

Rodin, A. E. and Overall, J. (1967). Statistical relationships of weight of the human pineal to age and malignancy. *Cancer* **20**, 1203–14.

Sack, R. L., Lewy, A. J., Erb, D. L., Vollmer, W. M., and Singer, C. M. (1986). Human melatonin production decreases with age. *Journal of Pineal Research* **3**, 379–88.

Silman, R. E., Leone, R. M., Hooper, R. J. L., and Preece, M. A. (1979). Melatonin, the pineal gland and human puberty. *Nature (London)* **282**, 301–3.

Sizonenko, P. C., Lang, U., Rivest, R. W., and Aubert, M. L. (1985). The pineal and pubertal development. In *Photoperiodism, melatonin and the pineal* (ed. D. Evered and S. Clark), pp. 208–30. Pitman, London.

Tamarkin, L. (1985). Discussion. In *Photoperiodism, melatonin and the pineal* (ed. D. Evered and S. Clark), pp. 225–30. Pitman, London.

Tamarkin, L., Reppert, S. M., Orloff, D. J., Klein, D. C., Yellon, S. M., Goldman, B. D. (1980). Ontogeny of the pineal melatonin rhythm in the Syrian (*Mesocricetus auratus*) and Siberian (*Phodopus sungorus*) hamsters and in the rat. *Endocrinology* **107**, 1061–4.

Tamarkin, L., Abastillas, P., Chen, H. C., McNemar, A., and Sidbury, J. B. (1982). The daily profile of plasma melatonin in obese and Prader-Willi syndrome children. *Journal of Clinical Endocrinology and Metabolism* **55**, 491–5.

Tang, F., Hadjiconstantinou, M., and Pang, S. F. (1985). Aging and diurnal rhythms of pineal serotonin, 5-hydroxyindoleacetic acid, norepinephrine, dopamine and serum melatonin in the male rat. *Neuroendocrinology* **40**, 160–4.

Tetsuo, M., Poth, M., and Markey, S. P. (1982). Melatonin metabolite excretion during childhood and puberty. *Journal of Clinical Endocrinology and Metabolism* **55**, 311–13.

Thomas, D. R., and Miles, A. (1988). The effect of age on melatonin secretion. *Biological Psychiatry*, in press.

Touitou, Y., *et al.* (1981). Age- and mental health-related circadian rhythms of plasma levels of melatonin, prolactin, luteinizing hormone and follicle-stimulating hormone in man. *Journal of Endocrinology* **91**, 467–75.

Touitou, Y., *et al.* (1984). Patterns of plasma melatonin with ageing and mental condition: stability of nyctohemeral rhythms and differences in seasonal variations. *Acta Endocrinologica* **106**, 145–51.

Vollrath, L. (1981). *The pineal organ.* Springer Verlag, Berlin.

Waldhauser, F. and Gisinger, B. (1986). The pineal gland and its development in human puberty. In *The pineal gland during development* (ed. D. Gupta and R. J. Reiter). Croom-Helm, London.

Waldhauser, F. and Steger, H. (1986). Changes in melatonin secretion with age and pubescence. *Journal of Neural Transmission* (Suppl.) **21**, 183–97.

Waldhauser, F. and Wurtman, R. J. (1983). The secretion and action of melatonin. In *Biochemical actions of hormones* (ed. C. Litwack), pp 187–225. Academic

Press, New York.

Waldhauser, F. and Wurtman, R. J. (1983). The secretion and action of melatonin. In *Biochemical actions of hormones* (ed. C. Litwack), pp. 187–225. Academic Press, New York.

Waldhauser, F., Weißenbacher, G., Frisch, H., Zeitlhuber, U., Waldhauser, M., and Wurtman, R. J. (1984). Fall in nocturnal serum melatonin during prepuberty and pubescence. *Lancet* i, 362–5.

Waldhauser, F., Weißenbacher, G., Tatzer, E., Gisinger, B., Waldhauser, M., Schemper, M., and Frisch, H. (1988). Alterations in nocturnal serum melatonin levels in humans with growth and aging. *Journal of Clinical Endocrinology and Metabolism* 66, 648–52.

Waldhauser, F., Frisch, H., Krautgasser-Gasparotti, A., Schober, E., and Bieglmayer, C. (1986). Serum melatonin is not affected by glucocorticoid replacement in congenital adrenal hyperplasia. *Acta Endocrinologica* 111, 355–9.

Wetterberg, L. (1979). Clinical importance of melatonin. *Progress in Brain Research* 52, 539–47.

Wurtman, R. J., Axelrod, J., and Barchas, J. D. (1964). Age and enzyme activity in the human pineal gland. *Endocrinology* 24, 299–301.

Yellon, S. M., Tamarkin, L., and Goldman, B. D. (1985). Maturation of the pineal melatonin rhythm in long- and short-day reared Djungarian hamsters. *Experientia* 41, 651–2.

Young, I. M., Francis, P. L., Leone, A. M., Stovell, P., and Silman, R. E. (1988). Constant pineal output and increasing body mass account for declining melatonin levels during human growth and sexual maturation. *Journal of Pineal Research* 5, 71–85.

9. Melatonin and antidepressant drugs: clinical pharmacology

Stuart A. Checkley and E. Palazidou

Introduction

The human pineal is a convenient clinical model for investigating the mechanism of action of antidepressant drugs, particularly upon noradrenergic mechanisms. These have been studied in detail through both *in vivo* and *in vitro* experiments and a model for the noradrenergic control of melatonin secretion has been established in the case of experimental animals (Fig. 9.1). It will be seen that clinical data suggest that each of the components of this model apply to man. Consequently, the model can be used to investigate the net effect of antidepressant drugs upon noradrenergic neurotransmission in the human.

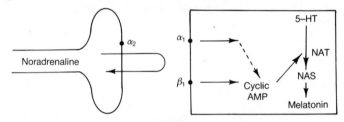

Fig. 9.1. Schematic diagram showing the role of noradrenalin, α_2-adrenoceptors, α_1-adrenoceptors and β_1-adrenoceptors in the regulation of melatonin synthesis. 5-HT=5-hydroxytryptophan, NAT=Nacetyltransferase, NAS=N acetylserotonin.

Basic and clinical pharmacology of melatonin in man

Influence of noradrenalin upon melatonin secretion in man

Considerable evidence from animal experiments suggests that the secretion of melatonin is dependent upon the noradrenergic innervation of the pineal (Klein 1978). Thus, the addition of noradrenalin to pineal cell culture *in vitro* increases the accumulation of cyclic AMP (Strada *et al.* 1972), the induction of N-acetyltransferase (NAT) (Deguchi 1973),

190

and the synthesis of melatonin (Axelrod *et al.* 1969). The neuro-endocrine regulation of melatonin secretion is reviewed in detail in Chapter 1.

The available clinical data are entirely consistent with these data from experimental animals. In patients with primary degeneration of the autonomic nervous system (Vaughan *et al.* 1979) and in patients with diabetic autonomic neuropathy (O'Brien *et al.* 1986) there is a failure of the normal nocturnal secretion of melatonin.

Conversely, the inhibition of noradrenalin uptake increases the secretion of melatonin in man (Checkley *et al.* 1985). The tricyclic antidepressant desipramine inhibits the uptake of noradrenlin 10–100 times more potently than the uptake of serotonin, and the drug oxaprotiline has been found to be an even more selective inhibitor of noradrenalin uptake. Furthermore, oxaprotiline has been resolved into two enantiomers of which only (+)-oxaprotiline inhibits noradrenalin uptake. In normal subjects we have shown that the secretion of melatonin is increased by desipramine and by (+)-oxaprotiline, but not by (−)-oxaprotiline (Fig. 9.2; Checkley *et al.* 1985).

The effect of desipramine on melatonin levels is likely to be due to an increase in secretion rather than a change in metabolism since the effects of desipramine upon the plasma concentrations of melatonin and 6-sulphatoxymelatonin were similar (Franey *et al.* 1986) in man. 6-sulphatoxymelatonin is the principal metabolite of melatonin and treatment with desipramine (100 mg orally at 09.00 hours) did not alter the urinary excretion of 6-sulphatoxymelatonin following the exogenous

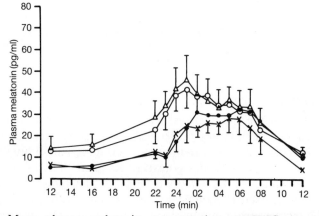

Fig. 9.2. Mean plasma melatonin concentrations (±SEM) in six healthy volunteers after treatment with 25 mg (+)-oxaprotiline t.d.s. (O—O) 25 mg (−)-oxaprotiline t.d.s., 50 mg desipramine t.d.s. (▲—▲), and placebo (●—●) for 24 hours each. [From Checkley *et al.* 1985, with permission.]

administration of 200 μg melatonin at 13.00 hours (Franey *et al.* 1986). Thus, in man as in animals, the secretion of melatonin is regulated by and dependent upon the integrity of the noradrenergic innervation of the human pineal (see Reiter, Chapter 1).

Influence of α_2-adrenoceptors on melatonin secretion in man

In both the peripheral and central nervous systems noradrenalin is thought to regulate its own release via negative feedback at pre-junctional α_2-adrenoceptors (Langer 1980). In the pineal also α_2-adrenoceptors regulate the release of noradrenalin (Pelayo *et al.* 1977).

α_2-adrenoceptors probably have a similar role in the human pineal as the α_2 agonist clonidine which inhibits the release of melatonin in man (Lewy *et al.* 1986), whereas the α_2 antagonist Org 3770 increases plasma melatonin concentrations (Fig. 9.3).

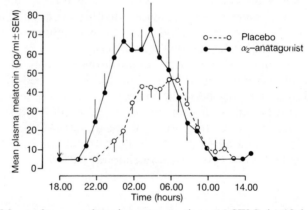

Fig. 9.3. Mean plasma melatonin concentrations ($\pm$ SEM) in 10 healthy male volunteers treated with α_2-antagonist (30 mg Org 3770) or placebo at 18.00 hours.

Influence of β_1-adrenoceptors

Binding sites for the β-antagonist L-alprenolol have been demonstrated on rat pineal cell membranes (Zatz *et al.* 1976). It is likely that l-alprenol binds to β_1- rather than β_2-adrenoceptors on these membranes since it is displaced by β_1- but not by β_2-agonists (Zatz *et al.* 1976). The β_1-agonist isoprenaline stimulates the activity of N-acetyltransferase (Brownstein *et al.* 1973) which itself is thought to regulate the synthesis of melatonin. The β-antagonist propranolol blocks the action both of isoprenaline (Brownstein *et al.* 1973) and of noradrenalin (Parfitt *et al.* 1976) on NAT.

In man the non-selective β-antagonist propranolol inhibits the nocturnal secretion of melatonin (Vaughan *et al.* 1976). The selective β_1-antagonist atenolol reduces plasma melatonin at midnight (Cowen *et al.* 1983*b*) and also the 24-hour urinary excretion of 6-sulphatoxy-melatonin (Arendt *et al.* 1985).

Thus, in man as in experimental animals, the melatonin-secreting pinealocytes have β_1-adrenoceptors whose activation is necessary for the synthesis of melatonin. In man, melatonin secretion cannot be increased by infusions of β-agonists (Lewy 1984*a,b*; Palazidou *et al.*, unpublished data), whereas in some species β-agonists do release melatonin (Lewy *et al.* 1980). This species difference may be due to the smaller doses of β-agonists which, for safety, must necessarily be given to human subjects.

Influence of cyclic AMP on melatonin in man

In the pineal as in other tissues the β-adrenoceptor linked to adenylate cyclase (Zatz *et al.* 1976). Cyclic AMP (cAMP) is thought to regulate the synthesis of melatonin in experimental animals since its derivative dibutyryl cyclic AMP increases pineal concentrations of melatonin (Berg and Klein 1972) and the activity of NAT (Klein *et al.* 1970) which is thought to regulate the synthesis of melatonin.

To determine whether cAMP also regulates melatonin production in man, normal volunteers have been treated with the phosphodiesterase inhibitor rolipram which increases the accumulation of cAMP in brain slices following the addition of noradrenalin (Schwabe *et al.* 1976). The effects of rolipram upon the urinary excretion of 6-sulphatoxymelatonin sulphate were then examined. The urinary excretion of 6-sulphatoxy-melatonin, as measured by RIA (Arendt *et al.* 1985), was significantly increased in the urine collected from midnight to 06.00 hours, but was significantly reduced in the collection from 06.00 hours to 12.00 hours (Fig. 9.4). A single dose of desipramine had a similar effect upon plasma melatonin which was increased during the first half of the night but reduced during the second half (Franey *et al.* 1986).

Influence of α_1-adrenoceptors on melatonin in man

Receptors with the characteristics of α_1-adrenoceptors have been demonstrated on rat (Sugden and Klein 1983) and ovine (Sugden *et al.* 1985) pineal cell membranes. The stimulation of these receptors with the α_1-agonist phenylephrine has little effect upon the activity of NAT (Vanecek *et al.* 1985). However, the combined stimulation of α_1-adreno-ceptors with phenylephrine and of β_1-adrenoceptors with isoprenaline results in greater changes in pineal cAMP and NAT activity (Vanecek *et al.* 1985) than are seen following the stimulation of α_1- or β_1-adreno-ceptors alone. Consequently, there is a functional synergism between the influences of α_1- and β_1-adrenoceptors upon the secretion of melatonin.

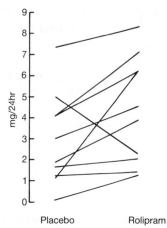

Fig. 9.4. The urinary excretion of sulphatoxy melatonin from 24.00 to 06.00 hours in 10 healthy male volunteers treated with a phosphodiesterase inhibitor (rolipram 1 mg t.d.s. for 24 hours) or placebo.

There are species differences in the relative influences of α_1- and β_1-adrenoceptors upon melatonin secretion. In the sheep the nocturnal secretion of melatonin has been reported to be inhibited by prazosin but not by propranolol (Sugden *et al.* 1985). On the other hand in the rat propranolol, but not prazosin, effects the activity of NAT (Klein 1978).

In man, α_1-adrenoceptors do influence the secretion of melatonin as the nocturnal secretion of melatonin is significantly reduced following treatment with prazosin (2 mg) at 19.00 hours (Fig. 9.5). Although this effect was statistically significant it was smaller than that of atenolol (Arendt *et al.* 1985) which completely abolishes the urinary excretion of 6-sulphatoxymelatonin. The cardiovascular effects of these drugs limit the dose–response studies that would be needed to compare the effects upon melatonin of blocking α_1- and β-adrenoceptors. In man both α_1- and β-adrenoceptors influence melatonin secretion, but the effect of β_1-adrenoceptors is dominant. Thus, it may be said in summary, that human melatonin secretion is:

(1) reduced by β_1 blockade;

(2) reduced by α_1 blockade;

(3) reduced by α_2 stimulation and increased by α_2 blockade;

(4) increased by noradrenalin uptake inhibition;

(5) increased by phosphodiesterase inhibition.

These data are consistent with the model (Fig. 9.1) which was developed on the basis of *in vivo* and *in vitro* experiments using the rat pineal (Klein 1985).

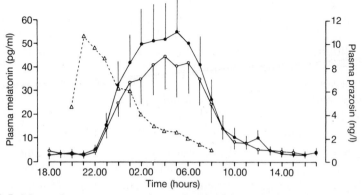

Fig. 9.5. Mean plasma concentrations of melatonin (± SEM) in 10 healthy male subjects treated with prazosin 1 mg (O—O) or placebo (●—●) at 16.00 hours. Mean plasma prazosin concentrations (△ – – – △) are also shown.

Effects of acute and chronic treatment with noradrenalin uptake inhibitors upon noradrenergic neurotransmission in the brains of experimental animals

Despite years of study it is still not known whether noradrenergic neurotransmission is increased or reduced following treatment with tricyclic antidepressant drugs. These drugs have different effects upon the different types of adrenoceptor present in the central nervous system and the net effect of these different influences upon noradrenergic neurotransmission is unresolved. However, the following observations have been recorded.

1. Acute treatment with tricyclic antidepressants increases intrasynaptic concentrations of noradrenalin as a result of inhibition of noradrenalin re-uptake (Iversen 1965); this effect will increase noradrenergic neurotransmission.

2. Some antidepressants such as desipramine block postsynaptic α_1-adrenoceptors following acute treatment (U'Pritchard *et al.* 1978); this effect will reduce noradrenergic neurotransmission.

3. Following *chronic* treatment, adaptive changes are seen at β_1-, α_1-, and α_2-adrenoceptors, with different influences upon noradrenergic neurotransmission.

4. Although a down-regulation at β-adrenoceptors will reduce noradrenergic neurotransmission, chronic treatment with antidepressant drugs results in other adaptive changes at adrenoceptors which will

tend to increase noradrenergic neurotransmission. Thus, chronic treatment with desipramine increases the functional responsiveness of postsynaptic α_1-adrenoceptors (Menkes *et al.* 1983). Under the same conditions there is a reduced functional status at presynaptic α_2-adrenoceptors with a resulting disinhibition of noradrenalin release (Svensson and Usdin 1978). Thirdly, the effect of tricyclic anti-depressants of noradrenalin re-uptake persists with chronic treatment and this effect together with the above changes in α_1- and α_2-adreno-ceptors will all tend to increase noradrenergic neurotransmission.

The most consistent changes occur at β_1-adrenoceptors. Almost every known antidepressant treatment (including tricyclic antidepressants, MAOIs, and ECT) decreases either the number of β-adrenoceptors or their coupling to the generation of the second messenger cAMP (Sulser 1984). This effect has been so consistent that Sulser has proposed that depression is due to an overactivity of noradrenergic neurotransmission and that it is the mechanism of action of chronic antidepressant drug treatment to reduce this overactivity (Sulser *et al.* 1978).

The pineal gland provides a convenient model for investigating the net effect of chronic treatment with antidepressant drugs upon noradrenergic neurotransmission. The secretion of melatonin is the physiological output of a typical noradrenergic sysem which, as shown in Fig. 9.1, includes α_1-, α_2-, and β_1-adrenoceptors. If noradrenergic neuro-transmission is increased by chronic antidepressant treatment then melatonin secretion will be increased, and vice versa. Such studies have been conducted in experimental animals, in normal human volunteers, and in depressed patients.

Effects of acute and chronic treatment with noradrenalin uptake inhibitors on melatonin secretion

Observations in experimental animals

Acute treatment with the relatively selective noradrenalin uptake inhibitor desipramine increases both pineal and plasma melatonin con-centrations (Parfitt and Klein 1977; Burns and Brown 1984). Three days of treatment with imipramine left pineal β-adrenoceptors unchanged, but caused a 15 per cent increase in pineal 5-HT, a 28 per cent increase in N-acetylserotonin, and a 32 per cent increase in pineal melatonin (Friedman *et al.* 1984). Thus, as expected, the *acute* effect of desi-pramine and imipramine is to increase noradrenergic neurotransmission in the rat pineal.

Chronic treatment with tricyclic antidepressants reduces all steps in the control of melatonin synthesis from the number and function of β-adrenoceptors, through the activity of the regulatory enzyme NAT, to the pineal concentrations of melatonin. The following observations may be listed.

1. Treatment with desipramine for 5, 7, and 10 days reduced the number of β-adrenoceptor binding sites in rat pineal (Moyer *et al.* 1979; Heydorn *et al.* 1982; Cowen *et al.* 1983*a*), and similar effects were seen after 21 days of treatment with imipramine (Heydorn *et al.* 1982).

2. After treatment with desipramine for 5, 7, and 10 days there was a reduced responsiveness of pineal β-adrenoreceptors as shown by an attenuated cAMP response to challenge with noradrenalin (Moyer *et al.* 1979) or isoprenaline (Heydorn *et al.* 1982; Cowen *et al.* 1983*a,b*).

3. Following treatment with imipramine for 21 days there was a 31 per cent reduction in the number of pineal β-adrenoceptors binding sites together with a 36 per cent reduction in NAT, a 25 per cent reduction in *N*-acetylserotonin and a 23 per cent reduction in pineal melatonin (Friedman *et al.* 1984). In this experiment, pineal 5-HT was unchanged and it would seem that a down-regulation at pineal β-adrenoceptors resulted in reduced noradrenergic neurotransmission and hence in a reduced synthesis of melatonin.

4. Similarly, chronic treatment with desipramine reduced nocturnal plasma and pineal concentrations of melatonin in one study (Heydorn *et al.* 1982). However, in another study the nocturnal secretion of melatonin was unchanged 10 hours after discontinuing chronic desipramine treatment, although 3 hours after stopping this treatment desipramine suppressed the daytime secretion of melatonin (Cowen *et al.* 1983*a,b*). The discrepancy between these last two reports is unexplained, but the consensus of all of these studies is that the secretion of melatonin is increased by *acute* treatment with tricyclic antidepressant drugs but is reduced by *chronic* treatment. In the case of the pineal the net effect of chronic antidepressant treatment is to reduce noradrenergic neurotransmission presumably as a consequence of the down-regulation of pineal β-adrenoceptors.

Observations in normal human volunteers

As has already been reviewed the secretion of melatonin in man is increased following treatment with desipramine and the highly selective noradrenalin uptake blocker (+)-oxaprotiline (Checkley *et al.* 1985). Furthermore, this effect is not due to a change in the metabolism of melatonin (Franey *et al.* 1986).

In contrast to the effects of acute treatment, chronic treatment for 7 and 21 days leaves plasma concentration unchanged (Thompson *et al.* 1985). Comparable data have been obtained by Cowen and associates (1985) who measured plasma melatonin at midnight after 0, 1, 4, 7, and 19 days of treatment with desipramine; there was a significant increase after 1 and 4 days of treatment, and then a reduction to almost baseline values after 19 days of treatment.

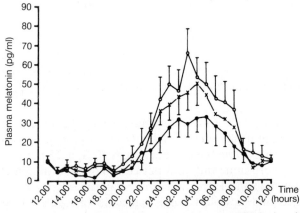

Fig. 9.6. Mean plasma melatonin concentrations (±SEM) in six depressed patients before treatment (●—●), and after 1 week (×—×) and 3 weeks (○—○) treatment with desipramine 2 mg/kg/day. [From Thompson *et al.* 1985, with permission.]

In normal volunteers the secretion of melatonin is increased following *acute* treatment with desipramine, but this effect is lost following *chronic* treatment. Presumably, the effects of adaptive changes at receptor level have cancelled out the functional effects of noradrenalin uptake inhibition. However, in contrast to the animal experiments, the adaptive changes do not reverse the primary drug effect.

Observations in depressed patients

Although the subject of an ongoing investigation no study has as yet updated the effect upon melatonin of a single dose of desipramine in depressed patients. However, after 7 and 21 days of treatment with desipramine there was a significant increase in plasma concentrations of melatonin in six depressed patients which is seen in Fig. 9.6 (Thompson *et al.* 1985). Similarly, in four depressed patients the urinary excretion of 6-sulphatoxymelatonin was increased following 21 days of treatment with desipramine (Sack and Lewy 1986).

From these preliminary data it would appear that desipramine has effects upon melatonin that differ between depressed patients and normal subjects. Chronic desipramine would appear to increase melatonin secretion in depressed patients, but not in normal volunteers. Although further studies are needed to determine whether or not depressed patients differ from normal subjects in the responsiveness of their pineals to antidepressant treatment several conclusions can be drawn from the above observations.

1. In depressed patients and in normal subjects desipramine increases the secretion of melatonin.

2. In all patients and volunteers investigated so far, chronic desipramine treatment has failed to reduce the secretion of melatonin as undoubtedly occurs in animals (Heydorn *et al.* 1982; Friedman *et al.* 1984).

It is therefore unlikely that desipramine reduces noradrenergic neuro-transmission as hypothesized by Sulser and colleagues (1978). Naturally, this conclusion only applies to the pineal which responds to desipramine in one way differently to the brain for, whereas chronic desipramine treatment reduces brain concentrations of noradrenalin, the opposite is seen in the pineal (Fuller and Perry 1977). Even though pineal pre-synaptic noradrenalin stores are increased by chronic treatment with desipramine in animals there is still a reduction in melatonin synthesis. In man no such reduction is seen. This is a further example from psycho-pharmacology of a crucial animal experiment not being applicable to man. Another example is the different neuroendocrine effects of electro-convulsive therapy (ECT) in animals and man (Checkley and Meldrum 1982); thus, in animals the growth hormone (GH) responses to clonidine and apomorphine are enhanced by ECT whereas in man they are not.

Effects of monoamine oxidase inhibitors (MAOIs) upon melatonin in animals and in man

Noradrenergic fibres in the pineal contain monamine oxidase (MAO) in the form MAO-A which preferentially metabolizes noradrenalin and serotonin. In contrast, pinealocytes contain MAO-B which has little effect on serotonin but which deaminates trace amines such as phenyl-ethylamine.

In monkeys chronic treatment with the MAO-B inhibitor pargyline had no effect upon plasma or CSF concentrations of melatonin although these were increased after treatment with the MAO-A inhibitor clorgyline (Garrick *et al.* 1985). Similar findings have also been reported in rodents (Oxenkrug *et al.* 1985). Increased daytime plasma melatonin concentrations were also reported in rats treated with the non-selective MAOI nialamide (Heydorn *et al.* 1982).

In depressed patients plasma melatonin concentrations were estimated between 20.00 and 08.30 hours before and after 3 weeks of treatment with the MAO-A inhibitor clorgyline, the MAO-B inhibitor pargyline, and the non-selective inhibitor tranylcypramine. Plasma melatonin was increased after treatment with clorgyline and tranylcypramine but not after treatment with pargyline (Murphy *et al.* 1986). These clinical and preclinical data show that the effect of MAOIs upon melatonin involves the inhibition of MAO-A presumably in noradrenergic fibres in the pineal. Whether this effect is mediated through the release of noradrenalin or serotonin is not known: the effect is blocked by 20 mg/kg propranolol (Heydorn *et al.* 1982) although such high doses of propranolol may have serotonin receptor blocking activity as well as β-blocking potency. It is likely that the effect of MAO-A inhibition is mediated through noradrenalin but further studies are needed to clarify this point.

Effects of 5-HT uptake inhibitors upon melatonin

Among the selective inhibitors of 5-HT uptake which are being investigated as potential antidepressants fluvoxamine has been reported to increase plasma concentrations of melatonin in man (Demisch *et al.* 1986). It is not known whether similar effects are seen in animals and produced by other selective inhibitors of 5-HT uptake. In this context it would also be necessary to investigate imipramine binding sites in pineal tissue since these are thought to be closely related to the 5-HT uptake mechanism (Langer *et al.* 1986).

Conclusion

The pineal gland is, in general, a good model for investigating drug effects particularly upon noradrenergic mechanisms. The role of serotonin is not certain and it is therefore wise to study selective noradrenalin uptake inhibitors such as desipramine, maprotiline, and oxaprotiline with which significant effects upon serotonin can probably be discounted. The pineal contains noradrenergic fibres which release noradrenalin under α_2-autoreceptor control. The melatonin secreting pinealocytes have α_1- and β_1-adrenoceptors which together influence melatonin secretion both in animals and in man. The clinical data are consistent with this model which was developed for the rat pineal (Klein 1985). In animals and in man acute inhibition of noradrenalin uptake increases melatonin secretion. In animals chronic antidepressant treatment down-regulates pineal β-adrenoceptors and it is presumably for this reason that melatonin secretion is reduced in animals treated chronically with tricyclic drugs. However, neither in normal human volunteers nor in depressed patients does desipramine reduce melatonin secretion. As desipramine is the drug which *par excellence* down-regulates β-adrenoceptors in animals these findings question the importance of down-regulation at β-adrenoceptors in the mechanism of action of antidepressant drugs.

Acknowledgements

This chapter includes unpublished data collected by Drs Franey and Palazidou for their respective Ph.D. theses which work was done in close collaboration with Dr Josephine Arendt.

References

Arendt, J., Bojkowski, C., Franey, C., Wright, J., and Marks, V. (1985). Immunoassay of 6-hydroxymelatonin sulphate in human plasma and urine: abolition of the 24 hour rhythm with atenolol. *Journal of Clinical Endocrinology and Metabolism* **60**, 1166–73.

Alphs, L. and Lovenberg, W. (1984). Modulation of rat pineal acetyl-Co A: arylamine N-acetyltransferase induction by α-adrenergic drugs. *Journal of Pharmacology and Experimental Therapeutics* **230**, 431–7.

Bittman, E. L. (1985). The role of rhythms in the response to melatonin. In *Photoperiodism melatonin and the pineal*, Ciba Foundation Symposium 117 (ed. R. L. Short), pp. 149–64. Pitman, London.

Axelrod, J., Shein, H. M., and Wurtman, R. J. (1969). Stimulation of ^{14}C melatonin synthesis from ^{14}C tryptophan by noradrenaline in rat pineal in organ culture. *Proceedings of the National Academy of Sciences USA* **69**, 2547–50.

Berg, G. R. and Klein, D. C. (1972). Norepinephrine increases the ^{32}P labelling of a phospholipid fraction of post-synaptic pineal membranes. *Journal of Neurochemistry* **19**, 2519–32.

Brownstein, U., Saavedra, J. U., and Axelrod, J. (1973). Control of pineal N-acetylserotonin by a β-adrenergic receptor. *Molecular Pharmacology* **9**, 605–11.

Burns, T. G. and Brown, G. M. (1984). The effects of acute and chronic desmethylimipramine treatment on pineal and serum melatonin and N-acetylserotonin. *Advances in the Biosciences* **53**, 25–30.

Checkley, S. A. and Meldrum, B. S. (1982). Studies on the mechanism of the antidepressant action of ECT. In *Neuropeptides: basic and clinical aspects* (ed. G. Fink and L. Whalley), pp. 216–26. Churchill Livingstone, Edinburgh.

Checkley, S. A., Thompson, C., Burton, S., Franey, C., and Arendt, J. (1985). Clinical studies of the effect of (+) and (−) oxaprotiline on noradrenaline uptake. *Psychopharmacology* **87**, 116–18.

Cowen, P. J., Fraser, S., Grahame-Smith, D. G., Green, A. R., and Stanford, C. (1983a). The effect of chronic antidepressant administration on β-adrenoceptor functions of the rat pineal. *British Journal of Pharmacology* **78**, 89–90.

Cowen, P. J., Fraser, S., Sammons, R., and Green, A. R. (1983*b*). Atenolol reduces plasma melatonin concentrations in man. *British Journal of Clinical Pharmacology* **15**, 579–81.

Cowen, P. J., Green, A. R., Graham-Smith, D. G., and Braddock, L. E. (1985). Plasma melatonin during desmethylimipramine treatment: evidence for changes in noradrenergic transmission. *British Journal of Clinical Pharmacology* **19**, 799–805.

Deguchi, T. (1973). Role of the β-adrenergic receptor in elevation of adenosine cyclic 3′, 5′-monophosphate and induction of serotonin N-acetyltransferase in rat pineal glands. *Molecular Pharmacology* **9**, 184.

Demisch, K., *et al.* (1986). Melatonin and cortisol increase after fluvoxamine. *British Journal of Clinical Pharmacology* **22**, 620–2.

Franey, C., Aldhous, M., Burton, S., Checkley, S. A., and Arendt, J. (1986). Acute treatment with desipramine stimulates melatonin and 6-sulphatoxymelatonin secretion in man. *British Journal of Clinical Pharmacology* **22**, 73–9.

Friedman, E., Yocca, F. D., and Cooper, T. B. (1984). Antidepressant drugs with varying pharmacological profiles alter pineal β-adrenergic mediated function. *Journal of Pharmacology and Experimental Therapeutics* **228**, 545–9.

Fuller, R. W. and Perry, K. W. (1977). Increase in pineal noradrenaline in rats by desipramine but not fluoxetine: implications concerning the specificity of these uptake inhibitors. *Journal of Pharmacy and Pharmacology* **29**, 710–11.

Garrick, N. A., Tamarkin, L., and Murphy, D. L. (1985). Marked enhancement of the nocturnal elevation of melatonin in rhesus monkeys by inhibitors of monoamine oxidase (MAO). *Pharmacologist* **27**, 196–9.

Heydorn, W. E., Brunswick, D. J., and Frazer, A. (1982). Effect of treatment of rats with antidepressants on melatonin concentration in the pineal gland. *Journal of Pharmacology and Experimental Therapeutics* **222**, 534–43.

Iversen, I. L. (1965). *The uptake and storage of noradrenaline in sympathetic nerves.* Cambridge University Press.

Jones, R. L., McGreer, P. L., and Greiner, A. C. (1969). Metabolism of exogenous melatonin in schizophrenic and non-schizophrenic volunteers. *Clinica Chimica Acta* **26**, 281–9.

Klein, D. C. (1978). The pineal gland: a model of neuro-endocrine regulation. In *The hypothalamus* (ed. S. Reichlin, R. J. Baldessarini, and J. B. Martin), pp. 303–27. Raven Press, New York.

Klein, D. C. (1985). Photoneural regulation of the mammalian pineal gland. In *Photoperiod melatonin and the pineal,* Ciba Foundation Symposium 117, pp. 38–51. Pitman, London.

Klein, D. C., Berg, G. R., and Weller, J. (1970). Melatonin synthesis: adenosine 3′, 5′ monophosphate and norepinephrine stimulate N-acetyltransferase. *Science* **168**, 979–80.

Langer, S. Z. (1980). Presynaptic regulation of the release of catecholamines. *Pharmacology Reviews* **32**, 337–62.

Langer, S. Z., Galzin, A. M., Lee, A. R., and Schoemaker, A. (1986). Anti-depressant binding sites in brain and platelets. In *Antidepressants and receptor function* (ed. D. Murphy), pp. 3–16. John Wiley and Sons, Chichester.

Lewy, A. J. (1984*a*). Human melatonin secretion: a marker for adrenergic function. In *Neurobiology of mood disorders* (ed. R. M. Post and J. C. Ballenger), pp. 207–14. Williams and Wilkins, Baltimore, Maryland.

Lewy, A. J. (1984*b*). Human melatonin secretion: a marker for the circadian system and the effects of light. In *Neurobiology of mood disorders* (ed. R. M. Post and J. C. Ballenger), pp. 215–26. Williams and Wilkins, Baltimore, Maryland.

Lewy, A. J., Tetsuo, M., Markey, S. P., Goodwin, F. K., and Kopin, I. J. (1980). Pinealectomy abolishes plasma melatonin in the rat. *Journal of Clinical Endocrinology and Metabolism* **50**, 204–5.

Lewy, A. J., Siever, L. J., Uhde, T. W., and Markey, S. P. (1986). Clonidine reduces plasma melatonin levels. *Journal of Pharmacy and Pharmacology* **38**, 555–6.

Menkes, D. B., Aghajanian, G. K., and Gallagher, D. W. (1983). Chronic antidepressant treatment enhances agonist affinity of brain alpha$_1$ adrenoceptors. *European Journal of Pharmacology* **87**, 35–41.

Moyer, J. A., Greenberg, L. H., Frazer, A., and Weiss, B. (1981). Subsensitivity of the β-adrenergic receptor-linked adenylate cyclase system of rat pineal gland following repeated treatment with desmethylimipramine and nialamide. *Biochemical Pharmacology* **19**, 187–93.

Murphy, D. L., Tamarkin, L., Sunderland, T., Garrick, N. A., and Cohen, R. M. (1986). Human plasma melatonin is elevated during treatment with monoamine oxidase inhibitors clorgyline and tranylcypromine but not deprenyl. *Psychiatry Research* **17**, 119–27.

O'Brien, I. A. D., Lewin, I. G., O'Hare, J. P., Arendt, J., and Corrall, R. J. M. (1986). Abnormal circadian Rhythm of melatonin in diabetic autonomic neuropathy. *Clinical Endocrinology* **24**, 359–64.

Oxenkrug, G. F., McCauley, R., McIntyre, I. M., and Filipowicz, C. (1985). Selective inhibition of MAO-A but not MAO-B activity increases rat pineal melatonin. *Journal of Neural Transmission* **61**, 265.

Parfitt, A. and Klein, D. C. (1977). Desmethylimipramine causes an increase in the production of ^{3}H-melatonin by isolated pineal glands. *Biochemical Pharmacology* **26**, 906–7.

Parfitt, A., Weller, J. L., and Klein, D. C. (1976). β-adrenergic blockers decrease adrenergically stimulated N-acetyltransferase activity in pineal glands in organ culture. *Neuropharmacology* **15**, 353–8.

Pelayo, F., Dubocovich, M. L., and Langer, S. Z. (1977). Regulation of noradrenaline release in the rat pineal gland through a negative feedback mechanism mediated by pre-synaptic α-adrenoceptors. *European Journal of Pharmacology* **45**, 317–18.

Sack, R. L. and Lewy, A. J. (1986). Desmethylimipramine treatment increases melatonin production in humans. *Biological Psychiatry* **21**, 406–9.

Schwabe, U., Miyake, U., Ohga, Y., and Daly, J. W. (1976). 4-(3-Cyclopentyloxy-4-methoxy)-2 pyrrolidone (2K 62711): a potent inhibitor of adenosine cyclic 3′, 5′ monophosphate phosphodiesterases in homogenates and tissue slices from rat brain. *Molecular Pharmacology* **12**, 900–10.

Strada, S., Klein, D. C., Weller, J. L., and Weis, B. (1972). Norepinephrine stimulation of cyclic adenosine monophosphate in cultured pineal gland. *Endocrinology* **90**, 1470–5.

Sugden, D. and Klein, D. C. (1983). Rat pineal α_1-adrenoceptors: identification and characterization using (^{125}I)iodo-2-[B-(4-hydroxyphenyl)-ethylamino-methyl]tetralone (^{125}I-HEAT). *Endocrinology* **114**, 435–40.

Sugden, D., Namboodiri, M. A. A., Klein, D. C., Pierce, J. E., Grady, R. K., Jr, and Mefford, I. (1985). Ovine pineal α_1-adrenoceptors: characterization and evidence for a functional role in the regulation of serum melatonin. *Endocrinology* **116**, 1960–7.

Sulser, F. (1984). Regulation and function of noradrenaline receptor systems in brain. Psychopharmacological aspects. *Neuropharmacology* **23**, 255–61.

Sulser, F., Vetulani, J., and Mobley, P. L. (1978). Mode of action of antidepressant drugs. *Biochemical Pharmacology* **27**, 257–61.

Svensson, T. H. and Usdin, T. (1978). Feedback inhibition of brain noradrenaline neurones by tricyclic antidepressants: α-receptor mediation. *Science* **202**, 1089–91.

Thompson, C., *et al.* (1985). The effect of desipramine upon melatonin and cortisol secretion in depressed and normal subjects. *British Journal of Psychiatry* **147**, 389–93.

U'Pritchard, D. C., Greenberg, D. A., Sheehan, P. P., and Snyder, S. H. (1978). Tricyclic antidepressants: therapeutic properties and affinity for α-adrenergic binding sites in the brain. *Science* **199**, 197–8.

Vanecek, J., Sugden, D., Weller, J., and Klein, D. C. (1985). Atypical synergistic α_1- and β_1-adrenergic regulation of adenosine 3′, 5′-monophosphate in cultured rat pinealocytes. *Endorcrinology* **116**, 2167–73.

Vaughan, G. M., *et al.* (1976). Nocturnal elevation of plasma melatonin and urinary 5-hydroxyindoleacetic acid in young men: attempts at modification by brief changes in environmental lighting and sleep and by autonomic drugs. *Journal of Clinical Endocrinology and Metabolism* **42**, 752–64.

Vaughan, G. M., McDonald, S. D., Jordan, R. M., Allen, J. P., Bell, R., and Stevens, E. A. (1979). Melatonin pituitary function and stress in humans. *Psychoneuroendocrinology* **4**, 351–62.

Wurtman, R. J. and Ozaki, Y. (1978). Physiological control of melatonin synthesis and secretion mechanisms generating rhythms in melatonin methoxytryptophol and arginine vasopressin levels, and effects on the pineal of endogeneous catecholamines, the oestrus cycle and environmental lighting. *Journal of Neural Transmission* (Suppl.) **13**, 59–64.

Zatz, M., Kebabian, J., Romero, J. A., Lefkowitz, R. J., and Axelrod, J. (1976). Pineal β-adrenergic receptor: correlation of binding of l-alprenolol with stimulation of adenylate cyclase. *Journal of Pharmacology and Experimental Therapeutics* **196**, 714–22.

10. Melatonin and major affective disorders

Robert L. Sack and Alfred J. Lewy

Introduction

Affective disorders are characterized by a persistent and pervasive alteration in mood (either depressed or elevated) associated with a constellation of symptoms (e.g. sleep and/or appetite disturbance, change in weight) that comprise a depressive or manic syndrome (American Psychiatric Association 1987). Some patients alternate between depression and elation (bipolar depressive disorder) while others have mood swings in one direction only (unipolar depression). The intensity of mood alteration can range from mild dysphoria (or elation) to severe psychosis. Affective disorders are quite common; the lifetime risk is 18 per cent for major depressive disorder and 1.2 per cent for bipolar disorder (Weissman and Meyers 1978).

Because of the varied manifestations of affective disorders, there has been an ongoing attempt to classify patients into meaningful subtypes. Patients have been classified as bipolar or unipolar, based on the presence or absence of mania or hypomania (Leonhard 1957); primary or secondary, depending on whether other diagnoses are present, e.g. alcoholism; typical or 'atypical', depending on the pattern of associated symptoms (West and Dally 1959); endogenous or reactive, depending on the apparent impact of precipitating stressors; and with or without melancholia (depression associated with prominent physiological symptoms).

'Winter depression', officially termed Major Depressive Disorder, Seasonal Pattern (American Psychiatric Association 1987), [also called 'Seasonal Affective Disorder' (Rosenthal *et al.* 1984)] is a new subtype of major depressive disorder. Patients with winter depression become depressed in the fall and recover in the spring. Associated symptoms are low energy, morning hypersomnia, and weight gain (see Thompson, Chapter 11). We and others have found that most winter depressive patients are unipolar (without manic or hypomanic symptoms), but other investigators (Rosenthal *et al.* 1984) have reported that many of these patients have bursts of hypomania and have thereby classified them as 'bipolar II' (Dunner *et al.* 1982). The issue of bipolarity in winter depression may well be resolved as greater consensus is developed regarding the definition of hypomania. Our group requires that an episode of elevated mood have an element of psychopathology, e.g. destructiveness or subsequent feeling of regret.

Because of the limitations of diagnosis based on phenomenology alone, great effort has been made to find biological 'markers', e.g. the dexamethasone suppression test (Carrol *et al.* 1981) or REM sleep latency test (Gillin *et al.* 1979; Kupfur 1976) that are distinctive signs of major affective disorder or of a particular diagnostic subtype. A biological marker does not have to be aetiologically related to an illness. Abnormalities present in both the symptomatic and asymptomatic phase of an illness are considered 'trait markers', while abnormalities present only during the symptomatic phase are considered 'state markers'.

As part of this chapter we will discuss four research strategies relating melatonin to major affective disorders, emphasizing the work carried out in our laboratory:

(1) measurement of the timing of melatonin production to assess circadian rhythm abnormalities in affective disorder;

(2) measurement of the suppression of melatonin production by light exposure to assess light sensitivity of the retinohypothalamic system;

(3) measurement of 24-hour melatonin production in depression to investigate 'melatonin deficiency' in depression;

(4) measurement of the effects of pharmacological challenge on melatonin production, using the pineal gland as an *in vivo* model of adrenergic synaptic function in humans.

For a view of the diagnostic significance of melatonin estimation in psychiatry, the reader is referred to Miles *et al.* Chapter 13.

The timing of melatonin production as a marker for circadian phase

Evidence for circadian rhythm abnormalities in affective disorder

Symptoms suggesting circadian rhythm disturbances are frequently seen in depression and mania, particularly in the sleep–wake cycle (Papousek 1975; Kripke *et al.* 1978; Wehr and Goodwin 1981, 1983). Melancholic depressed patients typically awaken before morning, while patients with 'atypical' and winter depression are very sleepy in the morning. Manic patients have a greatly decreased need for sleep. Depressive mood may have a diurnal variation, more intense in the morning than in the evening. Depression and mania may also manifest cyclic patterns of recurrence measured in weeks or months; as mentioned above, winter depression recurs each winter and remits each spring (Rosenthal *et al.* 1984; Thompson, Chapter 11). In animals, seasonal rhythms often result from circadian rhythms that interact with seasonal changes in daylength and darkness (Stetson *et al.* 1986). Patients with rapid-cycling bipolar illness may alternate between depression and mania every few weeks to months (Zis and Goodwin 1979). These longer cycles could be generated by two circadian oscillators that beat in and out of phase with each other (Halberg 1968).

While the rhythmic nature of mood symptoms has long been noted, this aspect of affective disorders has taken on increased significance in light of the growing understanding of circadian rhythm physiology. Because of the groundwork laid by basic animal and human studies, specific mechanisms can now be hypothesized to explain many of the cyclical manifestations of affective disorders. These hypotheses may be tested by measuring circadian rhythm markers in patients with mood disorders and/or treating patients with measures designed to correct hypothesized abnormalities and then assessing the impact on clinical symptomatology.

In our laboratory, we use these two strategies concurrently. The timing of melatonin production is employed to evaluate patients with affective (and sleep) disorders for possible rhythm abnormalities. Patients can be phase typed as phase advanced, phase delayed, or without phase abnormality (Lewy *et al.* 1984, 1985*b,c*). Patients may then be treated with appropriately timed bright light exposure to correct the phase abnormality. Light-induced phase shifts can be verified by repeat measurements of melatonin onset, with correlation of baseline and post-treatment circadian phase position with clinical symptomatology (Lewy *et al.* 1983, 1985*a,c*, 1987). Manipulation of the phase relationship between the melatonin rhythm and sleep (the circadian 'phase angle') can also be achieved by shifting the melatonin rhythm with appropriately timed, bright light exposure while holding sleep hours constant; alternatively sleep can be shifted while holding the light–dark cycle (and thus the melatonin rhythm) constant.

Using plasma melatonin to assess circadian phase position

The timing of melatonin production may be measured with a sensitive and specific gas chromatographic–negative chemical ionization mass spectrometric (GC–MS) assay for plasma melatonin (Lewy and Markey 1978), and several investigators, including ourselves, now believe melatonin secretion to be a very accurate indicator of circadian timing because it is relatively unaffected by stress, sleep, and other potential masking influences. Light exposure suppresses melatonin and can mask the melatonin rhythm but this problem is readily controlled by keeping patients in dim light conditions when drawing blood samples (Lewy *et al.* 1980). Other indicators of circadian phase (e.g. core body temperature, cortisol, sleep time) generate complex problems of interpretation because of masking effects. Animal experiments indicate that melatonin, core temperature, and cortisol rhythms are all generated within the suprachiasmatic nucleus (SCN) (Moore and Eichler 1972; Stephan and Zucker 1972; Inouye and Kawamura 1982).

In selecting a portion of the melatonin curve to use as an indicator of circadian phase, we have focused on the onset of production (normally occurring about 21.00 hours) since this surge in melatonin production is a very distinct event that is relatively unaffected by changes in substrate availability or variations in receptor sensitivity that might affect peak levels or the morning decline in production. Thus, for much of our clinical work, we measure the evening onset of melatonin production in dim light conditions. This is termed the 'dim light melatonin onset' or 'DLMO' (Lewy *et al.* 1985*b*). The DLMO is very convenient for research subjects since they can come to the clinic as out-patients in the evening and continue their daily routines without much interruption. Blood is typically sampled from 18.00 to 23.00 hours (4–5 ml every 30 minutes) through a plastic catheter which is kept patent with a dilute solution of heparin. The DLMO requires an assay which can reliably measure the transition from daytime (below 5 pg) to nocturnal plasma melatonin levels and so may not be appropriate for some of the radio-immunoassay (RIA) methods.

Circadian rhythms and winter depression

The discovery of bright light suppression of melatonin in humans (Lewy *et al.* 1980) suggested that humans might be sensitive to the solar light–dark cycle, even though the social and work day were extended with interior illumination; furthermore, it suggested that the biological effect of a summer photoperiod could be mimicked by extending the daylength with bright artificial light. In the winter of 1980, a patient who presented to the NIMH clinical centre with a history of regularly occurring depressions in the winter, hypomania in the spring and euthymia in the summer was treated by full-spectrum bright-light exposure in the morning (06.00–09.00 hours) and evening (16.00–19.00 hours) (Lewy *et al.* 1982). This treatment resulted in a prompt remission of his depression and stimulated further interest in bright light as a therapeutic modality.

The next light therapy study in the winter of 1981 involved a larger number of winter depressives and was designed to evaluate the possibility of a placebo effect. Using a randomized, cross-over design, Rosenthal and co-workers (1984) showed that bright white light was significantly more effective for winter depression than dim yellow light (both scheduled for 3 hours before dawn and 3 hours after dusk). No markers for circadian rhythms were measured in this study.

In 1983, we began to further evaluate the effects of light exposure in humans using the GC–MS melatonin assay as a marker for circadian phase position. In four normal volunteers studied for two weeks in the summer, dusk was advanced to 19.00 hours for the first week; then dawn was delayed to 09.00 hours the second week (Lewy *et al.* 1983). We found that exposing normal subjects to these long nights (simulating winter conditions) caused an initial unmasking of the underlying melatonin rhythm that was followed by apparent phase shifts. The results suggested that not only were humans similar to other animals regarding the suppressive effects of light, but that human circadian rhythms could be shifted by exposure to differing light schedules even if sleep and social cues were unchanged.

During the winter of 1984–85 we compared evening with morning light exposure in the treatment of winter depression for several reasons (Lewy *et al.* 1987). There was a question of patient convenience, i.e. did patients need to be under bright light both morning and evening as in earlier studies (6 hours total) or would light exposure at one time of the day be sufficient? However, there was also a more basic question: would the timing of light exposure produce differential effects as predicted by a phase response curve (PRC)? Such curves have been of fundamental importance in developing animal models of circadian rhythm physiology (DeCoursey 1960; Pittendrigh and Daan 1976). By exposing animals to single pulses of light (or other cues) in an otherwise constant dim light environment, lawful relationships have been described for the influence of environmental cues on biological rhythms. Light pulses near the animals subjective 'dawn' advance rhythms (cause them to occur earlier) while light pulses near around the time of subjective 'dusk' delay rhythms (cause them to occur later). Light pulses during the subjective day, have relatively little phase shifting effects. Hypothesizing a phase response curve for humans, we predicted that light in the evening would delay the DLMO and light in the morning would advance the DLMO. We were interested to see if these two times of light exposure would have different effects on winter depression symptoms.

Eight winter depressive patients and seven control subjects were enrolled in the winter months. We started each subject with an adaptation 'baseline' week of overnight dim light exposure (light intensity under 250 lux from 17.00 until 08.00 hours the following day) and a structured sleep schedule (22.00 to 06.00 hours). This lighting and sleep schedule was maintained through the duration of the study except that 2-hour 'pulses' of light exposure were introduced during the subsequent treatment weeks. In the second week, each subject was randomly assigned to either 2 hours of morning (06.00–08.00 hours) or evening (20.00–22.00 hours) light (2500 lux Vita-Lite). In the third week they were crossed over to the alternative schedule. In the fourth week, they received both morning and evening light exposure. Since each subject was exposed to the same total amount of light in weeks 2 and 3, the 'placebo' or non-specific effects of light exposure could be differentiated from phase shifting effects (which were in opposite directions). If total light exposure rather than timing were important, then the last week should produce the best clinical response. In summary, with this design, we hoped to see if the therapeutic benefits of light involved:

(1) an effect on circadian rhythms (correcting a phase abnormality);

(2) photoperiodic effects (related to lengthening of the daylight hours);

(3) effects unrelated to alterations in biological rhythms, i.e. were non-specific or placebo effects.

We found that the melatonin onset (DLMO) on the evening the study began was almost 2 hours later in the patients with winter depression than in the controls. Furthermore, morning light exposure advanced the melatonin onset (caused it to occur earlier) while evening light exposure delayed the melatonin onset (caused it to occur later); morning and evening light together caused the melatonin onset to occur at an intermediate time. Interestingly, winter depression patients showed a greater advance to morning light than controls, suggesting that patients had phase delayed PRCs as well. Antidepressant response to morning light treatment was superior to evening light; morning and evening light together (week 4) were less effective than morning light alone. This study provided evidence that winter depression was associated with a phase delay in circadian rhythms; treatment with morning light caused an advance (corrected the delay) and alleviated the symptoms (Fig. 10.1).

In the winter of 1985–86, we studied the duration of morning light exposure on clinical response and on melatonin onset in 14 patients with winter depression and six controls. Thirty minutes of morning light (06.00–06.30 hours) was compared to 2 hours (06.00–08.00 hours) of morning light. Sleep times and other light exposure were controlled in the same way as in the previous study. However, between each treatment week, a 'dim light' week was inserted. For most patients, 30 minutes of light exposure in the morning resulted in as much antidepressant response as 2 hours of exposure. Two hours of exposure caused greater phase advances in the melatonin onset and there was a linear relationship

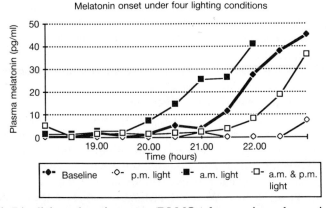

Fig. 10.1. Dim light melatonin onsets (DLMOs) from a winter depression patient. Each DLMO was measured at the end of a week of prescribed light exposure. AM light caused the DLMO to occur earlier while PM light caused the DLMO to occur later. AM and PM light together caused the DLMO to occur at an intermediate time, similar to the baseline condition.

between the degree of phase advance and clinical response. There were two patients that responded preferentially to 30 minutes, and four who responded preferentially to 2 hours. It may be that 2 hours of light advances some people too far. As in the previous study, winter depression patients show a greater advance to morning light than controls, suggesting that their PRCs are also phase delayed. Thus, we have begun to consider the importance, not only of the timing of melatonin production on a given night, but the phase shift in response to light (a test of PRC dynamics).

In 1987 we essentially replicated the study of 1984–85 on a group of 10 winter depressives and six control subjects, with the addition of a DLMO determination in the middle of each treatment week (Sack *et al.*, in preparation). Again the pre-baseline and baseline DLMOs were delayed in the depressed patients compared to the controls, and again, clinical response to morning light treatment was significantly better than to evening light. Morning light shifted the DLMO significantly earlier in the patients but had little effect on the controls. Evening light shifted the DLMO later in controls, suggesting that the PRC was delayed in the patients compared to the controls. Mid-week DLMOs (after 2 days of light exposure) were significantly shifted demonstrating that the phase-shifting effects of bright light occur within a few days after treatment has begun.

In summary, our current understanding of winter depression is that it is related to circadian rhythms that are abnormally phase delayed with respect to clock time and probably to sleep time as well. Morning light exposure corrects both the absolute phase disturbance as well as the phase angle relationship between the melatonin rhythm and sleep, since sleep is held constant.

Light therapy has become a very active area of clinical research and a number of other groups have conducted studies of light therapy in winter depression that are reviewed elsewhere in this volume (see Thompson, Chapter 11; Miles *et al.*, Chapter 13). There is a general consensus that light therapy is effective in winter depression, but the mechanism remains controversial. The Bethesda group has maintained that light treatment does not involve circadian mechanisms, but rather is mediated by some as yet undefined biological process (James *et al.* 1985; Rosenthal *et al.* 1985; Wehr *et al.* 1986). We have critically reviewed these and other studies elsewhere and feel that most of the evidence favours a mechanism involving circadian rhythms but, as in all other treatments in psychiatry, there are non-specific therapeutic factors as well, including a placebo effect and possibly an energizing effect from bright light exposure.

Phase delay sleep disorders

We have recently begun to focus our research strategies on the 'delayed sleep phase syndrome' (DSPS) (Weitzman *et al.* 1981) using our melatonin assay for evaluation and light exposure for intervention. Patients with DSPS have persistent difficulty falling asleep at the usual bedtime (e.g. before midnight) and subsequent difficulty awakening in the morning, i.e. their sleep time is delayed with respect to conventional norms. Patients with DSPS are of particular interest because they have a phase delay disorder similar to our winter depression subjects, but they are not depressed. Polysomnographic studies are usually unremarkable if the patient sleeps on his usual schedule; furthermore, subjective sleep complaints are minimal if the individual's social requirements do not involve being awake at the usual times (Wirz-Justice and Pringle 1987). Most patients have tried sedative-hypnotic drugs and many have tried extreme measures to awaken themselves in the morning (e.g. several alarm clocks, cold showers, etc.), but to little avail. Systematically delaying sleep by several hours each day until the patient is synchronized to a desired schedule ('chronotherapy') (Czeisler *et al.* 1981, 1986) has been employed with some success, but relapse is common if patients stay up late, even for a single night. A few cases of phase-advanced sleep disorder have also been described (Moldofsky *et al.* 1986).

We have treated a few patients with delayed sleep phase syndrome using bright light exposure in the morning. In pilot studies we have found that the melatonin onset prior to treatment is delayed (Singer, C. M. *et al.*, unpublished data). In several cases, bright light exposure in the morning has been successful in advancing sleep onset as well as melatonin onset. Light exposure may have to be very early (e.g. 05.00 hours) to be effective, suggesting a subsensitive phase response to morning light.

Phase angle disturbance between sleep and melatonin rhythms

Although both winter depressives and DSPS patients appear to be phase delayed, it may be that winter depressives have sleep that is relatively advanced with reference to the melatonin rhythm, while patients with DSPS have sleep that is either in normal phase relationship or delayed with respect to the melatonin rhythm, albeit in an abnormal phase relative to clock time. We have suggested that a phase angle abnormality may be depressogenic and have some preliminary evidence to support this hypothesis. In a 3-week protocol involving three subjects with winter depression, we delayed dawn by 5 hours for a week (holding sleep constant) and observed a worsening of depression (Sack, R. L. *et al.*, unpublished data). In the second week, we gradually delayed sleep until the patients were awakening to the light exposure; this resulted in improvement in mood. In the third week we gradually shifted sleep and light exposure back to the pre-treatment time; this resulted in continued clinical improvement.

Free-running melatonin rhythms in blind people: a model for testing hypotheses of circadian rhythm phase abnormalities and chronobiological effects of drugs

If light is an important time cue for human circadian rhythm regulation, then what happens to circadian rhythms in people who cannot perceive light? Initial studies of circadian rhythms in blind people found evidence for 'blunting' or decreased amplitude in core temperature, cortisol, and sleep rhythms (Migeon *et al.* 1956; Krieger and Rizzo 1971; Hollwich and Dieckhues 1972; D'Allesandro *et al.* 1974). In retrospect these findings probably resulted from averaging rhythms that were out of phase with each other. Miles *et al.* (1977) and then Orth *et al.* (1979) reported (in single case studies of totally blind subjects) that the cortisol rhythm was free-running with a period of about 24.5 hours, similar to the period of humans living in temporal isolation. Lewy and Newsome (1983), using the newly developed GC–MS assay, found that seven out of 10 blind subjects had abnormally timed melatonin production when evaluated on a single day. When studied longitudinally, one subject was found to have a free-running melatonin rhythm; another subject had a melatonin rhythm that was entrained but at an abnormal phase position. We have subsequently studied seven newly recruited totally blind subjects and have found a very consistent and stable free-running melatonin rhythm in four of the seven (Sack *et al.* 1987).

The period of the melatonin rhythm in the free-running blind subjects was remarkably precise. We were able to predict the phase of melatonin secretion (to within ±60 minutes) 2–4 weeks in advance of actual measurement. In two subjects, we had the opportunity to repeat the measurement of the circadian period (tau) of the free-running melatonin rhythm at 1 year (subject 1) and 2 years (subject 2) after the initial assessment. Tau was unchanged after this long time interval; thus, the circadian period appears to be a very stable individual trait (Fig. 10.2).

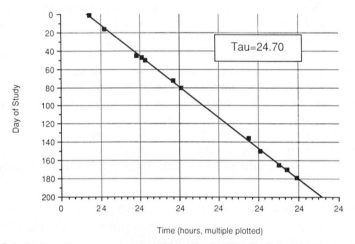

Tau=24.70

Day of Study

Time (hours, multiple plotted)

Fig. 10.2. Melatonin onsets measured in a totally blind subject for 6 months. The melatonin rhythm was very stable and precise.

One of the blind subjects kept a daily diary of sleep times, sleep quality, daytime sleepiness, mood, anxiety, and energy. He was attempting to maintain a conventional 24-hour schedule; consequently, his free-running melatonin rhythm (tau = 25.1 hours) was gradually beating in and out of phase with his desired sleep time. When his melatonin rhythm was maximally out of phase with his desired sleep time, he slept poorly at night, was very sleepy during the day, and felt depressed, tired, and anxious. When his melatonin rhythm came into synchrony with his desired sleep time, his sleep quality greatly improved and mood symptoms diminished. We eventually prescribed a sleep schedule that was synchronized to his melatonin rhythm (going to bed 1.1 hours later each day, 3 hours after his predicted melatonin onset), and his sleep quality and general well-being greatly improved.

It is of interest that the free-running melatonin rhythm was uninfluenced by sleep phase in these blind subjects; there was not even evidence of relative co-ordination. This finding is consistent with the one-oscillator model proposed by Daan and colleagues (1984), but not consistent with the predictions of a two-oscillator model of circadian rhythm generation as proposed by Kronauer and co-workers (1982).

It has been suggested that mood-active drugs exert their effect by influencing circadian rhythms. Free-running blind people can provide a model for testing the effects of drugs on the intrinsic circadian rhythm. For example, in a preliminary trial, lithium (900 mg at bedtime for 4 weeks) was found to lack any influence on the free-running melatonin rhythm in one blind subject (Sack *et al.*, unpublished). This is somewhat surprising since most animal studies have found that lithium slows circadian rhythms (Kripke and Wyborney 1980; McEachron *et al.* 1981, 1985).

We have also treated three subjects with exogenous melatonin administration in an attempt to entrain the free-running melatonin rhythm. Melatonin (5 mg orally) was administered at bedtime for 3 weeks; treatment was begun a few days before the individual's endogenous melatonin rhythm was approaching normal phase. Melatonin caused an 8-hour phase advance in one subject and a 12-hour advance in a second; in the third subject there was no phase shift. More subjects are being studied, but the preliminary evidence is consistent with the phase advance in endogenous melatonin secretion and fatigue rhythm observed by Arendt and co-workers in several sighted volunteers (Arendt *et al.* 1984; Wright *et al.* 1986). Melatonin has been shown to alleviate symptoms of jet-lag (Arendt *et al.* 1986). Furthermore, melatonin can entrain circadian rhythms in free-running rats (Redman *et al.* 1983) and produce phase shifts in lizards (Underwood 1986), both effects consistent with a PRC for melatonin administration. If the phase-shifting effects of melatonin in humans are analgous to the gonadal suppression in rodents, there may be a short (5-hour) period of sensitivity just before the dark period during which exogenous melatonin administration is chronobiologically active (Stetson *et al.* 1986). This prediction awaits further testing.

In summary, studies of melatonin rhythms in totally blind people are strongly supportive of an important role of light in the synchronization of human circadian rhythms. Blind volunteers with free-running rhythms are ideal subjects in whom to study the effects of drugs (used in common clinical dosages) on the endogenous pacemaker in humans.

Light suppression of melatonin in patients with major affective disorder

If light intensity is between 500 and 2500 lux, there is a fluence–response relationship between the intensity of light and the degree of suppression of melatonin: the brighter the light, the greater the suppression. Thus, by exposing individuals to light in this range at the time that melatonin is being actively produced (e.g. 02.00–04.00 hours), a measure of the retinohypothalamic sensitivity to light can be obtained. Using this paradigm, Lewy and co-workers (1981) reported apparent super-sensitivity to light in four bipolar patients, two manic and two depressed. Subsequently, light suppression of melatonin (500 lux from 02.00–04.00 hours) was measured in 11 euthymic bipolar patients who were off medications and compared to 24 normal controls (Lewy *et al.* 1985*a*). The patients averaged 61.5 ± 1.6 per cent suppression while the normal controls averaged 28.0 ± 5.8 per cent (mean $\pm$ SE); the difference was significant at the $p = 0.0003$ level. Supersensitivity to light may be a trait marker for bipolar disorder since it is present both during the symptomatic phase of the illness as well as during the euthymic phase. Recently, Nurnberger and co-workers (1987) have shown that the offspring (age 15–25 years) of manic-depressive individuals are also supersensitive to light compared to age-matched controls, further supporting this test as a potential trait marker.

Light suppression studies offer a new strategy for a dynamic assess-ment of the retinohypothalamic pineal axis. Discriminations may become finer when some of the technical and methodological issues are resolved. It would be helpful if an exact dose of light (intensity and colour) could be conveniently administered to subjects. Brainard and associates (1985) used a monochromatic fibre-optic light delivery system and a diffuser over the eye in order to establish fluence–response curves in normal subjects, but this technique requires considerable subject compliance and is difficult to apply to psychiatric patients. It would also be helpful to know whether the sensitivity to light varies over the night. It is possible that the lack of rebound observed by Beck-Friis and co-workers (1986) in depressed patients may represent a phase advance in the light sensitivity response.

It seems likely that the phase-shifting effects of light and the suppression of melatonin are related, but this has not yet been proven. Are patients who are supersensitive to the suppressive effects of light also supersensitive to the phase-shifting effects of light? The relationship of these two effects of lights remains to be studied directly. Miles and colleagues (Chapter 13, this volume) and Thompson (Chapter 11, this volume) have provided further discussions on the phenomenon of light-induced suppression of melatonin secretion, and its clinical significance.

Total overnight melatonin production in depression

A melatonin deficiency in depression?

Many studies have found reduced levels of melatonin production in depression (Wetterberg 1978; Beck-Friis *et al.* 1984, 1985; Claustrat *et al.* 1984; Nair *et al.* 1984; Brown *et al.* 1985*a,b*; Miles and Philbrick 1987, 1988; Frazer *et al.* 1986). Furthermore, Lewy *et al.* (1979) found melatonin levels to be lower in the depressed phase than in the manic episode of bipolar affective illness. Melatonin production could reflect adrenergic tone at the adrenergic–pineal junction; thus, these findings could be consistent with the classical 'catecholamine hypothesis' of depression (Schildkraut 1965).

While most studies have shown reduced levels of melatonin in depression, they must be interpreted with some caution. We and others have found significant reductions in melatonin production with ageing (Iguchi *et al.* 1982; Touitou *et al.* 1985; Sack *et al.* 1986; Nair *et al.* 1986). Since depressed patients tend to be older than control populations, age must be carefully controlled when comparing groups of subjects (see Waldhauser and Waldhauser, Chapter 8). In addition, a number of studies have sampled plasma levels either during the day (when levels are almost undetectable) or at a single time point during the night (Brown *et al.* 1985*a,b*) and not all studies have required that study patients be drug-free (Beck-Friis *et al.* 1984, 1985). Beta blockers and clonidine are known to potently suppress melatonin (discussed below). On the other hand, tricyclics may increase melatonin levels, obscuring a low melatonin level that might be present in the untreated state (discussed below). Patients with depression often suffer weight loss and this has been a confounding factor in evaluating adrenocortical function in depression (Carroll *et al.* 1981); however, body weight has not been found to have a consistent relationship with melatonin production (Sack *et al.* 1986). Finally, the specificity of low melatonin in depression has not been adequately evaluated; in one study, schizophrenics also manifested low melatonin levels (Ferrier *et al.* 1982; Miles and Philbrick 1987, 1988; Miles *et al.*, Chapters 12 and 13).

Plasma levels may not be ideal for studying the amplitude of melatonin production since plasma concentrations reflect metabolism as well as production. The ideal experiment to test the hypothesis of low melatonin levels in endogenous depression would involve measures of urinary 6-hydroxymelatonin (the major metabolite of melatonin), comparing depressed subjects with controls under comparable lighting conditions. Measurement of 24-hour urinary 6-hydroxymelatonin excretion effectively integrates the area under the curve of daily melatonin production. Boyce (1985) found low or absent 6-sulphatoxymelatonin excretion (the sulphated conjugate of 6-hydroxymelatonin) in the urine of three out of eight depressed subjects. We have measured 6-hydroxy-melatonin in patients with winter depression (unpublished data) and can find no significant difference from controls; group means between patients and controls are almost identical when melatonin production is corrected for age. Thus, although low melatonin levels may be associated with certain types of non-winter depression, they do not appear to be associated with winter depression.

Low melatonin in depression may be a 'state' variable rather than a 'trait' variable. In 11 euthymic bipolar patients who were undergoing light suppression studies, baseline plasma melatonin levels (measured at one point in the night) were almost identical to a control group (Lewy et al. 1985a). There is currently a need for a longitudinal study of nightly melatonin production in major endogenous depression, comparing patients in the symptomatic and euthymic phase of the illness. Miles and colleagues have discussed the potential diagnostic use (Chapter 13, this volume) and overall significance of lowered melatonin secretion in affective disorder (Miles and Philbrick 1987a,b).

Tricyclic antidepressant effects on melatonin production; the pineal as an *in vivo* model of adrenergic pharmacology

The pineal gland has yet another attraction for the psychiatrist interested in the biology of depression. The adrenergic innervation of the pineal gland has no antagonistic cholinergic innervation; thus, the pineal gland is a potentially useful model for the study of *in vivo* adrenergic function and pharmacology in humans (Lewy 1983). The total overnight production of melatonin is a useful means of estimating adrenergic stimulation of the pineal which is clearly influenced by adrenergically active drugs (Cowen et al. 1985a,b; Thompson et al. 1985; Lewy et al. 1986a,b; Murphy et al. 1986).

The down-regulation hypothesis for the action of tricyclic anti-depressants (TCA) proposes that chronic treatment (over several weeks) with these drugs causes reduced numbers of postsynaptic adrenergic receptors, presumably the result of chronically increased synaptic neuro-transmitter concentration (Vetulani and Sulser 1975). These receptor changes might be reflected in changes in melatonin production in patients undergoing chronic TCA treatment. In four subjects, we found that 6-hydroxymelatonin excretion increased over the first 4 weeks of treatment without any secondary decline (as might be predicted from down-regulation) (Sack et al. 1986). Our findings are consistent with reports from several other groups who found persistently increased melatonin production after TCA therapy (Thompson et al. 1985; Golden et al. 1984; but contrary to other investigators who have found no elevation after chronic treatment (Brown et al. 1985a; Cowen et al. 1985a,b). It is interesting to note that melatonin levels reverted to normal after three weeks of treatment with desipramine in non-depressed volunteers (Thompson *et al.* 1985) suggesting a difference between patients and normal controls in the effects of TCA administration. Checkley (Chapter 9, this volume) has reviewed in detail the effects of TCA administration. (See also Miles and Philbrick 1988.)

Discussion

Circadian rhythm pathology in major affective disorders

The average circadian period (tau) of humans maintained in temporal isolation is 25.0 ± 0.5 hours (Wever 1979). It is rare to find a human subject with a tau of less than 24 hours. Thus, in almost all cases, the human circadian clock must be advanced by an average of 0.5 hour each day in order to achieve entrainment. If tau is 24.3 hours, a person will have to advance 0.3 hours per day; if tau is 25.2 hours, a person will have to advance 1.2 hours per day (four times as much). Thus, the magnitude of net advance required for entrainment among normal people can vary 4–6-fold.

Normal entrainment results from the interaction of three factors: the intrinsic circadian period (tau), the intensity and timing of environmental time cues (e.g. light), and the shape of the PRC (the intrinsic sensitivity to time cues). Thus, it seems logical that pathology would be related to an abnormality in one (or more) of these three factors. Since the clock must be advanced each day in order to maintain entrainment, pathology of the entrainment mechanism is likely to be expressed in delayed or free-running rhythms. Pathology could also be expressed by the degree of coupling between rhythms; for example, sleep could be abnormally delayed or advanced with respect to other rhythms.

Melatonin rhythms are free-running in blind people because of the inaccessibility of light cues. Most patients with winter depression have phase-delayed melatonin rhythms. Since light sensitivity appears to be normal (suggesting a normally sensitive PRC), the syndrome may be due to the combination of a long tau and reduced light cues. In DSPS, not enough is known to speculate on the cause; a long tau, a diminished advance portion of the PRC, or subsensitivity to light could play a role. If the tendency to delay sleep relative to the other rhythms is very strong, a free-running rhythm may develop in a sighted person because such a person would be sleeping through the phase advance portion of their PRC, resulting in a daily net delay (Weber *et al.* 1980). Light cues may be sufficient, but falling on the 'dead zone' of the PRC.

It is quite possible that many people experience a degree of phase delay in the winter and that bright artificial light exposure in the morning may be reasonable as a matter of hygiene (as much as treatment) Kasper *et al.* (1987). Even in sunny climates, the exposure of modern white collar workers to sunlight is very limited (Savides *et al.* 1986).

It is rather ironic that interest in circadian rhythms and psychiatry began with a 'phase advance hypothesis' of depression. While much of the data is supportive of an advance in the sleep rhythm of endogenously depressed patients, evidence that the temperature, melatonin, and cortisol rhythms are advanced has not been robust. To some extent, it is more difficult to conceptualize mechanisms for a phase advance abnormality given the known properties of the human circadian system, i.e. the tendency for tau to be greater than 24 hours. It would be necessary to postulate a very short tau (close to or less than 24 hours) or a deficiency in delay mechanisms.

Is melatonin a mediator of circadian amplitude?

Major depressive disorders are characterized by a loss of circadian amplitude. For example, cortisol secretion is increased (Linkowski *et al.* 1985), but there is a 'flattening' of the circadian cortisol rhythm (Sachar *et al.* 1973) with an elevated nocturnal nadir (Miles and Philbrick 1988). Growth hormone secretion in the first half of the night is reduced whereas daytime levels are increased (Mendelwicz *et al.* 1985). The depth of sleep is reduced and REM sleep propensity is more evenly distributed across the night (Gillin *et al.* 1979). Nocturnal core body temperature is higher at night and there is less day–night variation (Avery *et al.* 1982). The elderly manifest many of the same diminished circadian amplitudes of sleep, cortisol secretion, and core body temperature as depressed patients.

Melatonin is quite consistently reduced in both depression and ageing. Treatment of depression with TCA or MAOI which increase nocturnal melatonin, also increases the amplitude of circadian rhythms in core temperature (Avery *et al.* 1982) and cortisol production (Sachar *et al.* 1973). Propranolol and clonidine (which decrease nocturnal melatonin production) can precipitate depressive symptoms.

Does melatonin play a role in mediating day–night differences in other rhythms or are the reduced levels of melatonin seen in depression and in the elderly an indication of reduced circadian amplitude caused by some more fundamental underlying cause? The answer to that question might be found by examining the effects of exogenous melatonin admininstration to individuals who manifest low endogenous production. When melatonin was given to depressed patients by Carmen *et al.* (1976), the results were generally negative and no further trials have been conducted. However, the doses employed undoubtedly produced plasma concentrations that were at least 100 times normal; furthermore, melatonin was given during the day rather than at in the evening, thus further blurring day–night differences. Therefore, further trials of melatonin in depression may be justifiable, especially for patients with low melatonin production (see Miles and Philbrick 1987, 1988, and Chapter 13).

Conclusions

The increasingly refined techniques for accurately measuring plasma melatonin in humans have provided the means for testing a number of important hypotheses regarding the pathophysiology of affective disorders. The timing of plasma melatonin secretion is an ideal marker for the hypothalamic circadian pacemaker and may become preferred to markers such as core temperature and cortisol widely used in human circadian studies to date. Perhaps measurements of plasma melatonin will provide the tool to bring the understanding of circadian physiology gained in the animal laboratory to the psychiatric clinic (Miles and Philbrick 1988, and Chapter 13).

References

American Psychiatric Association (1987). *Diagnostic and statistical manual of mental disorders*, 3rd edn. American Psychiatric Association, Washington, DC.

Arendt, J., Borbely, A. A., Franey, C., and Wright, J. (1984). The effects of chronic, small doses of melatonin given in the late afternoon on fatigue in man: A preliminary study. *Neuroscience Letters* **45**, 317–21.

Arendt, J., Aldhous, M., and Marks, V. (1986). Alleviation of jet lag by melatonin: Preliminary results of controlled double blind trial. *British Medical Journal (Clinical Research)*, **292**, 1170.

Avery, D., Wildschiodtz, G., and Rafaelson, O. J. (1982) Nocturnal temperature in affective disorders. *Journal of Affective Disorders* **4**, 61–71.

Beck-Friis, J., von Rosen, D., Kjellman, B. F., Ljunggren, G., and Wetterberg, L. (1984). Melatonin in relation to body measures, sex, age, season and the use of drugs in patients with major affective disorders and healthy subjects. *Psychoneuroendocrinology* **9**, 261–77.

Beck-Friis, J., *et al.* (1985). Serum melatonin in relation to clinical variables in patients with major depressive disorder and a hypothesis of a low melatonin syndrome. *Acta Psychiatrica Scandinavica* **71**, 319–30.

Beck-Friis, J., Borg, G., Mellgren, T., Unden, F., and Wetterberg, L. (1986). Nocturnal serum melatonin levels following evening bright light exposure. *Clinical Neuropharmacology* **9**, (Suppl. 4), 184–6.

Boyce, P. M. (1985). 6-Sulphatoxy melatonin in melancholia. *American Journal of Psychiatry* **142**, 125–7.

Brainard, G. C., *et al.* (1985). Effect of light wavelength on the suppression of nocturnal plasma melatonin in normal volunteers. *Annals of the New York Academy of Sciences* **453**, 376–8.

Brown, R. P., *et al.* (1985*a*). Nocturnal serum melatonin in major depressive disorder before and after desmethylimipramine treatment. *Psychopharmacology Bulletin* **21**, 579–81.

Brown, R. P., *et al.* (1985*b*). Differences in nocturnal melatonin secretion between melancholic depressed patients and control subjects. *American Journal of Psychiatry* **142**, 811–16.

Carmen, J., Post, R., Buswell, R., and Goodwin, F. (1976). Negative effects of melatonin on depression. *American Journal of Psychiatry* **133**, 1181–6.

Carroll, B. J., *et al.* (1981). A specific laboratory test for the diagnosis of melancholia. *Archives of General Psychiatry* **38**, 15–22.

Claustrat, B., Chazot, G., Brun, J., Jordan, D., and Sassolas, G. A. (1984). A chronobiological study of melatonin and cortisol secretion in depressed subjects: plasma melatonin, a biochemical marker in major depression. *Biological Psychiatry* **19**, 1215–28.

Cowen, P. J., Bevan, J. S., Gosden, B., and Elliott, S. A. (1985*a*). Treatment with beta-adrenoceptor blockers reduces plasma melatonin concentration. *British Journal of Clinical Pharmacology* **19**, 258–60.

Cowen, P. J., Green, A. R., Grahame-Smith, D. G., and Braddock, L. E. (1985*b*). Plasma melatonin during desmethylimipramine treatment: evidence for changes in noradrenergic transmission. *British Journal of Clinical Pharmacology* **19**(6), 799–805.

Czeisler, C. A., *et al.* (1981). Chronotherapy: Resetting the circadian clocks of patients with delayed sleep phase insomnia. *Sleep* **4**, 1–21.

Czeisler, C. A., *et al.* (1986). Bright light resets the human circadian pacemaker independent of the timing of the sleep–wake cycle. *Science* **233**, 667–71.

Daan, S., Beersma, D. G. M., and Borbely, A. A. (1984). Timing of human sleep: Recovery process gated by a circadian pacemaker. *American Journal of Physiology* **246**, R161–78.

D'Alessandro, B., Bellastella, A., Esposito, V., Colucci, C. F., and Montalbetti, N. (1974). Circadian rhythm of cortisol secretion in elderly and blind subjects. *British Medical Journal* **2**, 274.

DeCoursey, P. J. (1960). Daily light sensitivity rhythm in a rodent. *Science* **131**, 33–5.

Dunner, D. L., Russek, F. K., and Russek, B. (1982). Classification of bipolar affective disorder subtypes. *Comprehensive Psychiatry* **23**, 186–9.

Ferrier, I. N., Arendt, J., Johnstone, E. C., and Crow, T. J. (1982). Reduced nocturnal melatonin secretion in chronic schizophrenia: Relationship to body weight. *Clinical Endocrinology (Oxford)* **17**, 181–7.

Frazer, A., *et al.* (1986). Patterns of melatonin rhythms in depression. *Journal of Neural Transmission (Suppl.)* **21**, 269–690.

Gillin, J. C., Duncan, W., Pettigrew, K. D., Frankel, B., and Snyder, F. (1979). Successful separation of depressed, normal, and insomniac subjects by sleep EEG criteria. *Archives of General Psychiatry* **36**, 85–90.

Golden, R. N., Markey, S. P., and Potter, W. S. (1984). A new marker for adrenergic function in man. Presented at the American Psychiatric Association Annual Meeting, New Research Program, May 5–11, Los Angeles, California.

Halberg, F. (1968). Physiologic considerations underlying rhythometry with special reference to emotional illness. In *Cycles biologiques et psychiatrie* (ed. J. deAjuriaguerra), p. 73. Masson et Cie, Paris.

Hollwich, F. and Dieckhues, B. (1972). Circadian rhythm of cortisol level in normal subjects and in the blind. In *Contemporary ophthalmology* (ed. J. G. Gellows), pp. 472–5. Williams and Wilkins, Baltimore, Maryland.

Iguchi, H., Kato, K. I., and Ibayashi, H. (1982). Age-dependent reduction in serum melatonin concentrations in healthy human subjects. *Journal of Clinical Endocrinology and Metabolism* 55(1), 27–9.

Inouye, S. T. and Kawamura, H. (1982). Characteristics of a circadian pacemaker in the suprachiasmatic nucleus. *Journal of Comparative Physiology* **146**, 153–60.

James, S. P., Wehr, T. A., Sack, D. A., Parry, B. L., and Rosenthal, N. E. (1985). Treatment of seasonal affective disorder with light in the evening. *British Journal of Psychiatry* **147**, 424–8.

Jarret, D., Coble, P. A., and Kupfer, D. J. (1983). Reduced cortisol latency in depressive illness. *Archives of General Psychiatry* **40**, 506–11.

Kasper, S. F., Rogers, S., Schulz, P. M., Skwerer, R. G., and Rosenthal, N. E. (1987). The effects of bright light in normal subjects. Presented at the Annual Meeting, American Psychiatric Association, May 9–14, Chicago, Illinois.

Krieger, D. T. and Rizzo, F. (1971). Circadian periodicity of plasma 11-hydroxy-corticosteroid levels in subjects with partial and absent light perception. *Neuro-endocrinology* **32**, 165–79.

Kripke, D. F. and Wyborney, V. G. (1980). Lithium slows rat circadian activity rhythms. *Life Science* **26**, 1319–21.

Kripke, D. F., Mullaney, D. J., Atkinson, M., and Wolf, S. (1978). Circadian rhythm disorders in manic-depressives. *Biological Psychiatry* **123**, 335–51.

Kronauer, R. E., Czeisler, C. A., Pilato, C. A., Moore-Ede, M. C., and Weitzman, E. D. (1982). Mathematical model of the human circadian system with two interacting oscillators. *American Journal of Physiology* **242**, R3–17.

Kupfur, D. J. (1976). REM latency: a psychobiologic marker for primary depressive disease. *Biological Psychiatry* **11**, 159–74.

Leonhard, K. (1957). *Aufteilung der endogenen Psychosen*. Akademie Verlag, Berlin.

Lewy A. J. (1983). Human melatonin secretion (I): a marker adrenergic function. In *Biology of mood disorders,* (ed. R. M. Post and J. C. Ballenger), pp. 207–14. Plenum Press, New York.

Lewy, A. J. and Markey, S.P. (1978). Analysis of melatonin in human plasma by gas chromatography negative chemical ionization mass spectrometry. *Science* **201**, 741–3.

Lewy, A. J. and Newsome, D. A. (1983). Different types of melatonin circadian secretory rhythms in some blind subjects. *Journal of Clinical Endocrinology and Metabolism* **56**, 1103–7.

Lewy, A. J., Wehr, T. A., Gold, P., and Goodwin, F. K. (1979). Plasma melatonin in manic-depressive illness. In *Catecholamines: basic and clinical frontiers,* Vol. 2 (ed. E. Usdin, I. J. Kopin, and J. Barchas), pp. 1173–5. Pergamon Press, New York.

Lewy, A. J., Wehr, T. A., Goodwin, F. K., Newsome, D. A., and Markey, S. P.

(1980). Light suppresses melatonin secretion in humans. *Science* **210**, 1267–9.

Lewy, A. J., Wehr, T. A., Goodwin, F. K., Newsome, D. A., and Rosenthal, N. E. (1981). Manic-depressive patients may be supersensitive to light [letter]. *Lancet* **i**, 383–4.

Lewy, A. J., Kern, H. A., Rosenthal, N. E., and Wehr, T. A. (1982). Bright artificial light treatment of a manic-depressive patient with a seasonal mood cycle. *American Journal of Psychiatry* **139**, 1496–8.

Lewy, A. J., Sack, R. A., and Singer, C. L. (1984). Assessment and treatment of chronobiologic disorders using plasma melatonin levels and bright light exposure: The clock-gate model and the phase response curve. *Psychopharmacology Bulletin* **20**, 561–5.

Lewy *et al.*, (1985*a*). Supersensitivity to light: Possible trait marker for manic-depressive illness. *American Journal of Psychiatry* **142**, 725–7.

Lewy, A. J., Sack, R. L., and Singer, C. M. (1985*b*). Immediate and delayed effects of bright light on human melatonin production: shifting 'dawn and dusk': shifts the dim light melatonin onset·(DLMO). *Annals of the New York Academy of Sciences* **453**, 253–9.

Lewy, A. J., Sack, R. L., and Singer, C. M. (1985*c*). Treating phase typed chronobiological sleep and mood disorders using approximately timed bright artificial light. *Psychopharmacology Bulletin* **21**, 368–72.

Lewy, A. J., Sack, R. L., Singer, C. M., Miller, L. S., and Hoban, T. M. (1986*a*). Phase delay and hypersomnia [letter]. *American Journal of Psychiatry* **143**, 679–80.

Lewy, A. J., Siever, L. J., Uhde, T. W., and Markey, S. P. (1986*b*). Clonidine reduces plasma melatonin levels. *Journal of Pharmacy and Pharmacology* **38**, 555–6.

Lewy, A. J., Sack, R. L., Miller, L. S., and Hoban, T. M. (1987). Antidepressant and circadian phase-shifting effects of light. *Science* **235**, 352–4.

Linkowski, P., *et al.*, (1985). The 24-hour profile of adrenocorticotropin and cortisol in major depressive illness. *Journal of Clinical Endocrinology and Metabolism* **61**, 429–38.

McEachron, D. L., Kripke, D. F., and Wyborney, V. G. (1981). Lithium promotes entrainment of rats to long circadian light–dark cycles. *Psychiatric Research* **5**, 1–9.

McEachron, D. L., Kripke, D. F., Sharp, F. R.., Lewy, A. J., and McClellan, D. E. (1985). Lithium effects on selected circadian rhythms in rats. *Brain Research Bulletin* **15**, 347–50.

Mendlewicz, J., *et al.*, (1985). Diurnal hypersecretion of growth hormone in depression. *Journal of Clinical Endocrinology and Metabolism* **60**, 505–12.

Migeon, C. J., *et al.*, (1956). The diurnal variation of plasma levels and urinary excretion of 17-hydroxycorticosteroids in normal subjects, night workers and blind subjects. *Journal of Clinical Endocrinology and Metabolism* **16**, 622–33.

Miles, A. and Philbrick, D. R. S. (1987). Melatonin: Perspectives in laboratory medicine and clinical research. *CRC Critical Reviews in Clinical Laboratory Science* **25**, 231–53.

Miles, A. and Philbrick, D. R. S. (1988). Melatonin and psychiatry, a review. *Biological Psychiatry* **23**, 405–25.

Miles, L. E. M., Raynal, D. M., and Wilson, M. A. (1977). Blind man living in normal society has circadian rhythms of 24.9 hours. *Science* **198**, 421–3.

Moldofsky, H., Musisi, S., and Phillipson, E. A. (1986). Treatment of a case of advanced sleep phase syndrome by phase advance chronotherapy. *Sleep* **9**, 61–5.

Moore, R. Y. and Eichler, V. B. (1972). Loss of a circadian adrenal corticosterone rhythm following suprachiasmatic lesions in the rat. *Brain Research* **42**, 201–6.

Murphy, D. L., Garrick, N. A., Tamarkin, L., Taylor, P. L., and Markey, S. P. (1986). Effects of antidepressants and other psychotropic drugs on melatonin release and pineal gland function. *Journal of Neural Transmission* **21**, 291–309.

Nair, N. P., Hariharasubramanian, N., and Pilapil, C. (1984). Circadian rhythm of plasma melatonin in endogenous depression. *Progress in Neuropsychopharmacology and Biological Psychiatry* **8**, 715–18.

Nair, N. P., Hariharasubramanian, N., Pilapil, C., Isaac, I., and Thavundayil, J. X. (1986). Plasma melatonin—an index of brain ageing in humans? *Biological Psychiatry* **21**, 141–50.

Nurnberger, J., Berrettini, W., Tamarkin, L., and Gershon, E. (1987). Light suppression of melatonin in high-risk offspring of bipolar patients and age-matched controls. Presented at the Society of Biological Psychiatry Annual Meeting, May 6–10, Chicago, Illinois.

Orth, D. N., Besser, G. M., King, P. H., and Nicholson, W. E. (1979). Free-running circadian plasma cortisol rhythm in a blind human subject. *Clinical Endocrinology* **10**, 603–17.

Papousek, M. (1975). Chronobiological aspects of cyclothymia. *Fortschritte der Neurologie, Pscyhaitrie und Ihrer Grenzgebiete* **43**, 381–440.

Pittendrigh, C. S. and Daan, S. (1976). A functional analysis of circadian pacemakers in nocturnal rodents. I. The stability and lability of spontaneous frequency. *Journal of Comparative Physiology* **106**, 223–52.

Redman, J., Armstrong, S., and Ng, K. T. (1983). Free-running activity rhythms in the rat: entrainment by melatonin. *Science* **219**, 1080–1.

Rosenthal, N. E., *et al.* (1984). Seasonal affective disorder. A description of the syndrome and preliminary findings with light therapy. *Archives of General Psychiatry* **41**, 72–80.

Rosenthal, N. E., Sack, D. A., Carpenter, C. J., Parry, B. L., Mendelson, W. B., and Wehr, T. A. (1985). Antidepressant effects of light in seasonal affective disorder. *American Journal of Psychiatry* **142**, 163–70.

Sachar, E. J., Hellman, L., Roffwarg, H. P., Halpern, F. S., Fukushima, D. K., and Gallagher, T. F. (1973). Disrupted 24-hour patterns of cortisol secretion in psychotic depression. *Archives of General Psychiatry* **28**, 19–24.

Sack, R. L., Lewy, A. J., Erb, D. L., Vollmer, W. M., and Singer C. M. (1986). Human melatonin production decreases with age. *Journal of Pineal Research* **3**, 379–88.

Sack, R. L., Lewy, A. J., and Hoban, T. M. (1987). Free-running melatonin rhythms in blind people: phase shifts with melatonin and triazolam administration. In *Temporal disorder in human oscillatory systems* (ed. L. Rensing, U. an der Heiden, and M. C. Mackey), pp. 219–24. Springer-Verlag, Berlin.

Savides, T. J., Messin, S., Senger, C., and Kripke, D. F. (1986). Natural light exposure of young adults. *Physiology and Behaviour* **38**, 571–4.

Schildkraut, J. J. (1965). The catecholamine hypothesis of affective disorders: a review of supporting evidence. *American Journal of Psychiatry* **122**, 509–22.

Stephan, F. K. and Zucker, I. (1972). Circadian rhythms in drinking behaviour and locomotor activity of rats are eliminated by hypothalamic lesions. *Proceedings of the National Academy of Sciences USA* **69**, 1583–6.

Stetson, M. H., Sarafidis, E., and Rollag, M. D. (1986). Sensitivity to adult male Djungarian hamsters (*Phodopus sungorus sungorus*) to melatonin injections throughout the day: effects on the reproductive system and the pineal. *Biology of Reproduction* **35**, 618–23.

Thompson, C., *et al.* (1985). The effect of desipramine upon melatonin and cortisol secretion in depressed and normal subjects. *British Journal of Psychiatry* **147**, 389–93.

Touitou, Y., Fevre, M., Montange, M., Proust, J., Klinger, E., and Nakache, J. P. (1985). Age- and sex-associated modification of plasma melatonin concentrations in man. Relationship to pathology, malignant or not, and autopsy findings. *Acta Endocrinologica (Copenhagen)* **108**, 135–44.

Underwood, H. (1986). Circadian rhythms in lizards: phase response curve for melatonin. *Journal of Pineal Research* **3**, 187–96.

Vetulani, J. and Sulser, F. (1975). Action of various antidepressant treatments reduces reactivity of noradrenergic cyclic AMP-generating system in limbic forebrain. *Nature* **257**, 495–6.

Weber, A. L., Cary, M. S., Connor, N., and Keyes, P. (1980). Human non-24-hour sleep-wake cycles in an everyday environment. *Sleep* **2**, 347–54.

Wehr, T. A. and Goodwin, F. K. (1981). Biological Rhythms in Psychiatry. In *The American handbook of psychiatry VII*, 2nd edn (ed. S. Arieti), pp. 46–74. Basic Books, New York.

Wehr, T. A. and Goodwin, F. K. (1983). *Circadian rhythms in psychiatry*. Boxwood Press, Pacific Grove, California.

Wehr, T. A., Jacobsen, F. M., Sack, D. A., Arendt, J., Tamarkin, L., and Rosenthal, N. E. (1986). Phototherapy of seasonal affective disorder. Time of day and suppression of melatonin are not critical for antidepressant effects. *Archives of General Psychiatry* **43**, 870–5.

Weissman, M. M. and Meyers, J. K. (1978). Affective disorders in a U.S. urban community: the use of research diagnostic criteria in an epidemiological survey. *Archives of General Psychiatry* **35**, 1304–11.

Weitzman, E. D., *et al.* (1981). Delayed sleep phase syndrome: a chronobiologic disorder with sleep onset insomnia. *Archives of General Psychiatry* **38**, 737–46.

West, E. D. and Dally, P. J. (1959). Effects of iproniazid in depressive syndromes. *British Medical Journal* **1**, 1491–4.

Wetterberg, L. (1978). Melatonin in humans: physiological and clinical studies. *Journal of Neural Transmission* (Suppl.) **13**, 289–310.

Wever, R. A. (1979). *The circadian system of man. Results of experiments under temporal isolation.* Springer-Verlag, New York.

Wirz-Justice, A. and Pringle, C. (1987). The non-entrained life of a young gentleman at Oxford. *Sleep* **10**, 57–61.

Wright, J., Aldhous, M., Franey, C., English, C., and Arendt, J. (1986). The effects of exogenous melatonin on endocrine function in man. *Clinical Endocrinology* **24**, 375–82.

Zis, A. P. and Goodwin, F. K. (1979). Major affective disorder as a recurrent illness: A critical review. *Archives of General Psychiatry* **36**, 835–939.

11. Melatonin and seasonal affective disorder

Chris Thompson

Introduction

Seasonal Affective Disorder (SAD) is a recurrent condition in which regular depressive illnesses occur in the winter with remissions in the spring and summer (Rosenthal *et al.* 1984). Prominent symptoms are dysphoria, with sadness, anxiety, and irritability in almost equal measure, and pronounced fatigue and lethargy. The mood remains reactive to events although persistently low throughout the winter period. The vegetative symptoms of depression are those of reverse functional shift: hypersomnia, increased appetite, overeating, carbohydrate craving, and weight gain. Most of the sufferers are female, although the precise sex ratio varies between samples from 3:1 to 9:1. This may be due to differences in sampling frames.

Over half of the psychiatric diagnoses in all studies are bipolar according to Research Diagnostic Criteria (RDC), i.e. they have mania or hypomania in the spring or summer. The remainder are unipolar depressives. The proportion diagnosed as bipolar II (hypomania rather than mania) and unipolar varies between samples since the diagnosis of hypomania after remission of a long depression is difficult to distinguish from normal relief (Thompson and Isaacs 1988).

All of the cases have satisfied RDC for primary major depression as a condition of inclusion and the average number of episodes in the three main samples is around 15 (Thompson and Isaacs 1988; Wirz-Justice *et al.* 1986*a,b*; Rosenthal *et al.* 1984). About half have previously been treated with psychoactive drugs, and most have had difficulties at work and at home during winter, some losing their jobs because of the lethargy, hypersomnia, and irritability. For those with bipolar disorder the mania can also be troublesome.

Many studies have now shown that SAD is responsive to treatment with bright light. In these studies it has not yet proved possible to design a double-blind placebo controlled trial since it is always obvious to the patient that they are having light therapy. However bright light, except in

228

two studies (Isaacs *et al.* 1988; Wirz-Justice *et al.* 1986*b*) has proven superior to dim light as an antidepressant. Where phototherapy has been tried in non-seasonal affective disorder it has not been found to be effective, possibly suggesting a selective effect. These results suggest that SAD may be triggered by a lack of light in the winter and may be treated by artificial replacement light.

A condition which appears to be tied to the winter months clearly requires investigation of its pathogenesis. Many physiological functions vary with the seasons in mammals and in man (Aschoff 1981; Hildebrandt 1962) and any of these might prove to be the internal signal which passes on the zeitgeber's message that it is winter. Several of these physiological functions are known to be altered by light rather than temperature or any social variable that changes with the seasons. However, suspicion fell initially on melatonin as the critical variable intervening between winter lighting conditions and depression for several reasons. Firstly, melatonin is a strongly rhythmical signal which, in any individual, maintains a strict timing of daily secretion from one day to the next, and is remarkably free of the influence of masking variables such as sleep and temperature which might interfere with its role as a time cue for the circadian neuroendocrine system (Arendt 1985). Secondly, the secretion of melatonin varies not only with the day, but with the seasons. The pineal is active almost entirely in the hours of darkness and when the nights extend in the winter so the daily duration of secretion of melatonin extends by several hours. Again the yearly cycle of melatonin is remarkably regular and is known to govern the seasonal breeding rhythms of several species of mammals (Lincoln *et al.* 1981). Thirdly, melatonin is both entrained and directly suppressed by light of sufficient intensity. In normal human subjects this intensity is around 1500 lux, much greater than in rats, although there is evidence that in manic depressive patients on lithium (in an active phase of illness or in remission) the sensitivity is increased so that even 500 lux may suppress melatonin to some extent (Lewy *et al.* 1981; Sack and Lewy, Chapter 10). Fourthly, administering melatonin to normal subjects produces some of the symptoms of SAD, particularly fatigue (Arendt *et al.* 1985). Thus, the mechanism for the genesis of the illness and the response to light therapy would be almost within grasp if it could be shown that melatonin secretion was abnormal in SAD patients and that phototherapy modified it as a necessary condition for its effectiveness. To what extent have these hopes been fulfilled by the evidence?

Melatonin secretion in untreated SAD patients

A consistent alteration in melatonin secretion within SAD has not been convincingly demonstrated. However, Rosenthal *et al.* (1985*a*) have reported that the melatonin profiles in some SAD patients in the winter are abnormal although the data presented were very preliminary. Their data suggest that SAD patients may secrete more melatonin than normals in the late afternoon and early night up to 03.00 hours. However, this is so far based on published results for only four SAD patients and two normal controls. Lewy and Sack (1986) in nine SAD patients suggest that the onset of melatonin secretion may be delayed in the winter months, a finding at odds with Rosenthal and colleagues (1985*a*) reported above, who suggest a phase advance of secretory onset. In non-seasonal endogenous depression some workers have provided evidence for a reduction in nocturnal melatonin secretion (Beck-Friis *et al.* 1984; Claustrat *et al.* 1984; Miles *et al.*, Chapter 13). However, in at least one study where all relevant variables were carefully controlled, depressed patients secreted if anything slightly more melatonin than the matched normal group (Thompson *et al.* 1988).

Atenolol completely blocks melatonin secretion by β-receptor antagonism (Cowen *et al.* 1983). When 100 mg of atenolol was given to 18 SAD patients in the late afternoon there was no significant improvement of the symptoms of SAD (Rosenthal *et al.* 1985*b*) despite verification by radioimmunoassay of the drug-induced reduction in nocturnal melatonin secretion. Thus, unless it is the profile of melatonin which is important in generating SAD rather than simply the daily level of secretion, it appears that a reduction in melatonin by pharmacological means does not improve SAD.

Rosenthal and associates (1985*a*) treated eight SAD patients during the period of depression with phototherapy at an intensity of 2500 lux for 5–6 hours per day until they were almost entirely well. They then administered either 2.0–2.4 mg daily of melatonin or matching placebo tablets during the dark period for 1 week, in a double-blind cross-over protocol with 1 week of each condition.

Melatonin was measured hourly from 18.00 to 12.00 hours the following day before light treatment was commenced, during placebo treatment and during melatonin treatment. At baseline the nocturnal peak of melatonin varied from 40 to 128 pg/ml (mean: 69 pg/ml). The rise of melatonin appeared to begin rather early in the evening, although this was difficult to see since by 18.00 hours it was already at a mean level higher than daytime values. Following light treatment, melatonin levels between 18.00 hours and midnight were suppressed. During the melatonin phase of the cross-over there was a marked elevation of melatonin due to the exogenous compound.

The difficulty of interpreting the baseline melatonin values is a significant problem in this type of research. Interindividual variation in the peak values and timing of melatonin is very large, and normal data is sparse. There is a marked effect of age and menstrual status. Thus any comparison between normal and SAD groups needs to be very large and well controlled. It is therefore not possible to state whether or not the pre-treatment profile in these patients was abnormal.

The main part of the study was to determine whether melatonin would re-induce the symptoms of SAD in those successfully treated with light. During treatment with melatonin, the patients experienced a recrudescence of some of their symptoms of SAD, but not all. However, melatonin may produce some of these symptoms in non-seasonally depressed patients and normal individuals as well as in patients with SAD (Carman *et al.* 1976; Arendt *et al.* 1985). Melatonin may therefore induce these symptoms non-specifically.

Dietzel and co-workers (1986*a*), as part of a study of the effects of phototherapy on melatonin (see below), obtained an unexpected finding. In patients with Major Depressive Disorder diagnosed according to RDC, but not stated to be seasonal, the melatonin values were greater in the summer than in the winter, while in a control group of normal subjects the normal seasonal variation of greater winter values was found. This was significant at the $P < 0.05$ level. As these patients were successfully treated with phototherapy, they may well have suffered from SAD (see Kripke 1985) and the result is precisely the opposite of that predicted by the melatonin hypothesis of SAD, although it might be taken to suggest an abnormality of seasonal regulation of a less well specified sort.

In 24 samples of morning urine from patients in our series with SAD, the mean value was 0.34 ± 0.03 nM/l and the range was from 0.14 to 0.80 nM/l. This matches very closely the normal range as quoted by Wetterberg (1978). Morning urinary melatonin depends upon total nocturnal secretion, but correlates well ($r = 0.89$) with the 02.00 hour plasma value in a normal sample. In our sample there was thus no evidence of abnormal melatonin secretion.

Other neuroendocrine parameters also show no abnormality, which is consistent with the clinical picture of SAD as a usually, but by no means always, mild disorder with none of the classical endogenous symptoms (Thompson and Isaacs 1987). The dexamethasone suppression test in Rosenthal's first reported sample of 29 patients was normal in all those tested (Rosenthal *et al.* 1984). The protireline infusion test (TRH test) was also normal. Prolactin has been shown to be raised in SAD in one paper (Jacobson *et al.*, reported in Wirz-Justice 1986) and this may be due to serotoninergic overactivity. Serotonin is also important in feeding regulation and is an important consideration in future investigations of SAD.

Melatonin in the response to phototherapy

In phototherapy, the artificial light is given from a source which varies in size but is usually about 4 feet long and 2 feet wide positioned about 3 feet from the subject's eyes. Vita-lite or True-lite tubes are most often used, but any light which emits a relatively full spectrum appears to be effective. Indeed, the full spectrum may not itself be necessary. When subjects are covered from head to toe except for the eyes, the effect is still seen, but when they are exposed to the light over the full surface of the body with the eyes covered, the effect is lost (N. E. Rosenthal, personal communication). It is thus very likely that an intact optical system is necessary for the effect. Subjects sit in front of the lights glancing into them about every 30 seconds, but being otherwise occupied with television, reading, knitting, etc.

There have now been many studies on the response of depression, both seasonal and sporadic, to bright light. The light involved is sufficiently bright to have circadian effects (Czeisler *et al.* 1986) and to suppress melatonin, and it is therefore understandable that melatonin should be the focus of interest in its mechanism of action.

Lewy and colleagues (1982) first reported the successful treatment of a single male patient with a 13-year history of winter depressive episodes using bright artificial light (2000 lux), for 6 hours a day. This was followed by a report on nine patients which accompanied the first description of the syndrome of SAD (Rosenthal *et al.* 1984). In this study bright, full-spectrum, white light (2500 lux) was given for 6 hours daily, 3 hours before dawn and 3 hours after dusk. Dim yellow light (100 lux) was given as a control in a cross-over design. Each treatment lasted for 1 week. Bright light treatment led to a very significant improvement ($P < 0.001$) and yellow light had no effect.

This marked response to phototherapy in patients with SAD given to extend the day length was not repeated in studies on non-seasonal major depressive disorders by Kripke's group, some of which predated the SAD work (Kripke 1985). These authors were working on the hypothesis that non-seasonally depressed patients had disturbances of circadian rhythms, possibly with phase advance. They initially gave light as a 1-hour pulse in the morning with no significant effect. In addition, extending the period of treatment to several days and several hours a day also produced no significant effect. Although relatively few patients have been treated in this way, and generally for fewer hours than in SAD the evidence is so far in favour of the response to phototherapy being restricted to the seasonal type of affective disorder.

Several reasons were advanced for this effect in SAD.

1. The response depends on suppression of melatonin, which is itself in

some way connected with mediating the depression. This is the major hypothesis to be explored in this chapter.

Dissenting theories include:

2. The light therapy invariably involves a degree of sleep deprivation which may be the therapeutic factor.

3. The improvement is a placebo effect. It is unlikely that treatment expectations could have been equalized between the bright and dim light treatments so this cannot be ruled out. It is similar to comparing a large red tablet of active antidepressant with a small green placebo, the bigger more striking treatment will be at an advantage.

4. The light is operating as a zeitgeber (time cue) to restabilize abnormal circadian rhythms which are the mediators of the disorder.

Rosenthal and associates (1985c) reported a further 13 patients treated as in-patients (7) or out-patients (6) with bright light at 2500 lux or dim light (100 lux for in-patients or 300 lux for out-patients). In neither group was there a significant improvement with dim light, but both bright light conditions were significantly effective in both groups. In this study, sleep time was monitored, and was reduced equally in both bright and dim light conditions when compared to the baseline and withdrawal weeks, thus not accounting for the difference between bright and dim light.

In a subsequent study, James *et al.* (1985) found a significant improvement ($P < 0.01$) in nine patients treated with bright light for 5 hours in the evening alone, starting at 18.00 or 19.00 hours depending on convenience. There was no effect of dim light. This treatment involved no sleep deprivation, certainly not during the crucial early morning hours (Vogel *et al.* 1980), so the effects cannot be due to that mechanism.

An interesting subsidiary question is whether the therapeutic effects of sleep deprivation are at least partially mediated by exposure to bright light in the early morning. There has been only one rather inconclusive study of this possibility (Wehr *et al.* 1985), but it must clearly be taken into account in future studies of sleep deprivation. At any rate theory 2 above can be excluded.

Is suppression of melatonin necessary to the therapeutic effect? There are three ways of investigating this issue. First, melatonin can be measured during phototherapy to see if it is suppressed as the patients improve. Secondly, light given at such a time as not to influence melatonin secretion, i.e. during the day, can be investigated for its therapeutic potential. Thirdly, the threshold for therapeutic effect might not correspond to the threshold for suppression of melatonin. When considering these studies it should be born in mind that we already are aware that atenolol, which reduces melatonin secretion, does not improve the symptoms of SAD. This suggests that suppression of melatonin alone is insufficient for therapeutic effect.

Dietzel *et al.* (1986*b*) found that 15 patients with major affective disorders (probably SAD, but not specified as such) responded to light at 2800 lux from 06.00 to 09.00 hours and from 17.00 to 21.00 hours given for 1 day only. Melatonin was collected hourly through an in-dwelling cannula. This light treatment did not cause a reduction in melatonin, but the patients' depression scores did improve. This study is unusual as improvement occurred during the first day while in most other studies it came on gradually over the first 4 days of treatment. In addition, it is not stated that the patients actually suffered from SAD although this is very likely as they responded to light.

Rosenthal and associates (1985*a*) in the study described in the previous section were able to partially reverse the therapeutic effects of bright light with melatonin tablets but not with matching placebo tablets. The atypical symptoms of depression (characteristic of SAD) were those which were re-induced most clearly, suggesting that these symptoms may be more related to melatonin than the others, but also that melatonin cannot fully explain the antidepressant effects of light. Even those symptoms which worsened may have been due to non-specific effects of melatonin (Arendt *et al.* 1985).

We designed a treatment study to investigate whether photoperiod extension, artificially lengthening the day, was an essential part of the phototherapy effect (Isaacs *et al.* 1988). Eleven patients underwent a trial of three light treatments for 1 week each, with a 1 week withdrawal between each (Fig. 11.1).

1. Photoperiod extension for 2 hours before dawn and 2 hours after dusk with bright light (2500 lux).

2. Light augmentation in which the total dose of light was the same as in the photoperiod extension condition but was given within the bounds of natural daylight (10.00 to 14.00 hours).

3. Dim light (300 lux) given at the same time as the photoperiod extension, but below the physiological threshold.

We reasoned that if the photoperiod extension hypothesis, and thus the melatonin hypothesis, was correct then light augmentation should be no more effective than placebo, i.e. dim light. If, however, photoperiod extension is not essential then both of the bright light treatments should be equally effective and both better than dim light.

Patients were scored by blind raters using the Hamilton Rating Scale (HRS) for depression (first 17 items) and also an addendum to the HRS to cover atypical depressive symptoms (N. E. Rosenthal *et al.*, personal

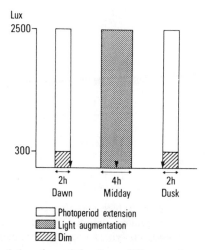

Fig. 11.1. Timing and duration of the three different phototherapeutic regimens. (For explanation see text.)

communication). The patients filled in a Weekly Mood Inventory (N. E. Rosenthal *et al.*, personal communication) which factors into four subscales, depressive mood and ideation, typical vegetative symptoms (weight loss, etc), atypical vegetative symptoms (weight gain, etc.), and hypomania euthymia items. Daily mood inventories were also completed by the patients.

The results were unexpected. Taking the results from the Hamilton Rating Scales first as being the most important because they were performed blind to treatment condition, the two bright light conditions together were not significantly more effective than the dim light treatment, which itself only had a modest effect (see Fig. 11.2). However, taking the bright light treatments separately, the light augmentation condition was significantly superior to the dim light and better than photoperiod extension, while photoperiod extension was no better than dim light. There were no differences between treatments for the self-rated scales of the weekly and daily mood inventories. However the daily mood inventory showed the same time-lag of effectiveness as in other studies, i.e. 4 days.

This result was first of all a failure to replicate the earlier findings of striking improvement in mood with bright light, and the reason for this is unclear. Secondly, the photoperiod extension hypothesis was decisively disproved by these findings, and with it the simple hypothesis that light acts by suppressing melatonin secretion in the morning and evening, bringing the profile of secretion back to a summer shape.

These findings are supported by a laboratory study of Wehr *et al.*

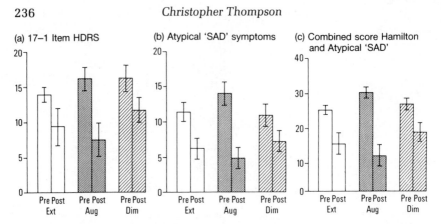

(a) 17–1 Item HDRS   (b) Atypical 'SAD' symptoms   (c) Combined score Hamilton and Atypical 'SAD'

Fig. 11.2. Hamilton rating scales before and after one week of phototherapy in each of three different regimens; photoperiodic extension (Ext), light augmentation (Aug) and dim light (Dim). (For explanation see text.)

(1986). In animals and probably in man, the signal that the season is changing is carried by the timing of dawn and dusk. This can be counterfeited by two brief pulses of light if the animal is kept in otherwise constant darkness or semi-darkness. The technique is called a skeleton photoperiod. Wehr *et al.* (1986) attempted to manipulate the skeleton photoperiods of patients with SAD. They admitted them to a research ward where they remained in semi-darkness (less than 120 lux) during the day, sleeping in full darkness between 23.00 and 07.00 hours every night. For summer conditions (Long Skeleton Photoperiod) they were exposed to bright light (2500 lux) from 07.30 to 10.30 hours and from 20.00 to 23.00 hours for 5 days. After a withdrawal week they were subjected to winter conditions (Short Skeleton Photoperiod) with bright light from 09.00 to 12.00 hours and from 14.00 to 17.00 hours for 5 days. The treatments were given in a cross-over design. The bright light treatment was significantly antidepressant, but equally so in both Skeleton Photoperiod conditions. However, nocturnal melatonin secretion was 22 per cent lower in the summer than the winter conditions as predicted, showing again that suppression of melatonin by extension of the photoperiod is not essential to therapeutic effect. One problem with the interpretation of this study is that no dim light control treatment was used so that the improvements seen with these two treatments may have been no better than dim light itself could have produced.

Thus, studies which have used the timing of light therapy to investigate the importance of the suppression of melatonin have all failed to demonstrate a clear involvement of this hormone. Is there a threshold of light intensity or duration below which phototherapy is ineffective, and does this correspond to the threshold for the suppression of melatonin?

Wirz-Justice *et al.* (1986*a*) treated 12 patients with 2500 lux white light and 250 lux yellow light for 4 hours a day (2 hours in the morning and evening) for at least 5 days each, with a 1-week withdrawal between treatments. There were no differences between the two treatment regimens, both causing significant improvements in clinical state. However, when only the atypical symptom score was taken into account bright light was a slightly superior treatment to dim.

The first interpretation of this result which one is led to consider is that the efficacy of the intervention with bright light was no greater than the placebo condition. This was the way the authors clearly intended to interpret the results during the design of the study, and it is how others have done so. However, the authors argued against this explanation on several grounds. First, the antidepressant effect was lost on withdrawal of yellow light. Secondly, the delay in response was the same as in other studies, i.e. 3–4 days. Thirdly, repeated use of the lights led to sustained improvement. None of these reasons is especially convincing and it remains possible that the effect of phototherapy in this study was simply a placebo. In a further study (Wirz-Justice *et al.* 1986*b*) these workers have shown that as little as half an hour of bright morning light is effective, and that 1 or 2 hours appear to be more effective, but equal to each other.

In a further study of 19 SAD patients, 2 hours of morning light was compared to half-an-hour of morning light (Lewy and Sack 1986). There was no difference between the treatments, supporting to some extent the findings of Wirz-Justice (1986*b*), but raising once again the possibility that, in this study at least, neither treatment was better than placebo.

From these studies it appears that bright light in the morning may be effective even if only 1 hour is given, but half-an-hour is probably not enough. However, even dim light may be effective. The authors argue that this may be because depressed patients are more sensitive to light than normals (Lewy *et al.* 1981), although this alone would not explain the differences between the Bethesda and the Basle studies. In addition, the supersensitivity to light in depressed patients has so far only been shown in manic-depressives without a seasonal cycle. It is likely that the therapeutic effect is made up of several factors, one of which may well be a marked placebo effect.

The simple theory of melatonin suppression by phototherapy would thus seem no longer tenable. Indeed, melatonin suppression does not occur during successful experiments with light, and light at a time of day which does not affect melatonin secretion can be clinically effective. In addition, light at an intensity lower than the physiological threshold for melatonin suppression may also be effective.

Are there any alternative mechanisms which involve melatonin?

There are two possibilities. There may be an abnormality of end organ sensitivity (Rosenthal *et al.* 1985*d*) which may itself vary seasonally. At present there is no information bearing on this hypothesis, especially as the site of action of melatonin remains unknown. The second and more interesting possibility is that the time-cueing properties of melatonin, rather than the nightly volume of secretion, are important. We have seen that the timing of phototherapy appears not to be crucial to the success of treatment, since evening, morning, and midday light are all effective, However, only two studies have compared morning with evening light, which would be the most discriminating way to examine an effect of circadian timing on the efficacy of phototherapy. Lewy and colleagues (1985) have proposed that patients with major depressive disorders have phase-advanced circadian rhythms, including the rhythm of melatonin. In contrast, patients with SAD are hypothesized to have phase delay of circadian rhythms. There is little evidence for this hypothesis and the preliminary data of Rosenthal and associates (1985*a–d*) appear to show an early rather than a late rise in melatonin secretion.

Czeisler *et al.* (1986) have shown that in humans light can reset the circadian clock during temporal isolation, light in the morning causing a phase advance relative to the environment. Thus, if SAD patients are phase-delayed as Lewy suggests, then morning light should either be the only time for which a therapeutic effect is shown (which it is not) or at least it should be more effective than evening light. This was, in fact, the result obtained by Lewy in a study of eight SAD patients treated with either 2 hours of bright light in the morning or the evening and seven normal controls (Lewy and Sack 1986). The SAD patients at the start of treatment had significantly later melatonin onset times than the controls, suggesting on this basis that they may be phase-delayed. Morning light led to an improvement in depression and an advance of the melatonin onset. Evening light led to no improvement and a further delay in melatonin onset. This then is some support for melatonin acting not as a pathogenic agent, but as a time cue for circadian abnormality. A third condition of light in the morning and evening produced intermediate improvements and intermediate shifts in melatonin secretion.

Hellekson *et al.* (1986) obtained contrary results. They found no differences between 2 hours of bright light given in the morning or the evening, or 1 hour of bright light split between morning and evening. Only six patients completed this study so there may have been a type II statistical error.

The phase-typing hypothesis, however, might not be supported by Lewy's findings (Lewy and Sack 1986) since there may be a circadian rhythm detectable in retinal sensitivity to light, with the point of maximum sensitivity at dawn (Reme and Wirz-Justice 1985). Thus, patients treated at dawn would have received more light than those treated at dusk. When this factor is taken into account the Lewy studies (Lewy and Sack 1986) would support the hypothesis that the total dose of light is the most important factor in the antidepressant effect rather than its timing, and would agree with the studies of Isaacs *et al.* (1988), Wehr *et al.* (1986), Hellekson *et al.* (1986), and Wirz-Justice *et al.* (1986*a,b*).

Conclusions

Thus far, no convincing evidence has been presented to demonstrate that melatonin is positively involved in the syndrome of SAD or in the response to phototherapy. If melatonin is involved, it is through a mechanism which is not yet understood. However, it is possible that a role for melatonin in circadian timing is the active ingredient rather than overall levels of secretion.

If melatonin itself is not involved, then what is? Discounting the possibility that the response to light is a placebo effect, serotonin involvement may be suspected. In rodents, eating behaviour is controlled by the serotoninergic system and the symptoms of overeating and weight gain may reflect an abnormality of this system in SAD patients. Serotoninergic mechanisms also appear to be involved in the stabilization of circadian rhythms (Martin and Redfern 1984). Interesting new work suggests that in rats, constant light conditions lead to alterations in brain serotonin and beta receptor function similar to those seen on long-term treatment with antidepressants (Cox *et al.* 1986). This mechanism would be consistent with the response to light being due to the total dose rather than the timing.

Additional support is derived from evidence that the syndrome of SAD is related not so much to the photoperiod, but to the daily sunshine since most subjects appear to be sensitive to short-term changes in weather patterns (Rosenthal *et al.* 1986). The photoperiodic change with the seasons would then act to induce SAD by reducing the number of hours available in each day to receive light of sufficient intensity to prevent depression. This may be why the syndrome appears mainly in winter, although varying in onset and offset depending on shorter-term weather conditions in spring and autumn. The model fits both the variability in the timing of the syndrome and the response to phototherapy rather better than the previous models based on photoperiod alone and suggests that the response to phototherapy will depend on a mechanism other than melatonin suppression, possibly a direct effect upon central serotonin and beta-receptors.

Acknowledgements

I am grateful to Professor L. Wetterberg who kindly performed the melatonin assays on the urine samples in our patients.

References

Arendt, J. (1985). The Pineal Hormone melatonin in seasonal and circadian rhythms. In *Circadian rhythms in the CNS* (ed. P. H. Redfern, K. F. Martin, and I. Campbell), pp. 15–28. MacMillan, Oxford.

Arendt, J., *et al.* (1985). Some effects of melatonin and the control of its secretion in humans. In *Photoperiodism, melatonin and the pineal*, Ciba Foundation Symposium 117, pp. 266–83. Pitman, London.

Aschoff, J. (1981). Annual rhythms in man. In *Handbook of behavioural neuro-biology* (ed. J. Aschoff), Vol. 4, pp. 475–87. Plenum Press, New York.

Beck-Friis, J., von Rosen, D., Kjellman, B. F., Ljungren, J. G., and Wetterberg, L. (1984). Melatonin in relation to body measures, sex, age, season and the use of drugs in patients with major affective disorders and healthy subjects. *Psycho-neuroendocrinology* **9**, 261–77.

Carman, J. S., Post, R. M., Buswell, R., and Goodwin, F. K. (1976). Negative effects of melatonin on depression. *American Journal of Psychiatry* **133**, 1181–6.

Claustrat, B., Chazot, G., Brun, J., Jordan, D., and Sassolas, G. (1984). A chronobiological study of melatonin and cortisol secretion in depressed subjects: plasma melatonin, a biochemical marker in major depression. *Biological Psychiatry* **19**, 1215–28.

Cowen, P. J., Fraser, S., Sammous, R., and Green, A. R. (1983). Atenolol reduces plasma melatonin concentrations in man. *British Journal of Clinical Pharmacology* **15**, 579–81.

Cox, C. M., Mason, R., Neal, A., and Parker, T. L. (1986). Altered 5-HT sensitivity and synaptic morphology in rat CNS induced by long term exposure to continuous light. *British Journal of Pharmacology* **89**, 528P.

Czeisler, C. A., *et al.* (1986). Bright light resets the human circadian pacemaker independent of the timing of the sleep-wake cycle. *Science* **233**, 667–71.

Dietzel, M., Lesch, O. M., and Reschenhofer, F. (1986a). Seasonal variation of melatonin differs in depressed patients and healthy controls. *Clinical Neuropharmacology* (Suppl.) **4**, 199–201.

Dietzel, M., Waldhauser, F., Lesch, O. M., Musalek, M., and Walter, H. (1986b). Bright light treatment success not explained by melatonin. *Conference on recent advances in affective disorders*, Basle, Summer 1986 (Abstract).

Hellekson, C. J., Kline, J. A., and Rosenthal, N. E. (1986). Phototherapy for Seasonal Affective Disorder in Alaska. *American Journal of Psychiatry* **143**, 1035–7.

Hildenbrandt, G. (1962). Biologische Rhythmen und ihre Bedeutung fur die Bader- und Klimaheilkunde. In *Handbuch der Bader- und Klimaheilkunde* (ed. A. Amelung and A. Evers), pp. 730–85. FK Schattauer verlag, Stuttgart.

Isaacs, G., Stainer, D. S., Sensky, T. E., Moor, S., and Thompson, C. (1988).

Phototherapy and its mechanisms of action in seasonal affective disorder. *Journal of Affective Disorders* **14**, 13–19.

James, S. P., *et al.* (1985). Treatment of seasonal affective disorder with evening light. *British Journal of Psychiatry* **147**, 424–8.

Kripke, D. F. (1985). Therapeutic effects of bright light in depressed patients. *Annals of the New York Academy of Science* **453**, 270–81.

Lincoln, G. A., Almeida, O. F. X., and Arendt, J. (1981). Role of melatonin and circadian rhythms in seasonal reproduction in rams. *Journal of Reproductive Fertility* (Suppl.) **30**, 23–31.

Lewy, A. J., Wehr, T. A., Goodwin, F. K., Newsome, D. A., and Rosenthal, N. E. (1981). Manic-Depressive patients may be supersensitive to light. *Lancet* **i**, 383–4.

Lewy, A. J., Kern, H. E., Rosenthal, N. E., and Wehr, T. A. (1982). Bright artificial light treatment of a manic-depressive patient with a seasonal mood cycle. *American Journal of Psychiatry* **139**, 1496–8.

Lewy, A. J., Sack, R. L., and Singer, C. M. (1985). Immediate and delayed effects of bright light on human melatonin production: shifting dawn and dusk shifts the dim light melatonin onset (DLMO). *Annals of the New York Academy of Sciences* **453**, 253–9.

Lewy, A. J. and Sack, R. L. (1986). Melatonin physiology and light therapy, *Clinical Neuropharmacology* (Suppl.) **4**, 196–8.

Martin, K. F. and Redfern, P. H. (1982). The effects of Clomipramine on the 24 hour variation of 5-HT and tryptophan concentration in the rat brain. *British Journal of Pharmacology* **76**, 288.

Reme, C. and Wirz-Justice, A. (1985). Circadian rhythms of the retina and their pharmacological modulation. In *Circadian rhythms in the CNS* (ed. P. Redfern, K. Martin, and I. Campbell), pp. 135–46. MacMillan, Oxford.

Rosenthal, N. E., *et al.* (1984). Seasonal Affective Disorder. *Archives of General Psychiatry* **41**, 72–80.

Rosenthal, N. E., *et al.* (1985*a*). Seasonal affective disorder and phototherapy. *Annals of the New York Academy of Science* **453**, 260–9.

Rosenthal, N. E., *et al.* (1985*b*). The role of melatonin in seasonal affective disorder and phototherapy. In *Melatonin in humans* (ed. R. J. Wurtman and F. Waldhauser), pp. 72–80. Cambridge, Massachusetts.

Rosenthal, N. E., Sack, D. A., Carpenter, C. J., Parry, B. L., Mendelson, W. B., and Wehr, T. A. (1985*c*). Antidepressant response to light in seasonal affective disorder. *American Journal of Psychiatry* **142**, 163–70.

Rosenthal, N. E., Sack, D. A., Jacobson, F. M., Parry, B. L., James, S. P., Tamarkin, L., Arendt, J., and Wehr, T. A. (1985*d*). Consensus and controversy in seasonal affective disorders and phototherapy. Paper presented to the IVth World Congress of Biological Psychiatry, Philadelphia.

Rosenthal, N. E., Sack, D. A., Jacobson, F. M., Skwerer, R. G., and Wehr, T. A. (1986). Seasonal affective disorder and light: past, present and future. *Clinical Neuropharmacology* (Suppl.) **4**, 193–5.

Thompson, C. and Isaacs, G. (1988). Seasonal affective disorder; a British sample—symptomatology in relation to mode of referral and diagnostic subtype. *Journal of Affective Disorders* **14**, 1–13.

Thompson, C., Franey, C., Arendt, J., and Checkley, S. A. (1988). A comparison of melatonin secretion in normal subjects and depressed patients. *British Journal of Psychiatry* **152**, 260–6.

Vogel, G. W., Vogel, F., McAbee, R. S., and Thurmond, A. J. (1980). Improvement of depression by REM sleep deprivation: new findings and a theory. *Archives of General Psychiatry* **37**, 247.

Wehr, T. A., Rosenthal, N. E., Sack, D. A., and Gillin, J. C. (1985). Antidepressant effects of sleep deprivation in bright and dim light. *Acta Psychiatrica Scandinavica* **72**, 161–5.

Wehr, T. A., Jacobsen, F. M., Sack, D. A., Arendt, J., Tamarkin, L., and Rosenthal, N. E. (1986). Timing of phototherapy and it's effect on melatonin secretion do not appear to be critical for its antidepressant effect in seasonal affective disorder. *Archives of General Psychiatry* **43**, 870.

Wetterberg, L. (1978). Melatonin in humans, physiological and clinical studies. *Journal of Neural Transmission* (Suppl.) **13**, 289–310.

Wirz-Justice, A. (1986). Light therapy for depression: present status, problems and perspectives. *Psychopathology* **19** (Suppl. 2), 136–41.

Wirz-Justice, A., Bucheli, C., Graw, P., Kielholz, P., Fisch, H-U., and Woggon, B. (1986*a*). Light treatment of seasonal affective disorder in Switzerland. *Acta Psychiatrica Scandinavica* **74**, 193–204.

Wirz-Justice, A., Bucheli, C., Graw, P., Kielholz, P., Fisch, H-U., and Woggon, B. (1986*b*). How much light is antidepressant? *Psychiatry Research* **17**, 75–7.

12. Melatonin and schizophrenia— a biochemical link?

Andrew Miles and Joseph E. Grey

Introduction

A historical association of the pineal with psychiatric disorder has existed since at least three centuries before the birth of Christ when Herophilas of Alexandria (325–280 BC) first mentioned the pineal body as an entity functionally involved in controlling the 'stream of thoughts'. The ancient Greeks also showed interest in the pineal proposing this organ as the seat of the soul (see Vesalius 1555), a startling concept which was later extended by Descartes (1662) with a philosophy that designated the pineal as the organ which unites the immortal soul with the body. In the seventeenth and eighteenth centuries many physicians positively associated the pineal with psychosis. King (1686) had published a case report of a 75-year-old academic with acute psychosis (probably schizophrenic) who upon post-mortem examination had shown pineal calcification. King reported, 'The glandula pinealis was firm and fair, and of ordinary size. Feeling it and finding it harder than ordinary (and talking of a gentleman then present of Descartes' opinion that it was the seat of the soul) I pressed it and found in it a stone in a film, or rather a petrified gland in a film. I took out this stone and kept it as a great rarity. I do not remember I ever heard of such a thing before. I am sure of all the brains I have dissected (and I may say that I have dissected more than one hundred) I never saw such a one'. Two hundred years later, pineal extracts were being used in the experimental therapy of *Dementia praecox* (Becker 1920) more as part of the then almost fashionable belief in the therapeutic efficacy of glandular preparations, than on any real scientific basis. Indeed, melatonin itself was not 'discovered' until 1958 (Lerner *et al.* 1958) and the many trials with pineal extracts at the turn of the present century were so unsatisfactorily conducted that we have omitted their description here. Such reports are, nevertheless, early contributions to the modern literature and in this regard the interested reader is referred to the study of Kitay and Altschule (1954) which presents a detailed listing of these early experiments.

243

The modern literature

Pineal extracts in schizophrenia

Altschule (1957) conducted the first serious clinical investigations of pineal extracts in schizophrenic patients using beef pineal preparations as experimental agents for this psychosis. The findings were published in the *New England Journal of Medicine* in 1957, and reported the effects of aqueous extracts of acetone-dried beef pineal substance in chronic schizophrenia. The effects reported were, essentially, improvement followed by refractoriness.

The preparations for administration by injection were produced by solubilization of 1 g acetone-dried beef pineal substance with 10 ml of distilled water followed by filtering of the product and daily injection of the equivalent of 1 g beef pineal substance to eight patients with chronic schizophrenia. The initial clinical picture was improvement soon followed by a return to the orginal state. Altschule (1957) discussed the state of affairs by presenting clinical data from three of the eight patients under study. The first case quoted was that of a 30-year-old female who, prior to treatment, exhibited bizarre postures, hallucinations, delusions, and was incoherent. Following a course of several injections of the extract she improved rapidly—hallucinations were attenuated and some degree of insight into illness was observed. This degree of clinical improvement was maintained for nearly 3 weeks until a slow symptomatic relapse was recorded. A similar response was noted in a 23-year-old male who, prior to treatment, had shown apathy and had experienced hallucinations. Following 1 week of treatment, the patient became cheerful and active with disappearance of hallucinations. Considerable insight into illness was recorded, but after 3 weeks of therapy clinical relapse was seen. A more rapid onset of clinical improvement was observed in a 32-year-old female who had been psychotic for 5 years, grimacing and gesturing bizarrely. She was frequently dishevelled and incontinent. Following 6 days of treatment she spoke sanely and coherently, and expressed disgust at her condition with a desire for cosmetic improvement. For a period of 3 days the improvement was sustained, but relapse occurred on Day 9 of therapy.

Interestingly, Altschule (1957) noted that patients given the extract for 2–3 weeks lost refractoriness in 30–60 days, but those in whom treatment was continued over 4 weeks developed an apparently irreversible refractoriness. Furthermore, the effects described were prevented by concurrent administration of 75 mg cortisone every 6 hours. When cortisone was discontinued, clinical improvement was seen and the course of events followed that described in the patients receiving extract alone.

Altschule (1957) also conducted experiments using an alkaline protein-free extract. This preparation was produced by solubilizing 1 g of the acetone-dried beef pineal extract described above in 10 ml distilled water at 50–55°C at pH 8–9 for 4 hours with chemical reduction (HCl) of pH to neutrality. The preparation was then cooled and four parts of acetone added, left for 48 hours at −10°C, and then filtered at the same temperature. Acetone was removed by gentle heating and residual protein removed by centrifugation. The extract prepared by this means was administered to three patients in 1 g equivalents. A clinical improvement was noted that continued for more than 1 week following the final injection, but relapse then followed. Altschule (1957) prepared similar extracts at 70°C and these were equipotent to those prepared at 50–55°C. In addition to these experiments, about 30 others were performed with extracts prepared from organic solvent extraction or aqueously at various acidities. Such extracts were found to be either completely inert or markedly less potent than the alkaline extract described. The author concluded that

protein-free pineal extracts made at a pH of about 8.5 and at temperatures between 55 and 70°C reverse biochemical abnormalities of schizophrenia and cause clinical improvements

and these were indeed ambitious statements. This study preceded the isolation and characterization of melatonin (Lerner *et al.* 1958) and represented a considerable improvement on earlier studies with pineal extracts where results had been contradictory and where chronic cases of schizophrenia were not always used thus making suspect any clinical improvements that might represent spontaneous remissions.

The next significant contribution that extended the work of Altschule (1957) was from Eldred and colleagues (1960) and was similarly published in the *New England Journal of Medicine.* Eldred and associates (1960) studied eight females with chronic schizophrenia under impressively well standardized conditions in a placebo-controlled double-blind study using the pineal extract described by Altschule (1957). A statistically significant clinical improvement following injection of pineal extract was noted, but the degree of improvement was not outstanding. Although the findings of Eldred and colleagues (1960) do not constitute definitive evidence for a therapeutic efficacy of the pineal extract, the findings are noteworthy since the patients who showed improvement had proved refractory to a multiplicity of psychiatric treatment regimes over an average period of 20 years, and the mean age of the patients was 52. Also interesting was the initial worsening of the patients during the first 4 weeks of treatment, although by the end of the study only minimal statistically significant improvement was recorded.

Eldred and colleagues (1960) concluded their study by stating that

> although the findings are not such as to determine the future place of pineal extract in the treatment of schizophrenia, they are sufficient to encourage and give direction to further studies of its effects with a larger, younger and less chronically ill population.

The suggestion of Eldred and his associates was certainly not immediately adopted by other workers in the field, and little research in this area was published until 1974 when Bigelow (1974) reported his study of the effects of the 'Altschule extract' in chronic schizophrenia in a placebo-controlled double-blind study.

The patients selected for Bigelow's investigation (1974) had a clear history of schizophrenic or schizoaffective illness of at least 3 years duration. Patients with a history of illness exceeding 15 years duration, or who displayed affective blunting were excluded. The pineal extract was essentially that previously described by Altschule (1957) and Eldred *et al.* (1960). The placebo employed was a dilute liver extract of similar appearance to the pineal extract. All patients received at least one course each of extract and placebo lasting 3 weeks, and, if therapeutic response appeared during the initial trial of extract, the individual was subjected to further cross-over experiments. The clinical status of each patient was assessed by a 28-item nurses' rating scale designed to measure a broad range of psychopathology (Wyatt and Kupfer 1968).

The results of the study confirmed the earlier reports and demonstrated the aqueous pineal extract capable of producing beneficial effects in five of the 10 patients investigated. However, the clinical improvement was apparently completely dependent upon continued concurrent antipsychotic medication since, in two of the patients (nos. 4 and 9 of the study) demonstrating clinical improvement, relapse was produced by reduction in maintenance dosage of antipsychotic drugs. Bigelow (1974) concluded that the overall improvement in the patients was 'small'.

It is important to note that the extract employed in the investigations reviewed was 'free of assayable melatonin' (Bigelow 1974), but changes in the metabolism and physiology of melatonin that may be induced by injection of the pineal extract have not been assessed. Indeed, it would be interesting, in addition, to screen the extract for the presence of melatonin with our present sophisticated analytical facility (see Miles *et al.*, Chapter 13). It is probably true to say that there remains a lot more to be learned about the nature and exact effects of the pineal extract or perhaps of one of its many constituent compounds. Indeed, with current biochemical interest in the pineal extending beyond melatonin to other methoxyindoles and pineal peptides, etc., it is quite possible that discoveries in the future will relate to results from the past.

Melatonin and schizophrenia: metabolic aspects

A small, but significant quantity of research has been published which has specifically studied the metabolism and effects of melatonin in schizophrenic individuals. McIsaac (1961) was probably the first investigator to suggest an aetiological role of melatonin in the schizophrenic syndrome, deriving his hypothesis from the observation of a marked similarity between the hallucinogenic agent harmine and the then newly characterized pineal hormone melatonin. The suggestion that melatonin or one of its precursors may under abnormal conditions be metabolized into compounds of aetiological significance in schizophrenia remains an interesting possibility.

Smythies (1976) later suggested that abnormal O- or N-methylation of endogenously synthesized indoleamines or catecholamines might well contribute aetiologically to schizophrenic illness. The logic of this hypothesis was based on the finding that hydroxyindole, O-methyltransferase (HIOMT), the pineal enzyme directly responsible for melatonin synthesis (see Fig. 12.1) has the ability to synthesize the psychotogen N-acetyl-3,4-dimethoxyphenylethylamine (NADMPEA) from the abnormal substrate N-acetyl, 4-hydroxy, 3-methoxyphenylethylamine. Indeed, Hartley and Smith (1973) have shown that there is a variety of psychotomimetic compounds that have the capacity to stimulate HIOMT activity. The same authors further suggested that if HIOMT became out of phase with its normal substrate, it would

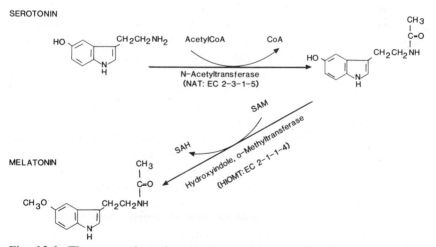

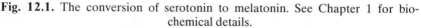

Fig. 12.1. The conversion of serotonin to melatonin. See Chapter 1 for biochemical details.

theoretically be free to act on abnormal substrates (depending on their availability) producing methylated compounds of possible importance in the aetiology of the schizophrenic psychosis. The neuroleptic butyro-phenone haloperidol is highly concentrated by pinealocytes where it blocks HIOMT activity and thus the capacity of this enzyme to act on any abnormal substrates that may be available (Hartley *et al.* 1977). This neuroleptic might thus, in addition to its well documented effects on dopaminergic function, block synthesis of abnormal methylated metabolites by HIOMT in schizophrenic illness. Clearly, further research into these hypotheses is required before we can assess the relative importance of such a metabolic lesion in the whole syndrome.

Further possibilities with regard to abnormal metabolism of melatonin in schizophrenia were rationalized by McIsaac (1961) already mentioned, who had suggested that melatonin could be metabolized to the harmala alkaloid and psychotogen 10-methoxyharmalan. Since 10-methoxyharmalan and its derivatives can potently inhibit the activity of monoamine oxidase (Udenfriend *et al.* 1958), disruption of normal serotoninergic neurotransmission would be a logical consequence with a shunting of accumulated serotonin into 5-methoxyharmalan synthesis (McIsaac 1961).

These studies, based largely on hypothetical biochemistry, are not made convincing by studies such those of Owen and colleagues (1983). These investigators could show no significant differences in the activity of HIOMT in post-mortem specimens of schizophrenic pineal when compared to control tissue. There remains, nevertheless, a striking similarity between melatonin and several compounds of known psychoto-mimetic potency (Fig. 12.2).

A further metabolic abnormality, this time of a catabolic type, has been reported to occur in schizophrenic illness. Jones and associates (1969) have documented a slightly different profile of excreted meta-bolites in psychotic individuals with several *unknown* metabolites being detected. No observations on any clinical change in the patients' symptomatology following melatonin administration were made in this study, but data do exist which report related effects. Indeed, melatonin administration to patients with schizophrenic illness in remission apparently results in relapse and a florid exacerbation of psychosis appearing *de novo* (Altshcule, reported in Carman *et al.* 1976). In this study, 300 mg melatonin was administered intravenously in a peak dose to two schizophrenic individuals who at the time of study were in remission. The clinical relapse appeared shortly after infusion, and hallucinosis was maintained for 18–36 hours.

The effect of melatonin administration in these two schizophrenics, though not of statistical significance, is nevertheless of interest since

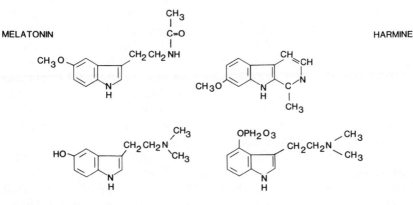

Fig. 12.2. The structural similarity between melatonin and some psycho-tomimetic agents.

lysergide, mescaline, harmine, and dimethyltryptamine all have the capacity to induce pineal HIOMT (Hartley and Smith 1973; Lynch *et al.* 1973; Holtz *et al.* 1974) thus increasing melatonin synthesis and the potential for generation of abnormal compounds through inappropriate methylation reactions catalysed by this enzyme. Indeed, amphetamine, L-dopa and cocaine are all capable of generating a syndrome that closely resembles functional psychosis, and all increase melatonin synthesis (Backstrom and Wetterberg 1973; Lynch *et al.* 1973; Holtz *et al.* 1974). In addition, Winters and colleagues (1973) claim that the functional integrity of the pineal gland is essential for the manifestations of hallucinogen-induced behaviour, and melatonin itself is not devoid of direct dopaminergic actions that could be related to the schizophrenic disease process (Wendel *et al.* 1974; Bradbury *et al.* 1983; Zisabel and Laudon 1983).

Melatonin and schizophrenia: clinical aspects

A significant literature has accrued that argues against abnormal metabolism of melatonin in schizophrenic illness. Hanssen and his colleagues (1980) have reported a low 08.00 hour and high midnight plasma melatonin level in schizophrenic individuals, but in this same study no abnormality in the circadian organization of secretion could be demonstrated. Significant to the findings was that no difference could be shown between the mean resting morning melatonin level of the three schizophrenic individuals studied and their non-schizophrenic siblings. Four years later, Beckmann and associates (1984) investigated melatonin levels in the CSF of 13 drug-free paranoid schizophrenics, 15 neuroleptic-treated schizophrenic subjects, and 16 healthy volunteers. In this study, no significant differences in CSF melatonin levels could be identified between the groups, although a valid criticism of this study would be that the time of CSF collection was 09.00–10.00 hours and thus a time not optimum for investigation. The authors concluded that their study did not support a contention for abnormal melatonin metabolism in schizophrenia, but that an abnormality in circadian organization of secretory profile could not be ruled out.

The most recent and well conducted study at the time of writing, which involved study of melatonin secretion in schizophrenic individuals, has been reported by Steiner and Brown (1985). These authors investigated the circadian profile and level of secretion of melatonin in 10 schizophrenic subjects on admission to hospital, and no abnormalities of these parameters could be detected. However, abnormal melatonin secretion may be present in chronic schizophrenics with affective components to their illness. Wetterberg and co-workers (1982) have suggested that such a phenomenon was present in a group of chronic schizophrenics studied by Ferrier and colleagues (1982). It is important to consider the possibility that abnormal melatonin secretion in schizophrenia may have no direct relevance to the schizophrenic aetiology, but perhaps only to the development of affective pathology with which melatonin abnormalities are more strongly associated (see Chapters 9–12, this volume).

Conclusions

The existing literature does not support a positive involvement of melatonin in schizophrenic illness or abnormalities in its secretion secondary to the onset of illness. Studies have been relatively few in number, and on the whole have not been exemplary in control and interpretation. Indeed, until well-controlled, long-term circadian assessments of melatonin secretion have been conducted in precisely defined diagnostic subgroups of schizophrenia, we must continue to ask the question: 'Melatonin and schizophrenia—a biochemical link?'

Acknowledgments

The authors thank Professor Peter McGuffin and Dr David M. Shaw for their advice, Dr Alan Wardrop and Mr Daniel M. Barrow for their interest and helpful discussions, and Mrs Lillian Lewis for excellent secretarial assistance.

References

Altschule, M. D. (1957). Some effects of aqueous extracts of acetone dried beef pineal substance in chronic schizophrenia. *New England Journal of Medicine* **257**, 919–22.

Backstrom, M. and Wetterberg, L. (1973). Increased N-acetylserotonin and melatonin formation induced by *d*-amphetamine in rat pineal gland organ culture via a *β*-adrenergic receptor mechanism. *Acta Physiologica Scandinavica* **87**, 113–20.

Becker, W. J. (1920). Epiglandol bei Dementia Praecox. *Therapeutische Halbmonatscher* **34**, 667–8.

Beckmann, H., Wetterberg, L., and Gattaz, W. F. (1984). Melatonin immuno-reactivity in the cerebrospinal fluid of schizophrenic patients and healthy controls. *Psychiatry Research* **11**, 107–10.

Bigelow, L. B. (1974). Effects of aqueous pineal extract in chronic schizophrenia. *Biological Psychiatry* **8**, 5–15.

Bradbury, A. J., Kelly, M. E., and Smith, J. A. (1985). Melatonin action in the midbrain can regulate dopamine function both behaviourally and biochemi-cally. In *The pineal gland, endocrine aspects* (ed. G. M. Brown and S. D. Wainwright), pp. 327–32. Pergamon Press, Oxford.

Carman, J. S., Post, R. M., Buswell, R., and Goodwin, F. K. (1976). Negative effects of melatonin on depression. *American Journal of Psychiatry* **133**, 1181–6.

Descartes, R. (1662, posthumous). *De Homine, Figuris et Latinitate donatus a F. Schuyl*, Lugduni Batavorum.

Eldred, S. H., Bell, N. W., and Lewis, L. J. (1960). A pilot study comparing the effects of pineal extract and a placebo in patients with chronic schizophrenia. *New England Journal of Medicine* **263**, 1330–5.

Ferrier, I. N., Arendt, J., Johnstone, E. C., and Crow, T. J. (1982). Reduced nocturnal secretion of melatonin in chronic schizophrenia; relationship to body weight. *Clinical Endocrinology* **17**, 181–7.

Hanssen, T., Heyden, T., Sundberg, J., Alfredsson, G., Nyback, H., and Wetterberg, L. (1980). Propranolol in schizophrenia: clinical, metabolic and pharmacological findings. *Archives of General Psychiatry* **37**, 685.

Hartley, R. and Smith, J. A. (1973). The activation of pineal hydroxyindole, O-methyltransferase by psychotomimetic drugs. *Journal of Pharmacy and Pharmacology* **25**, 751–2.

Hartley, R., Padwick, D., and Smith, J. A. (1977). The inhibition of pineal hydroxyindole, O-methyltransferase by haloperidol and fluphenazine. *Journal of Pharmacy and Pharmacology* **24**, 100–3.

Holtz, R. W., DeGuchi, T. C., and Axelrod, J. (1974). Stimulation of serotonin N-acetyltransferase in pineal organ culture by drugs. *Journal of Neurochemistry* **22**, 205–9.

Jones, R. L., McGeer, P. L., and Greiner, A. C. (1969). Metabolism of exogenous melatonin in schizophrenic and non-schizophrenic subjects. *Clinica Chimica Acta* **26**, 281–6.

King, E. (1686). A relation of a petrified Glandula Pinealis, lately found in the dissection of a brain. *Philosophical Transcripts of the Royal Society of London* **185**, 228–31.

Kitay, J. I. and Altschule, M. D. (1954). *The pineal gland: a review of the physiological literature.* Harvard University Press, Cambridge, Massachusetts.

Lerner, A. B., Case, J. D., Takahashi, Y., Lee, T. H., and More, W. (1958). Isolation of melatonin, the pineal factor that lightens melanocytes. *Journal of the American Chemical Society* **80**, 2587.

Lynch, H. J., Wang, P., and Wurtman, R. J. (1973). Increase in rat pineal melatonin content following L-dopa administration. *Life Science* **12**, 145–51.

McIsaac, W. M. (1961). A biochemical concept of mental disease. *Postgraduate Medicine* **30**, 111–18.

Owen, F., Ferrier, I. N., and Poulter, M. (1983). Hydroxyindole, O-methyltransferase activity in human pineals: a comparison of controls and schizophrenics. *Clinical Endocrinology* **19**, 313–17.

Smythies, J. R. (1976). Recent progress in schizophrenia research. *Lancet* **ii**, 136–9.

Steiner, M. and Brown, G. M. (1985). Melatonin–cortisol ratio and the dexamethasone suppression test in newly admitted psychiatric inpatients. In *The pineal gland: endocrine aspects* (ed. G. M. Brown and S. D. Wainwright), pp. 347–53. Pergamon Press, Oxford.

Udenfriend, J., Witkop, B., Redfield, R. G., and Weissbach, H. (1958). Studies with reversible inhibitors of monoamine oxidase; harmaline and related compounds. *Biochemical Pharmacology* **1**, 160–5.

Vesalius, A. (1555). *De Humani Corporis Fabrica,* Libri Septem, 2nd edn. Basiliae.

Wendel, O. T., Waterbury, L. D., and Pearce, L. A. (1974). Increase in monoamine concentrations in rat brain following melatonin administration. *Experientia* **30**, 1167–8.

Wetterberg, L., *et al.* (1982). Melatonin and cortisol levels in psychiatric illness. *Lancet* **ii**, 100.

Winters, W. D., Alcaraz, M., Cervantes, M. Y., and Flores-Guzman, C. (1973). The synergistic effect of reduced visual input on ketamine action—the possible role of the pineal gland. *Neuropharmacology* **12**, 407–16.

Wyatt, R. J. and Kupfer, P. J. (1968). A fourteen symptom behaviour and mood rating scale for longitudinal patient evaluation by nurses. *Psychological Reports* **23**, 1331–6.

Zisapel, N. and Laudon, M. (1983). Inhibition by melatonin of dopamine release from rat hypothalamus. *Brain Research* **272**, 378–81.

13. Melatonin—a diagnostic marker in laboratory medicine?

Andrew Miles and D. Roger Thomas

Introduction

During the last 40 years the correlation of specific biochemical and histological abnormalities with specific disease processes has led to a precise branch of clinical science: laboratory medicine. Within this discipline it is recognized that some markers actually represent integral components of the pathological lesion under investigation, whereas others are often indirectly involved, representing aetiologically unrelated though secondarily precipitated phenomena. Within this discipline it is also true to say that an observed abnormality in a biological parameter does not have to be aetiologically involved in the disease process to be of clinical and diagnostic utility. The quantitation of melatonin in various biological fluids has not yet any well established role in this regard, but evidence is accumulating which suggests an important utility for melatonin assay in laboratory medicine and clinical research. This chapter reviews the state of the art in assay methodology and discusses the available data which suggest a role for melatonin as a tool in the laboratory diagnosis and investigation of disease.

Melatonin estimation

Radioimmunoassay (RIA) of melatonin is the most powerful relatively uncomplicated and comparatively inexpensive technique in routine use for the sensitive and accurate quantitation of analyte in plasma, serum, cerebrospinal fluid (CSF), and saliva (Arendt *et al.* 1975; Arendt 1978, 1981; Miles and Philbrick 1987). The first assay techniques, which from our present RIA position we see as almost qualitative rather than quantitative, were the bioassay and fluorometric techniques (Arendt 1978), but these soon became replaced by the first RIAs of which at least four had been published by 1976 (Arendt *et al.* 1975; Levine and Riceberg 1975; Rollag and Niswender 1976; Pang *et al.* 1976). Following the published description of these techniques, all of which had used relatively specific antisera of high titre and good sensitivity, a steady flow of other RIA systems appeared with increasing improvements in characteristics of specificity, sensitivity, accuracy, and precision. Among such systems were those of Arendt and Wilkinson (1978), Moore and colleagues (1979), Ferrier *et al.* (1982*a*), Iguchi *et al.* (1982*a*), Wetterberg *et al.* (1982), Tamarkin *et al.* (1982), Waldhauser *et al.* (1984), Claustrat *et al.* (1984*a*), and Vakurri *et al.* (1984*a*).

The advent of non-solvent extraction or direct RIAs in laboratory medicine was welcomed by biochemists, and the most prominent of these for plasma melatonin assay has been described by Fraser and associates (1983). This technique represents a rapid RIA system of excellent specificity, good sensitivity and has coefficient of variability characteristics well within recognized limits. A similar method for the estimation of the major melatonin metabolite, 6-hydroxymelatonin, directly from serum and urine has been described by Arendt and colleagues (1985). Most recently, Sleghart and associates (1987) have published a new radioimmunoassay procedure for serum melatonin. The technique shows an intra-assay coefficient of variation of 7.3 per cent and inter-assay coefficient of variation of 9.1 per cent with excellent recovery of analyte. The procedure involves extraction of melatonin from serum on Baker reverse phase C-18 columns, and attendant column preparation and maintenance procedures are necessitated. Such methodology increases the complexity and cost of melatonin estimation, and direct immunoassay of melatonin from serum may be preferred (Fraser *et al.* 1983).

The practical benefits of an RIA system need no description here, but RIAs are certainly not without their technical and related problems. Indeed, since our present knowledge of pineal endocrinology is heavily based on studies that have employed various RIA systems for melatonin estimation, the importance of a complete validation of any RIA system for use cannot be over-emphasized (Rollag 1981). Many of the assays currently popular have shown a zero value in plasma from subjects who have undergone pinealectomy proving that they can measure at least something from pineal tissue (Markey and Buell 1982; Miles and Philbrick 1987). This fact alone, however, does not exclude the possibility that such an assay system is measuring something else from the pineal in addition to melatonin. Furthermore, there can be no guarantee that an RIA will measure melatonin in all samples, since unsuspected interference and modification of system efficiency might arise with the presence of drugs, haemoglobin, lipids, or anticoagulants in plasma (Yalow and Berson 1971; Johansson *et al.* 1985; Wetterberg *et al.* 1984*b*; Waldhauser *et al.* 1984). All new assays must therefore describe full characteristics and these should be closely reproduced in the laboratory that employs them (Lynch 1983). A recommended approach in the initial setting-up of a new RIA system is to compare results from the same set of samples with a different method, and with different antiserum in the same method, and then to compare all of these results with those from a completely different method such as gas chromatography–mass spectrometry (GC–MS) (Fellenberg *et al.* 1980; Francis *et al.* 1987; Markey *et al.* 1985; Tetsuo *et al.* 1980; Young *et al.* 1985). Attention must always be paid to the initial extraction of melatonin from samples if the assay is not 'direct', and to the final separation technique. Indeed, using the same antiserum it is possible through modification of the extraction procedure (e.g. choice of solvent, time of extraction, etc.) to greatly improve initial recoveries. In addition, it should be remembered that the efficiency of any assay system depends heavily on the skill of the operator. In the hands of one technician a melatonin RIA may show impressive assay characteristics and quality control, whereas in another's hands these parameters may be unsatisfactory. A permanent international quality control scheme for melatonin RIA would be of considerable benefit in the detection and solving of technical difficulties, and this is something that investigators in the field must develop.

An interesting immunological technique that assays melatonin non-isotopically using enzyme-labelled antibodies has been described by Ferrua and Masseyeff (1985). Plasma samples are first chloroform extracted, and then melatonin and a methoxytryptamine hemisuccinate–human serum albumin conjugate (physically adsorbed on to a poly-styrene sphere) compete for a limited and fixed quantity of peroxidase-labelled antimelatonin IgG. Following incubation and wash procedures, the enzymatic activity bound to the sphere is measured with a chromogenic substrate. The method can reportedly detect quantities of melatonin in the 22 fmol region with good precision. This technique is an example of a non-isotopic system and in this differs from the vast majority of current immunoassays which employ ^{3}H-melatonin as a radiotracer (of which there are at least four specific activities currently commercially available). Iodinated melatonin has been prepared and characterized (Vakkuri *et al.* 1984*a,b*; Tiefenauer 1984), but γ-emitting melatonin labels have yet to receive use in most pineal research laboratories. The use of such labels within validated immunoassays is attractive since detection and counting of γ-radiation as the final step in analytical technique is less expensive, easier and 'neater' than tritium counting (Miles and Philbrick 1987).

Plasma assay of melatonin is essential to the intensive characterization of circadian rhythms, but absolute excretion as assessed by urinary

6-hydroxymelatonin quantitation, is often a valuable index. Various methods exist for estimation of this analyte including GC–MS (see Miles and Philbrick 1987, for refs) and RIA. Currently, the most popular RIA available for urinary 6-hydroxymelatonin assay is that described by Arendt and her collegues (1985), and this particular technique has the added advantage of an ability to measure serum metabolite prior to its urinary clearance. 6-hydroxymelatonin sulphate—the conjugated metabolite—is extremely stable in urine for at least 24 hours at 4°C, and for at least 6 months at −20°C (Bojkowski 1985).

In addition to plasma, serum, urine, ovarian follicular fluid (see Reiter, Chapter 1), and CSF (Hedlund et al. 1977; Rollag et al. 1978; Reppert et al. 1980) melatonin is also present in human saliva (Miles et al. 1985a,b, 1987, 1988; Vakkuri 1985) and this observation is of immediate relevance with regard to new methodology and practicability of experimental approach. The general practical relevance of salivary analyte RIAs in laboratory medicine may be easily appreciated, and has been fairly usefully discussed in a review from the Tenovus Institute, United Kingdom (Riad-Fahmy et al. 1982) of the applicability of these methods in the investigation of endocrine function. In recognizing the current progress and level of knowledge in pineal science, one must remain aware of the limitations often encountered. One limitation, and indeed a major one, has been the practical difficulties associated with critical longitudinal assessment of the circadian characteristics of human melatonin secretion. Such difficulties have been most manifest to investigators of paediatric pineal endocrinology and also to those in psychiatry research where long-term studies of the plasma melatonin profile are limited by both ethical and practical considerations. Such difficulties led Miles and his colleagues (1985a) to develop a direct RIA for salivary melatonin based on some principles of the direct plasma melatonin RIA previously described by Fraser et al. (1983). Extensive investigation of salivary melatonin profiles have been performed (Miles et al. 1985a,b, 1987a,b) and the most consistent findings indicate that salivary melatonin concentrations represent a mean 30 per cent of the corresponding plasma value to which they remain highly correlated at all times ($r = 0.98$, $P > 0.001$). The results were consistent with a report by Cardinali and co-workers (1972) that melatonin is around 70 per cent bound to plasma albumin, suggesting the fraction in saliva to be the unbound plasma quantity and thus the fraction of immediate functional relevance (Riad-Fahmy et al. 1982). McIntyre and associates (1987) have recently confirmed these findings, and the work of these authors has been discussed previously in this volume (see Reiter, Chapter 1).

Arato and colleagues (1985) have shown that despite interindividual variation in timing, profile, and level of melatonin secretion, secretion within a given individual shows remarkable stability. This makes alterations in melatonin secretion within the individual those of most immediate biological significance. Such intra-individual modifications in secretory characteristics are reported to occur in bipolar affective disorders (Lewy *et al.* 1979; Miles and Philbrick 1987, 1988). Monitoring of these changes and assessment of their clinical and diagnostic significance from longitudinal community-based studies may become more practicably possible using salivary melatonin analysis as one methodological approach. Indeed, the longitudinal profile of salivary melatonin within the individual has been shown remarkably stable (Miles *et al.* 1987*b*). A second potential clinical utility of salivary melatonin estimation may be seen in cases of melatonin-secreting pineal tumours (Miles *et al.* 1985*c*; Miles and Philbrick 1987, 1988). Surgical pinealectomy or functional pinealectomy produced by irradiation removes the site of origin of detectable endogenous melatonin (Neuwelt and Lewy 1983; Vorkapic *et al.* 1988) and the appearance of melatonin in the saliva of patients treated for this tumour type may indicate tumour regeneration or metastatic complication and has use in long-term out-patient follow up. Figure 13.1 shows the superimposed profiles of plasma and salivary melatonin levels in 10 volunteer subjects over a 24-hour cycle. The longitudinal stability of the salivary melatonin rhythm in male subjects is shown in Fig. 13.2.

A multiplicity of analytical techniques is thus available for the quantitation of melatonin in a diverse range of body fluids. The next step in immunological technique research must be the development of a monoclonal antibody based chemiluminescent assay for melatonin in plasma. Such a technique will provide the classical convenience and practicability of the conventional RIA system, but has the potential for a greatly enhanced level of specificity, sensitivity, precision, and accuracy.

Melatonin in clinical investigation and diagnosis

Introduction

The potential usefulness of melatonin estimation as a diagnostic marker in laboratory medicine was probably first suggested by Wetterberg and associates (1979) who proposed that assay of melatonin in conjunction with cortisol estimation could provide a melatonin–cortisol ratio that was diagnostically specific for major affective disorder. Subsequent studies have cast doubt on the clinical utility of melatonin estimation in *simple* form. Indeed, disturbed melatonin secretion appears not confined to major affective disorder, but has been recorded in a multiplicity of other clinical categories. Given observations such as these, all potential

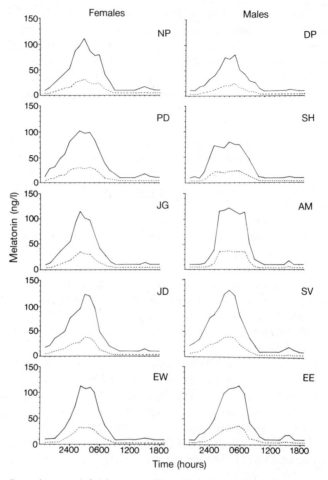

Fig. 13.1. Superimposed 24-hour profiles of plasma and salivary melatonin in volunteer subjects. Each individual profile represents hourly estimation of plasma and salivary melatonin in five males (19–33 years) and five females (20–25 years). [Reproduced from Miles *et al.* 1987*a*, with permission.]

diagnostic applications of melatonin estimation must be carefully evaluated (Miles and Philbrick 1987, 1988).

Melatonin levels in neoplastic disease

In most investigations of melatonin secretion in cancer patients, estimations have been performed following establishment of disease or in its advanced stages. In terms of the pathobiology of neoplasia, the results from such studies have not really meant very much. However, alterations in melatonin secretion in individuals with neoplastic disease need not

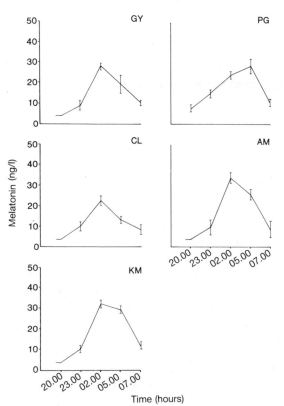

Fig. 13.2. Nocturnal profiles of salivary melatonin levels in volunteer males. Saliva was collected at 08.00, 23.00, 02.00, 05.00, and 07.00 hours on Monday, Wednesday, and Friday of a week for four consecutive weeks. Individual profiles therefore represent individual assessments of longitudinal salivary melatonin profiles (mean ± SD). [Reproduced from Miles *et al.* 1987*b*, with permission.]

have to be in any way of pathogenetic significance to be of clinical utility. Indeed, even if the altered secretion is purely a secondary phenomenon, or a 'pineal reaction' to the presence of neoplastic tissue, it might well have clinical significance. Investigations must involve consideration of what abnormalities in particular are specific to the tumour and its type, that is, what aspect of the circadian profile must be analysed with diagnostic applications in mind. As Drs Blask and Hill point out in Chapter 7 of this volume, the phase relationships of the circadian profile may be important in some neoplasms whereas in others the amplitude of the nocturnal rise or the presence of elevated daytime concentrations may be important. However, longitudinal studies of melatonin secretion from the beginning or ideally *before* the genesis of malignancy of any kind, have not yet been reported in a statistically viable population of patients. Until such studies appear we must deduce what we can from the existing studies with their attendant limitations in this regard, if comment is to be made on the potential clinical and diagnostic relevance of melatonin analysis in neoplastic processes. The importance of melatonin in the pathobiology of neoplasia has already been discussed in excellent detail by Drs Blask and Hill earlier in this volume, and the reader is referred to Chapter 7 for that information.

Reports of melatonin levels in neoplastic disease have, to date, been uncommon and rather contradictory. Raikhlin and colleagues (1980) reported elevated melatonin levels in the serum of oncological patients which were apparently independent of tumour classification, localization, and type, whereas other workers have reported in contrast generally lowered levels (Bartsch *et al.* 1981; Tamarkin *et al.* 1982). Thus, either elevated or reduced melatonin secretion might be related to oncological status. In addition, some cancer patients may have essentially normal levels. Lissoni and associates (1986*a*) concluded, after a study of pineal function in 35 oncological patients, that two subpopulations of patients exist in relation to altered melatonin secretion; one with essentially normal levels and the other with elevated output. These authors interpreted the high levels of melatonin in the latter category to represent a functional increase in pineal activity in an attempt to increase the secretion of an oncostatic agent(s) which may be melatonin itself or a related factor—the secretion of which from the pineal is *marked* by melatonin.

Although generation of melatonin by the tumour itself or altered hepatic metabolism of hormone are important considerations, a *true* alteration in pineal activity may be occurring in these patients. Indeed, Lapin and Frowein (1981) have observed true decreases of secretion of rat pineal melatonin in parallel with tumour growth and neoplastic progression in these animals. Relkin (1976) had similarly put forward the suggestion that during the early stages of tumour growth, the pineal gland becomes hyperactive in an effort to secrete oncostatic agents, and that this continued hyperactivity results in a final degeneration of the gland and consequently reduced melatonin levels. This hypothesis has been supported in more recent years by the studies of Bartsch and colleagues (see below).

Melatonin secretion in prostatic and mammary carcinoma

Bartsch *et al.* (1985) have studied melatonin secretion in males with varying types of prostate tumours. These investigators observed increased melatonin levels in patients with incidental prostatic carcinoma (where malignant cells can be demonstrated, but where no marked proliferation has occurred), essentially normal levels in benign prostatic hypertrophy, but markedly lowered levels in established prostatic carcinoma. Such findings are of differential diagnostic potential since benign disease cannot at present be differential from malignant disease by other notable biochemical indices of general prostatic pathology such as prostatic acid phosphatase or prostate-specific antigen measurement. However, Birau (1981) has reported daytime levels of melatonin in patients with prostatic carcinoma as either undetectable, normal, or very high (> 100 ng/l), and Touitou *et al.* (1985) have documented normal daytime plasma melatonin in two elderly males hospitalized for prostate carcinoma. Such findings do not necessarily contradict the conclusions of Bartsch *et al.* (1985) or the general hypothesis of gradually decreasing pineal function with neoplastic progression. They do, however, point to the critical need for replication of the findings of Bartsch *et al.* (1985) using essentially the same experimental design, but with long-term longitudinal monitoring of the circadian profile of melatonin in patients with these prostatic pathologies. A full description of melatonin secretion during the course of the disease may then be possible, and details from it could be correlated with clinically determined diagnosis, stage, and prognosis. The possible effects of drugs must be carefully considered in view of at least one report (Schloot *et al.* 1981) that methyloestradiol and related agents may attenuate melatonin secretion. Damian and colleagues (1987) have documented interesting modulatory effects of pineal extracts on prostatic acid phosphatase activity. A similar potential role for melatonin in differentiating organ-specific tumour pathologies may be present in patients with mammary carcinoma. Drs Blask and Hill have earlier in this volume discussed the available data implicating melatonin aetiologically with mammary carcinoma, and only data pertinent to diagnostic potential will be discussed here.

From investigations so far reported, melatonin levels have been fairly consistently described as reduced or essentially normal in patients with mammary carcinoma, although there have been two reports of elevated daytime melatonin levels in breast cancer patients (Birau 1981; Schloot *et al.* 1981). Again, such a finding is not inconsistent with current hypotheses and understanding, and has been discussed in relation to prostatic carcinomas in males. Probably, the most important observation of concern here is the apparent association of oestrogen and pro-gesterone receptor status of breast cancer and melatonin secretion both in terms of circulating levels and in terms of biochemical effects (Blask and Hill, this volume; Miles and Philbrick 1987). Bartsch and associates (1981) have reported that, in 10 menopausal women with advanced stages of mammary carcinoma, 24-hour urinary melatonin excretion was 31 per cent lower than in age-matched controls, but the difference did not reach statistical significance. However, the patients with oestrogen-receptor-positive (ER+) tumours had urinary melatonin levels lower than those patients with oestrogen-receptor-negative (ER−) tumours, although the authors did not evaluate this apparent association statistically. Tamarkin *et al.* (1982) have shown that, although a normal circadian rhythm may be seen in both breast cancer patients and control subjects, when the patients were divided into ER+ and ER− categories, those patients with ER+ tumours showed a 41 per cent reduction in amplitude of the nocturnal peak compared to the amplitude of secretion in ER− patients which was essentially normal. Thus, tumour oestrogen-receptor concentrations are apparently inversely correlated with the nocturnal peak plasma concentration of melatonin. Danforth and colleagues (1985) have more recently confirmed these early results, and have additionally reported the same correlation between progesterone-receptor status of breast tumour and nocturnal peak plasma concentrations of melatonin. Interestingly, this correlation between sex steroid hormone-receptor concentration and melatonin is specific, and no correlation between melatonin levels and glucocorticoid receptors could be demonstrated.

The apparent ability of melatonin assay to differentiate ER+ from ER−, and PR+ from PR− mammary tumours indicates a potential diagnostic utility for melatonin assay in this regard. A 24-hour secretory rhythm profile with estimation of absolute excretion of melatonin may well prove a useful adjunct in the routine differential diagnosis of mammary tumours. This potential ability is an important one since both prognosis and therapeutic approach are assessed by determination of the sex steroid hormone-receptor status of biopsied breast carcinoma (King *et al.* 1980).

Neurodiagnostic value of melatonin analysis: pineal tumours

Up to 1 per cent of all intracranial neoplasms fall into the diagnostic category of pineal tumour. The histology of these tumours has been fairly recently reviewed elsewhere (see Miles and Philbrick 1987 for refs). The potential utility of melatonin analysis in the differential biochemical diagnosis of these histologically heterogeneous tumours remains incompletely resolved, but our current knowledge suggests at least some diagnostic possibilities within a broader clinical investigation.

Barber and associates (1978) documented considerably elevated daytime melatonin levels in patients with pineal tumours. In five patients studied, analysis of melatonin showed plasma levels of 140, 142, 182, 183, and 131 ng/l, with a corresponding control value of 20 ± 3 ng/l. The authors suggested that, on this basis, melatonin analysis could be a useful diagnostic tool in initial investigations of the patient. However, Kennaway and colleagues (1979) were unable to demonstrate any elevation of melatonin secretion in two patients with pineal tumours, and Tapp (1978) had reported a plasma melatonin value of 185 ng/l in a 30-year-old male in whom histological diagnosis had shown not a *true* pinealoma, but rather a pineal associated tumour. More recently, Neuwelt and Lewy (1983) reported an essentially normal circadian profile of melatonin in a young male with a well encapsulated pineal tumour with elements of low-grade astrocytoma and pinealoblastoma.

True tumours of the pinealocytes might be expected to result in elevated levels of melatonin, and non-pinealocyte or pineal-associated tumours to result in normal or lowered levels through local destruction of melatonin synthesizing cells. However, Quay (1980) has pointed out that, even in tumours which show great increases in pinealocytes genetically equipped for melatonin biosynthesis, the enzymatic capacity for synthesis may be only 4–7 per cent of normal pinealocytes depending on the tumour cellular differentiation characteristics. In cases of pinealoblastomas, melatonin secretion may be normal upon initial examination (Neuwelt and Lewy 1983), but with time following neoplastic progression levels may drop or become elevated. Indeed, as Fetell and Stein (1986) have discussed, even non-parenchymal tumours might sometimes result in increased melatonin secretion through local interference with the regulatory mechanisms of the pineal gland. Such results bring the specificity and utility of melatonin assay as a neurodiagnostic tool into question (Arendt 1978).

If a pineal tumour can be demonstrated through melatonin assay to be melatonin-secreting, then monitoring melatonin levels within the patient may well prove considerably useful. Miles and colleagues (1985a) have reported the case of a 17-year-old female neurosurgical patient who presented with an intracranial growth located in the left cerebral hemisphere. The patient had been treated with irradiation over 2 years previously for a primary pinealoma. Analysis of 11.30 hours plasma melatonin revealed a value of 120 ng/l with an age- and sex-matched control value of 14 ng/l. Irradiation therapy of the hemispheric tumour resulted in its rapid disappearance, and 11.30 hours melatonin analysis at this stage showed consistently undetectable levels. The authors hypothesized that the high level of melatonin recorded in the patient was derived from the hemispheric tumour which might thus have represented a single or closely associated group of metastases from the original pinealoma. This report demonstrates a potentially important role for melatonin assay in the evaluation of such patients. Not only could melatonin assay be employed on patient presentation to determine whether or not the pineal tumour was hormone-secreting, but it might be employed in the long-term out-patient follow-up of the patient to screen for tumour regeneration or metastatic complication. Such a screening may be ideally carried out in saliva samples as previously described above.

Clearly, further reports of melatonin secretion in cases of pineal tumour should be accompanied by a histological report to allow a more confident interpretation of the diagnostic value of melatonin analysis. One such report, and the most extensive to date, has been presented by Vorkapic and colleagues (1988). Nine patients with pineal region tumours diagnosed with computer assisted tomography (CAT) were studied to assess the neurodiagnostic application of melatonin analysis. The histological diagnosis accompanied with corresponding pre- and post-operative serum melatonin levels are summarized in Table 13.1. Prior to surgery, elevated melatonin levels could not be demonstrated in any of the nine patients, despite the presence of two cases of pinealocytomas. In three of the subjects (1–3, Table 13.1) serum melatonin remained undetectable throughout the entire period of study. One patient (case 4, Table 13.1) presented with extremely low melatonin levels and an absence of circadian organization of secretion. The remaining five patients (5–9, Table 13.1) exhibited a normal circadian pattern and level of secretion. Examination following tumour resection showed that all but one of the eight patients (case 3 died from brainstem oedema on third post-operative day) had undetectable or extremely low melatonin levels (see Table 13.1). The exceptional patient (case 9, Table 13.1), who had

Table 13.1. Pre- and post-operative serum melatonin concentrations in histologically diagnosed pineal region tumours. Data is summarized from Vorkapic *et al.* (1988). For discussion see text.

Case number	Histological diagnosis	02.00 hours serum melatonin (ng/l)	
		Pre-operative	Post-operative
1	Germinoma	< 10	< 10
2	Meningeoma	< 10	< 10
3	Pinealoblastoma	11	DIED
4	Teratoma	18	11*
5	Pinealocytoma	36	< 10
6	Pinealoblastoma	100	< 10
7	Teratoma	84	14*
8	Pinealocytoma	46	18*
9	Pilocytic astrocytoma	80	130

*Values obtained through radioimmunoassay may here represent 'assay noise', but incompleteness of pinealectomy should nevertheless always be considered.

presented with a pineo-mesencephalic pilocytic astrocytoma dislocating, but not involving the pineal itself, exhibited a normal pattern of secretion. In this patient, resection was possible without damage of pineal tissue.

From this study of Vorkapic and colleagues (1988) it would appear that, on patient presentation, reduced or normal plasma melatonin rather than elevated concentrations of hormone are the consistent findings. Following tumour resection the absence of melatonin secretion also proved useful as an index of the completeness of pinealectomy. Vorkapic's study (1988) represents a considerable advance in our understanding of melatonin secretion in pinealoma and its modification by the presence of neoplastic infiltration, and its experimental design is an excellent one which subsequent investigators should adhere to. Final comments on diagnostic utility must be reserved until such studies have been performed and presented.

Melatonin secretion in psychiatric disorder

Much current data suggest a role for melatonin analysis as a clinical tool and diagnostic marker in psychiatry (Miles and Philbrick 1987, 1988). This potential role has been explored in areas such as study of the interactions of antidepressant and antipsychotic drugs with melatonin and the pineal, investigation of the apparent pineal dysfunction of affective disorders, evaluation of the role of melatonin in light and light therapy induced effects in depressive patients, and investigation of pineal involvement in depressive hypercortisolaemia. Additional potential may be seen in the evaluation and management of mania and acute schizophrenia.

Melatonin—a diagnostic tool in major affective disorder?

Jimerson and colleagues (1977) were among the first investigators to assess melatonin secretion in patients with major affective disorder, and at the time of their study could not demonstrate any significant alterations in pineal function assessed by this means. Subsequent studies employing more sophisticated methodology demonstrated marked changes in maximal level of secretion (Lewy *et al.* 1979; Wetterberg *et al.* 1981, 1984*a*; Beck-Friis *et al.* 1988), a failure to exhibit in some cases normal circadian characteristics (Mendelwicz *et al.* 1979), phase advances in secretory onset (Beck-Friis *et al.* 1981; Lewy 1984) and a supersensitivity to high-intensity light induced suppression of nocturnal melatonin secretion (Lewy *et al.* 1981; Lewy 1983). These phenomena are now regarded as the cardinal features of the apparent pineal dysfunction of affective disorder.

The reduction in circulating melatonin levels in depressed patients has been extensively documented (Lewy *et al.* 1979; Mendelwicz *et al.* 1979; Wirz-Justice and Arendt 1979; Wetterberg *et al.* 1979, 1981, 1984*a*; Beck-Friis *et al.* 1981, 1985; Wetterberg 1983; Claustrat *et al.* 1984*b*; Lewy 1984; Brown *et al.* 1985; Boyce 1985; Steiner and Brown 1985; Frazer *et al.* 1986; McIntyre *et al.* 1986), and has been included into a formally described 'low melatonin syndrome' (Beck-Friis *et al.* 1985). However, there are still important questions which have not yet been completely answered; Is there a section of the 'normal' population with melatonin levels as low as those in depressed patients, and if so, can low melatonin still be regarded as a *trait* or *state* marker for depressive illness? There is general agreement on the first half of this question, if not on the second. Some 'normals' do have low melatonin levels, but they could, after all, be more susceptible to depressive disorder than other 'normals' with higher levels. What appears significant is that many studies have shown a distinct association between lowered melatonin levels and melancholic depressive subtype (Boyce 1985). Brown and colleagues (1985) have recorded lower nocturnal melatonin secretion in depressed patients with melancholia than in depressive patients without melan- cholia. Similarly, Boyce (1985) has shown an increased association of reduced urinary 6-hydroxymelatonin excretion with melancholic subtype of depressive illness than in those depressives who show no element of melancholia in their symptomatology. Such findings, if replicated more extensively, may prove useful in the differentiation of depressive sub- types.

The biosynthesis and secretion of melatonin are under noradrenergic control, and this means that lowered levels of melatonin in depression may reflect not a generalized pineal dysfunction *per se*, but only a reduction in the central nervous system noradrenergic tone (Lewy *et al.* 1979; Claustrat *et al.* 1984*b*). In addition, lowered secretion of melatonin could conceivably result from disturbances in the central sero- toninergic systems also. The pinealocyte synthesizes melatonin from pineal serotonin which is itself derived from peripherally available tryptophan (Klein and Weller 1970). A decrease in systemic tryptophan availability could result in a deficiency of pineal serotonin for melatonin synthesis by a noradrenergic mechanism which in depression has long been considered impaired (Schildkraut 1965). Furthermore, tryptophan hypercatabolism is likely to occur in the hypercortisolaemic depressive through a cortisol-mediated induction of hepatic tryptophan pyrrolase synthesis (Curzon and Green 1968).

Factors other than the presence of psychiatric disturbance may influence melatonin levels, and these require careful assessment when considering any biochemical parameter as a biological marker. Light (Wetterberg 1978; Lewy *et al.* 1983), age (Touitou *et al.* 1981, 1985;

Iguchi *et al.* 1982*a*), body weight (Arendt *et al.* 1982; Ferrier *et al.* 1982*a*) body height (Beck-Friis *et al.* 1984), the use of glasses (Erickson *et al.* 1983), drugs (Hanssen *et al.* 1977; Lissoni *et al.* 1986*a,b*; Checkley, this volume), and genetic variation (Wetterberg *et al.* 1983) have been reported to influence circulating melatonin levels. Not all studies have, however, been in agreement on the presence of hypo-melatoninaemia in major affective disorder (Thompson *et al.* 1988).

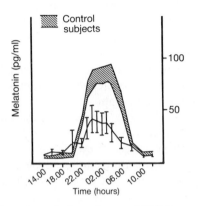

Fig. 13.3. Plasma melatonin variations (mean ± SEM) over a 24-hour period in controls and depressed patients. The typical 'depressive hypomelatoninaemia' referred to in the text may be clearly seen here. [Reproduced from Claustrat *et al.* 1984*b*, with permission.]

Melatonin as a marker for biological rhythms and during phototherapy

The timing of melatonin secretion can be useful in assessing circadian phase positions. This is best determined by screening for the time of secretory onset which is a discrete event that can be determined with high accuracy.

Lewy and colleagues (1979) were among the first investigators to demonstrate an early onset of melatonin secretion in depressed patients, with this onset of secretion occurring even earlier in mania (Lewy *et al.* 1979; Lewy 1984). Speculations for the presence of a possible dysfunction in the retino-pineal tract somehow involved in the circadian shift appeared soon after. Such an abnormality was first documented by Lewy and colleagues (1981) who showed that manic-depressive patients were supersensitive to the suppressive effects of high-intensity light on nocturnal melatonin secretion. This response may well be a useful *trait* marker. In a study of 11 euthymic manic-depressive patients not taking medication, 500 lux suppressed melatonin secretion twice as much as in 24 healthy volunteers (Lewy *et al.* 1985*a*).

Lewy and colleagues have treated winter depressives by exposure to bright light between 06.00 and 09.00 hours, and between 16.00 and 19.00 hours, and have found that most of these patients respond more favourably to morning light than to evening light alone. Furthermore, the response to morning light alone is greater than the response to light at both times (Lewy *et al.* 1984, 1985*b*). Their results were consistent with the contention that such patients had phase-delayed circadian rhythms (Lewy *et al.* 1984, 1985*b*) in contrast to a phase advance which characterizes most other endogenous depressives who have early morning waking. The endogenous depressives were observed to respond preferentially solely to evening light rather than morning light (Lewy *et al.* 1983, 1984, 1985*b*). From their results it appears that morning bright light exposure is therapeutically active in patients with delayed sleep phase syndrome, and evening light of efficacy in those patients with advanced sleep phase syndrome. Given these data, Lewy has recommended that all patients showing evidence of chronobiological sleep or mood disorders be 'phase typed' (Lewy *et al.* 1984, 1985*b,c*). Such a clinical and diagnostic procedure must involve assay of melatonin secretory onset and, in quantitative terms, level of secretory onset (see Sack and Lewy, Chapter 10).

It is noteworthy that the light required to produce any antidepressant effects must be of an intensity adequate to suppress melatonin synthesis and secretion (Rosenthal *et al.* 1986; Kripke 1986). If phototherapy has to suppress melatonin secretion in order to achieve clinical result, then whether or not the effect is biologically or coincidentally involved in the process is of little relevance to a diagnostic use of melatonin analysis. Indeed, melatonin assay is then capable of primarily determining the presence of light sensitivity, and also in determining the dose of light necessary to achieve a percentage suppression critical to therapeutic effect. Furthermore, if supersensitivity to light can be shown a specific response of depressives by testing for the presence of this response in other psychiatric categories, then it has obvious diagnostic selectivity.

Rebound in nocturnal melatonin release after exposure to light is lacking in depression

Beck-Friis and associates (1988) have reported evening light exposure capable of reducing the nocturnal rise in melatonin secretion of healthy subjects. The suppression is followed by a rebound increase in nocturnal melatonin release in 'normals', but not in some manic-depressives. The biochemical mechanisms responsible for this failure are unclear, although a failure to sensitize the receptor mechanism responsible for the rebound increase, or a deficiency in the accumulation of melatonin or a precursor, or both are immediate hypothetical possibilities. If this failure to show 'rebound' in melatonin secretion is consistently reported, then another potential diagnostic test for major depressive disorder will receive use. The ability of depressives to convert 5-hydroxytryptophan (5-HTP) to melatonin at various times of the 24-hour cycle, compared to healthy subjects, could form the basis of another diagnostic test (Namboodiri *et al.* 1983), although recent evidence suggests no serum melatonin response to 5-HTP administration in healthy children and adults (Cavallo *et al.* 1987).

Melatonin—value in the management of schizophrenia and mania?

Although there are some well known historical associations of the pineal with schizophrenia (Descartes 1662; Becker 1920; Kitay and Altschule 1954; Miles and Philbrick 1988) little diagnostic and clinical value for melatonin assay in this psychosis may be described. The possible link between the pineal and schizophrenia has been discussed in the previous chapter of this volume, and it is clear that melatonin levels and rhythms have been found essentially normal by several independent groups of investigators (Hanssen *et al.* 1980; Owen *et al.* 1983; Beckmann *et al.* 1984; Steiner and Brown 1985).

A reduced melatonin secretion may be present in chronic schizophrenia with affective pathology (Ferrier *et al.* 1982*a,b*; Miles and Philbrick 1987, 1988) and it is interesting to speculate whether or not a reduction in melatonin secretion parallels the appearance of affective pathology, and might thus be of use in differentiating schizoaffective psychoses from uncomplicated schizophrenia. In this regard it is important to note that personality disorders with affective pathology may have lowered melatonin secretion (Steiner and Brown 1985). Also worthy of further study in respect of diagnostic utility is the interesting observation of Jones and colleagues (1969) that melatonin administration to schizophrenic subjects results in a minor difference in the type and ratio of excreted compounds when compared to psychiatrically normal individuals.

Drugs of documented therapeutic use in acute schizophrenia can effectively block melatonin secretion through direct blockade of pineal noradrenergic receptors and by inhibition of the biosynthetic pathway (Hartley *et al.* 1972; Hanssen *et al.* 1977). Monitoring of melatonin secretion during institution of neuroleptic therapy may thus be useful in assessing the therapeutic outcome with these drugs since suppression of melatonin secretion by these drugs is, in many cases, likely to parallel the interaction of these agents with dopamine receptors. The same potential clinical utility may be present in mania where neuroleptics are valuable in rapid control and management of acute symptoms, and where melatonin secretion may be initially elevated (Lewy *et al.* 1979), thus increasing the sensitivity of the procedure.

Other diseases

Birau (1981) has studied melatonin secretion in a wide variety of clinical syndromes with important findings. Markedly disturbed secretion of melatonin was reported in Klinefelter's syndrome (KS), Turner's syndrome (TS), sarcoidosis, psoriasis vulgaris, and spina bifida occulta (SBO), in individuals studied over 31 hours. Patients with KS exhibited depressed melatonin secretion with an absence of circadian rhythm when compared to healthy controls. However, a minor increase in secretion was detected at 20.00 hours, and the 08.00 and 14.00 hours serum melatonin values were significantly higher than in the control group. Similar results were recorded for patients with TS.

Study of melatonin secretion in 32 patients with SBO (eight children, eight early pubertal children, 16 adult patients) demonstrated marked differences in level and profile of secretion between the subgroups. Of particular interest is the marked differences in the 08.00 hours melatonin level between healthy male children (30 ng/l) and male children with SBO (> 200 ng/l), and healthy female children (37 ng/l) and female children with SBO (> 240 ng/l). Furthermore, in the children with SBO melatonin levels were consistently 100 per cent increased or greater at all times above controls (Fig. 13.4). If such results can be shown independent of orthopaedic and drug treatments employed in the management of SBO, then melatonin analysis will prove of diagnostic significance in this condition. Indeed, if the pre-partum amniotic fluid levels of melatonin can be shown clearly increased in mothers carrying children with SBO, analysis of melatonin could form an interesting adjunct in the screening for, and prenatal diagnosis of the condition.

Reduced melatonin secretion has also been reported in patients with Alzheimer's disease and suggestions have been put forward that melatonin may be a useful index of human brain ageing (Nair *et al.* 1985) and that melatonin analysis may be useful in the monitoring of presenile dementias.

Conclusions

Advances in the methodology of pineal research have made possible our recognition of the presence of altered and disturbed melatonin secretion within a diverse range of diseases, and some of these observations are proving to be of diagnostic and clinical relevance. The correlation of melatonin levels with sex steroid receptor concentrations of mammary

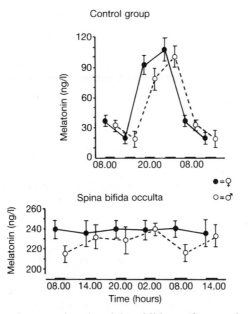

Fig. 13.4. Melatonin secretion in eight children (four males, four females; 6.2 ± 1.7 years) with spina bifida occulta. Time points are mean ± SD, with age- and sex-matched control children. [Reproduced from Birau 1981, with permission.]

carcinomas should prove useful in confirming radioligand differential diagnosis of sex steroid positive and negative tumours, and could play an important part in the approach to therapy and formulation of prognosis. A similar differential diagnostic potential for melatonin analysis may be seen in prostate medicine where melatonin assay may differentiate benign prostatic hypertrophy from prostatic carcinoma, and could reveal incidental carcinoma of the prostate. This might become an important application because conventional diagnostic markers in prostate medicine cannot differentate in this way. Melatonin assay is also of potential use in the diagnosis of a melatonin secreting pineal tumour, and in the long-term follow-up of patients treated for such pathology. In psychiatry, melatonin assay may be employed in the assessment of central noradrenergic receptor function in depressive illness, in assessing the biological potency of phototherapy, in phase typing depressive illnesses, in the management of acute schizophrenia and mania, and in the general investigation of psychiatric disturbance.

The alteration of melatonin secretion in other disease categories apart from psychiatric and oncological areas, together with the relative paucity of data on the exact effects of age, sex, weight, height, and drugs, etc., means that laboratory medicine cannot yet make definitive statements on the diagnostic use of melatonin assay, and for the moment we must remain content with constructive speculation. However, current data at least indicate a plausible utility of melatonin analysis in diagnostically orientated clinical investigation and future research in this regard can only clarify our existing knowledge.

Acknowledgements

The authors thank Professor Peter McGuffin and Dr David M. Shaw for their advice, Dr Alan Wardrop and Messrs Daniel M. Barrow and Geoffrey Yeoman for their interest, and Mrs Lillian Lewis for excellent secretarial assistance.

References

Arato, M., Grof, E., Grof, P., Laszlo, I., and Brown, G. M. (1985). Reproducibility of the overnight melatonin secretion pattern in healthy men. In *The pineal; endocrine aspects*, Advances in the Biosciences, No. 53 (ed. G. M. Brown and S. D. Wainwright), pp. 277–82. Pergamon Press, Oxford and New York.

Arendt, J. (1978). Melatonin assays in body fluids. *Journal of Neural Transmission (Supplement)* **13**, 265–78.

Arendt, J. (1978). Melatonin as a tumour marker in a patient with pinealoma. *British Medical Journal* **2**, 635.

Arendt, J. (1981). Current status of assay methods for melatonin. *Advances in the Biosciences* **29**, 3–7.

Arendt, J. and Wilkinson, M. (1978). Melatonin. In *Methods of hormone radio-immunoassay*, 2nd edn (ed. B. M. Jaffe and H. R. Behrmann), p. 101. Academic Press, New York.

Arendt, J., Paunier, L., and Sizonenko, P. C. (1975). Melatonin radio-immunoassay. *Journal of Clinical Endocrinology and Metabolism* **40**, 347–50.

Arendt, J., Hampton, S., English, J., Kwasowski, P., and Marks, V. (1982). The 24 hour profiles of melatonin, cortisol, insulin, c-peptide and gip following a meal and subsequent fasting. *Clinical Endocrinology* **16**, 89–95.

Arendt, J., Bojkowski, C., Franey, C., Wright, J., and Marks, V. (1985). Immunoassay of 6-hydroxymelatonin sulphate in human plasma and urine; abolition of the 24 hour rhythm with atenolol. *Journal of Clinical Endocrinology and Metabolism* **60**, 1166–72.

Barber, S. G., Smith, J. A., Cove, D. R., Smith, S. C. H., and London, D. R. (1978). Marker for pineal tumours. *Lancet* **ii**, 372.

Bartsch, C., Bartsch, J., Jain, A. K., Laumas, K. R., and Wetterberg, L. (1981). Urinary melatonin levels in human breast cancer patients. *Journal of Neural Transmission* **52**, 281–94.

Bartsch, C., Bartsch, H., Fluchter, S. B., Atlanasio, A., and Gupta, D. (1985). Evidence for modulation of melatonin secretion in men with benign and malignant tumours of the prostate. *Journal of Pineal Research* **2**, 121–32.

Becker, W. J. (1920). Epiglandol bei Dementia Praecox. *Therapeutische Halbmonastcher* **34**, 667–8.

Beck-Friis, J., *et al.* (1981). Hormonal changes in acute depression. In *Biological psychiatry* (ed. C. Perris, G. Struwe, and B. Jansson), pp. 1244–8. Biomedical Press BV, Amsterdam.

Beck-Friis, J., von Rosen, D., Kjellman, B. F., Ljunggren, J. G., and Wetterberg, L. (1984). Melatonin in relation to body measures, sex, age, season and the use of drugs in patients with major affective disorders and healthy subjects. *Psychoneuroendocrinology* **9**, 261–77.

Beck-Friis, J., Borg, G., and Wetterberg, L. (1988). Rebound increase of nocturnal melatonin levels following evening suppression by bright light exposure in healthy men; relationship to cortisol levels and morning exposure. *Conference on medical and biological effects of light*, October 31–November 2, 1984. The New York Academy of Sciences (in press).

Beck-Friis, J., *et al.* (1985). Serum melatonin in relation in clinical variables in patients with major depressive disorders; a hypothesis of a low melatonin syndrome. *Acta Psychiatrica Scandanavica* **71**, 319–30.

Beckmann, H., Wetterberg, L., and Gattaz, W. F. (1984). Melatonin immunoreactivity in the cerebrospinal fluid of schizophrenic patients and healthy controls. *Psychiatry Research* **11**, 107–10.

Birau, N. (1981). Melatonin in human serum; Progress in screening investigation and clinic. In *Melatonin: current status and perspectives*, Advances in the Biosciences, No. 29 (ed. N. Birau and W. Schloot), pp. 297–326. Pergamon Press, New York.

Bojkowski, C. (1985). Unpublished data; reported in Arendt *et al.* 1985.

Boyce, P. M. (1985). 6-sulphatoxymelatonin in melancholia. *American Journal of Psychiatry* **142**, 125–7.

Brown, R. P., *et al.* (1985). Differences in nocturnal melatonin secretion between melancholic depressed patients and control subjects. *American Journal of Psychiatry* **142**, 811–16.

Cardinali, D. P., Lynch, H. J., and Wurtman, R. J. (1972). Binding of melatonin to human and rat plasma proteins. *Endocrinology* **91**, 1213–16.

Cavallo, A., Richards, G. E., Meyer, W. J., and Waldrop, R. D. (1987). Evaluation of 5-hydroxytryptophan administration as a test of pineal function in humans. *Hormone Research* **27**, 69–73.

Claustrat, B., Harthe, C., Vitte, P. A., Cohen, R., and Chazot, G. (1984*a*). Melatonin radioimmunoassay—an improved method. *Journal of Steroid Biochemistry* **20**, 1447.

Claustrat, B., Chazot, G., Brun, J., Jordan, D., and Sassolas, G. (1984*b*). A chronobiological study of melatonin and cortisol secretion in depressed subjects; plasma melatonin; a biochemical marker in major depression. *Biological Psychiatry* **19**, 1215–28.

Curzon, G., and Green, A. R. (1968). Effect of hydrocortisone on rat brain 5-hydroxytryptamine. *Life Science* **7**, 657.

Damian, E., Ionas, O., and Bodescu, I. (1987). Effect of melatonin-free pineal extract on acid phosphatase activity in the castrated and testosterone treated rat. *Endocrinologie* **25**, 9–14.

Danforth, D. N., Tamarkin, L., Mullvihill, J., Bagley, C. S., and Lippman, M. (1985). Plasma melatonin and the hormone dependency of human breast cancer. *Journal of Clinical Oncology* **3**, 941–8.

Descartes, R. (1662, posthumous). De Homine, Figuris et Latinitate donatus a F. Schuyl, Lugduni Batavorum.

Erikson, C., Kuller, R., and Wetterberg, L. (1983). Non visual effects of light. *Neuroendocrinology Letters* **5**, 412.

Fellenberg, A. J., Phillipou, G., and Seamark, R. F. (1980). Specific quantitation of urinary 6-hydroxymelatonin sulphate by GC-MS. *Biomedical Mass Spectrometry* **7**, 84–7.

Ferrier, I. N., Arendt, J., Johnstone, E. C., and Crow, T. J. (1982*a*). Reduced nocturnal melatonin secretion in chronic schizophrenia; relationship to body weight. *Clinical Endocrinology* **17**, 181–7.

Ferrier, I. N., Johnstone, E. C., Crow, T. J., and Arendt, J. (1982*b*). Melatonin-cortisol ratio in psychiatric illness. *Lancet* **i**, 1070.

Ferrua, B. and Masseyeff, R. (1985). Immunoassay of melatonin with enzyme labelled antibodies. *Journal of Immunoassay* **6**, 79–94.

Fetell, M. R. and Stein, B. M. (1986). Neuroendocrine aspects of pineal tumours. *Neurological Clinics* **4**, 877–905.

Francis, P. L., Leone, A. M., Young, I. M., Stovell, P., and Silman, R. E. (1987). GC-MS assay for 6-hydroxymelatonin sulphate and 6-hydroxymelatonin glucuronide in urine. *Clinical Chemistry* **33**, 453–7.

Fraser, S., Cowen, P., Franklin, M., Franey, C., and Arendt, J. (1983). Direct radioimmunoassay for melatonin in plasma. *Clinical Chemistry* **29**, 396–7.

Frazer, A., *et al.* (1986). Patterns of melatonin rhythms in depression. *Journal of Neural Transmission (Suppl.)* **21**, 269–90.

Hanssen, T., Heydorn, T., Sundberg, I., and Wetterberg, L. (1977). Effect of propranolol on serum melatonin. *Lancet* **ii**, 309.

Hanssen, T., Heydorn, T., Sundberg, J., Alfredsson, G., Nyback, H., and Wetterberg, L. (1980). Propranolol in schizophrenia; clinical, metabolic and pharmacological findings. *Archives of General Psychiatry* **37**, 685–90.

Hartley, R., Padwick, D., and Smith, J. A. (1972). The inhibition of pineal HIOMT by haloperidol and fluphenazine. *Journal of Pharmacy and Pharmacology* **24**, 100–3.

Hedlund, L., Lischko, N. M., Rollag, M. D., and Niswender, G. D. (1977). Melatonin: daily cycle in plasma and cerebrospinal fluid of calves. *Science* **195**, 686.

Iguchi, H., Kato, K., and Ibayashi, H. (1982). Age dependent reduction in serum melatonin concentration in healthy subjects. *Journal of Clinical Endocrinology and Metabolism* **55**, 27–9.

Jimerson, D. C., Lynch, H. J., and Post, R. M. (1977). Urinary melatonin rhythms during sleep deprivation in depressed patients and normals. *Life Science* **20**, 1501–8.

Johansson, G. E. K., Ho, A. K., Chik, C. L., and Brown, G. M. (1985). Interference in melatonin radioimmunoassay by heparin preparations. In *The pineal: endocrine aspects*, Advances in the Biosciences, No. 53 (ed. G. M. Brown and S. D. Wainwright), pp. 47–52. Pergamon Press, Oxford.

Jones, R. L., McGeer, P. L., and Greiner, A. C. (1969). Metabolism of exogenous melatonin in schizophrenic and non-schizophrenic volunteers. *Clinica Chimica Acta* **26**, 281–6.

Kennaway, D. J., McCulloch, G., Mathews, C. D., and Seamark, R. F. (1979). Plasma melatonin, LH, FSH, prolactin and corticoids in two patients with pinealoma. *Journal of Clinical Endocrinology and Metabolism* **49**, 144–5.

King, R. J. B., Hayward, J. L., Masters, J. B. W., Willis, R. R., and Rubens, R. D. (1980). Steroid receptor assays as prognostic aids in treatment of breast cancer. In *Steroid receptors and hormone dependent neoplasia* (ed. J. L. Wittliff and O. Dapunt), pp. 249–54. Masson, New York.

Kitay, J. I. and Altschule, M. D. (1954). *The pineal gland.* Harvard University Press, Cambridge, Massachusetts.

Klein, D. C. and Weller, J. L. (1970). Indole metabolism in the pineal gland; a circadian rhythm in NAT activity. *Science* **169**, 1093–5.

Kripke, D. F. (1986). The therapeutic effect of bright light in depressed patients. *Annals of the New York Academy of Science* **453**, 270–81.

Lapin, V. and Frowein, A. (1981). Effect of growing tumours on pineal melatonin levels in male rats. *Journal of Neural Transmission* **52**, 123–36.

Lerner, A. B., Case, J. D., Takahashi, Y., Lee, T. H., and More, D. (1958). Isolation of melatonin, the pineal factor that lightens melanocytes. *Journal of the American Chemical Society* **80**, 2587.

Levine, L. and Riceberg, L. J. (1975). Radioimmunoassay for melatonin. *Research Communications in Chemical Pathology* **10**, 693.

Lewy, A. J. (1983). Effects of light on human melatonin production and the human circadian system. *Progress in Neuropsychopharmacology and Biological Psychiatry* **7**, 551–6.

Lewy, A. J. (1984). Human melatonin secretion: a marker for the circadian system and the effects of light. In *Neurobiology of mood disorders* (ed. R. N. Post and J. C. Ballender), pp. 215–26. William and Wilkins, Baltimore, Maryland.

Lewy, A. J. and Markey, S. P. (1978). Analysis of melatonin in human plasma by gas chromatography negative chemical ionisation mass spectrometry. *Science (Washington DC)* **201**, 741–3.

Lewy, A. J., Wehr, T. A., Gold, P. W., and Goodwin, F. K. (1979). Plasma melatonin in manic-depressive illness. In *Catecholamines: basic and clinical frontiers*, Vol. **2** (ed. E. Usdin, I. J. Kopin, and J. Barchas), pp. 1173–5. Pergamon Press, Oxford.

Lewy, A. J., Wehr, T. A., Goodwin, F. K., Newsome, D., and Rosenthal, N. E. (1981). Manic-depressive patients may be supersensitive to light. *Lancet* **i**, 383–4.

Lewy, A. J., Sack, R. L., Fredrickson, R. H., Reave, S. M., Denney, D., and Zielske, D. R. (1983). The use of bright light in the treatments of chronobio-logical sleep and mood disorders: the phase response curve. *Psychopharmaco-logy Bulletin* **19**, 523–5.

Lewy, A. J., Sack, R. L., and Singer, C. L. (1984). Assessment and treatment of chronobiological disorders using plasma melatonin levels and bright light exposure: the clock–gate model and the phase response curve. *Psycho-pharmacology Bulletin* **20**, 561–5.

Lewy, A. J., Nurnberger, J. I., and Wehr, T. A. (1985*a*). Supersensitivity to light—a possible trait marker for manic-depressive illness. *American Journal of Psychiatry* **142**, 725–7.

Lewy, A. J., Sack, R. L., and Singer, C. M. (1985*b*). Treating phase-typed chronobiological sleep and mood disorders using appropriately timed bright artificial light. *Psychopharmacology Bulletin* **21**, 368–72.

Lewy, A. J., Sack, R. L., and Singer, C. M. (1985*c*). Immediate and delayed effects of bright light on human melatonin production: shifting 'dawn' and 'dusk' shifts the dim light melatonin onset (DLMO). *Annals of the New York Academy of Science* **453**, 253–9.

Lissoni, P., *et al.* (1986*a*). A clinical study of the pineal gland activity in onco-logical patients. *Cancer* **57**, 837–42.

Lissoni, P., *et al.* (1986*b*). Effect of Tetrahydrocannabinol on melatonin secretion in man. *Hormone and Metabolic Research* **18**, 77–8.

Lynch, H. J. (1983). Assay methodology. In *The pineal gland* (ed. R. M. Relkin), pp. 129–50. Elsevier-North Holland, New York.

Markey, S. P. and Buell, P. E. (1982). Pinealectomy abolishes 6-hydroxymelatonin excretion by male rats. *Endocrinology* **11**, 425–6.

Markey, S. P., Higa, S., Shih, M., Danforth, D. N., and Tamarkin, L. (1985). The correlation between human plasma melatonin levels and urinary 6-hydroxy-melatonin excretion. *Clinica Chimica Acta* **150**, 221–5.

McIntyre, I. M., Judd, F., Norman, T. R., and Burrows, G. D. (1986). Plasma melatonin concentrations in depression. *Australian and New Zealand Journal of Psychiatry* **20**, 381–3.

McIntyre, I. M., Norman, T. R., Burrows, G. D., and Armstrong, S. M. (1987). Melatonin rhythm in human plasma and saliva. *Journal of Pineal Research* **4**, 177–83.

Mendelwicz, J., Linkowski, P., and Branchey, L. (1979). Abnormal 24 hour pattern of melatonin secretion in depression. *Lancet* **ii**, 1362.

Miles, A. and Philbrick, D. R. S. (1987). Melatonin: current perspectives in laboratory medicine and clinical research. *CRC Reviews in Clinical Laboratory Sciences* **25**, 231–53.

Miles, A. and Philbrick, D. R. S. (1988). Melatonin and psychiatry, a review. *Biological Psychiatry* **23**, 405–25.

Miles, A., Philbrick, D. R. S., Tidmarsh, S. F., and Shaw, D. M. (1985*a*). Direct radioimmunoassay of melatonin in saliva. *Clinical Chemistry* **31**, 1412–13.

Miles, A., Philbrick, D. R. S., Shaw, D. M., Tidmarsh, S. F., and Pugh, A. J. (1985*b*). Salivary melatonin estimation in clinical research. *Clinical Chemistry* **31**, 2041–2.

Miles, A., Tidmarsh, S. F., Philbrick, D. R. S., and Shaw, D. M. (1985*c*). Diagnostic potential of melatonin analysis in pineal tumour. *New England Journal of Medicine* **313**, 329–30.

Miles, A., Philbrick, D. R. S., Thomas, D. R., and Grey, J. E. (1987*a*). Diagnostic and clinical implications of plasma and salivary melatonin assay in laboratory medicine. *Clinical Chemistry* **33**, 1295–7.

Miles, A., Thomas, D. R., and Grey, J. E. (1987*b*). Longitudinal profiles of salivary melatonin in adult males. *Clinical Chemistry* **33**, 1957–9.

Moore, D. C., Paunier, L., and Sizonenko, P. C. (1979). Effect of adrenergic stimulation and blockade on melatonin secretion in the human. *Progress in Brain Research* **52**, 517–21.

Nair, N. P. V., Hariharasubramanian, N., Pilapil, C., Isaac, I., and Thavundayil, J. X. (1985). Plasma melatonin—an index of brain ageing in humans? *Biological Psychiatry* **21**, 141–50.

Namboodiri, M. A. A., Sugden, K., Klein, D. C., and Mefford, I. N. (1983). 5-hydroxytryptophan elevated serum melatonin. *Science (Washington DC)* **221**, 659–61.

Neuwelt, E. A. and Lewy, A. J. (1983). Disappearance of plasma melatonin after removal of a neoplastic pineal gland. *The New England Journal of Medicine* **308**, 1132–5.

Owen, F., Ferrier, I. N., and Poulter, M. (1983). HIOMT activity in human pineals, a comparison of controls and schizophrenics. *Clinical Endocrinology* **19**, 313–17.

Pang, S. F., Brown, G. M., Grota, L. J., and Rodman, R. L. (1976). Radio-immunoassay of melatonin in pineal glands, Harderian glands, retinas and sera of rats and chickens. *Federal Proceedings of the American Society for Experimental Biology* **35**, 691–7.

Quay, W. B. (1980). Experimental and spontaneous pineal tumours: findings relating to endocrine and oncogenic factors and mechanisms. *Journal of Neural Transmission* **48**, 9.

Raikhlin, N. T., Kvetnoy, I. M., and Tyurin, E. S. (1980). Melatonin in the blood serum of oncological patients. *Clinical Medicine (Moscow)* **58**, 77.

Relkin, R. (1976). The pineal and human disease. In *Annual research reviews of the pineal* (ed. R. Relkin), p. 76. Loundsdale House, Lancaster, England.

Reppert, S. M., Perlew, M. J., and Klein, D. C. (1980). Cerebrospinal fluid melatonin. In *Neurobiology of cerebrospinal fluid* (ed. J. H. Wood), pp. 579–89. Plenum Press, New York.

Riad-Fahmy, D., Read, G. F., Walker, R. F., and Grinbits, K. (1982). Steroids in saliva for assessing endocrine function. *Endocrine Reviews* **3**, 367–95.

Rollag, M. D. and Niswender, G. (1976). Radioimmunoassay of serum concentrations of melatonin in sheep exposed to different lighting regimes. *Endocrinology* **98**, 482–9.

Rollag, M. D. (1981). Methods for measuring pineal hormones. In *The pineal gland, anatomy and biochemistry* (ed. R. J. Reiter) Vol. 1, pp. 273–302. CRC Press, Boca Raton, Florida.

Rollag, M. D., Morgan, R. J., and Niswender, G. D. (1978). Route of melatonin secretion in sheep. *Endocrinology* **102**, 1.

Rosenthal, N. E., *et al.* (1986). Seasonal affective disorder and phototherapy. *Annals of the New York Academy of Science* **453**, 260–9.

Schildkraut, I. J. (1965). The catecholamine hypothesis of affective disorders, a review of supporting evidence. *American Journal of Psychiatry* **122**, 509–12.

Schloot, W., Dubbels, R., and Birau, N. (1981). Genetics of melatonin. In *Melatonin—current status and perspectives* (ed. N. Birau and W. Schloot), pp. 269–84. Pergamon Press, Elmsford.

Sleghart, W., Ronca, E., Drexler, G., and Karall, S. (1987). Improved radioimmunoassay of melatonin in serum. *Clinical Chemistry* **33**, 604–5.

Steiner, M. and Brown, G. M. (1985). Melatonon-cortisol ratio and the dexamethasone suppression test in newly admitted psychiatric patients. In *The pineal gland; endocrine aspects* Advances in the Biosciences No. 53 (ed. G. M. Brown and S. D. Wainwright), 347–53. Pergamon Press, New York.

Tamarkin, L., *et al.*(1982). Decreased nocturnal plasma melatonin peak in patients with oestrogen receptor positive breast cancer. *Science* **216**, 1003–5.

Tapp, E. (1978). Melatonin as a tumour marker in pinealoma. *British Medical Journal* **2**, 636.

Tetsuo, M., Markey, S. P., and Kopin, I. J. (1980). Measurement of 6-hydroxy-melatonin in human urine and its diurnal variations. *Life Science* **27**, 105–9.

Thompson, C., Franey, C., Arendt, J., and Checkley, S. A. (1988). A comparison of melatonin in secretion in normal subjects and depressed patients. *British Journal of Psychiatry* **152**, 260–6.

Tiefenauer, L. X. (1984). Prevention of bridge binding effects in haptenic immunoassay systems exemplified by an iodinated radioimmunoassay for melatonin. *Journal of Immunological Methods* **74**, 293.

Touitou, Y., Fevre-Montagne, M., and Lagoguey, M. (1981). Age and mental health related circadian rhythms of plasma levels of melatonin, prolactin, LH and FSH in man. *Journal of Endocrinology* **9**, 467–75.

Touitou, Y., Fevre-Montagne, M., Proust, J., Klinger, E., and Nakache, J. P. (1985). Age and sex associated modification of plasma melatonin concentrations in man. Relationship to pathology, malignant or not, and autopsy findings. *Acta Endocrinologica* **108**, 135–44.

Vakkuri, O., Leppaluoto, J., and Vuolteenaho, O. (1984a). Development and validation of a melatonin radioimmunoassay using radioiodinated melatonin as tracer. *Acta Endocrinologica* **106**, 152–7.

Vakkuri, O., Lamsa, E., Rahkamaa, E., Ruotsalainer, H., and Lappaluoto, J. (1984b). Iodinated melatonin: preparation and characterisation of the molecular structure by mass and ^{1}H-NMR spectroscopy. *Analytical Biochemistry* **142**, 284–9.

Vakkuri, O. (1985). Diurnal rhythm of melatonin in human saliva. *Acta Physiologica Scandanavica* **23**, 151–4.

Vorkapic, P., Waldhauser, F., Bruckner, R., Biegesmayer, C., Schmidbauer, M., and Pendl, G. (1988). Serum melatonin levels—a neurodiagnostic tool in pineal region tumours. *Neurosurgery* (in press).

Waldhauser, F., Weissenbacher, G., Zeitlhuber, U., Waldhauser, M., Frisch, H., and Wurtman, R. J. (1984). Fall in nocturnal serum melatonin levels during prepuberty and pubescence. *Lancet* **i**, 362–5.

Wetterberg, L. (1978). Melatonin in humans. Physiological and clinical studies. *Journal of Neural Transmission (Suppl.)* **13**, 289–310.

Wetterberg, L., Beck-Friis, J., Aperia, B., Kjellman, B. F., Ljunggren, J. G., Petterson, U., and Sjolin, A. (1979). Melatonin–cortisol in depression. *Lancet* **ii**, 1361.

Wetterberg, L. (1983). The relationship between the pineal gland and the pituitary-adrenal axis in health, endocrine and psychiatric conditions. *Psycho-neuroendocrinology* **8**, 75–80.

Wetterberg, L., *et al.*(1981). Pineal–hypothalamic–pituitary function in patients with depressive illness. In *Steroid hormone regulation of the brain* (ed. K. Fuxe, J. A. Gustafsson, and L. Wetterberg), pp. 397–403. Pergamon Press, Oxford.

Wetterberg, L., Aperia, B., Beck-Friis, J., Kjellman, B. F., and Ljunggren, J. G. (1982). Melatonin and cortisol levels in psychiatric illness. *Lancet* **ii**, 100.

Wetterberg, L., Iselius, L., and Lindsten, J. (1983). Genetic regulation of melatonin excretion in urine. *Clinical Genetics* **24**, 403–6.

Wetterberg, L., Beck-Friis, J., Kjellman, B. F., and Ljunggren, J. G. (1984*a*). Circadian rhythms in melatonin and cortisol secretion in depression. In *Frontiers in biochemical and pharmacological research in depression* (ed. E. Usdin, H. Asberg, L. Bertilsson, and F. Sjoqvist), pp. 197–205. Raven Press, New York.

Wetterberg, L., Saaf, J., Noren, R., Waldenlind, E., and Friberg, Y. (1984*b*). Interference in melatonin radioimmunoassay by dimethylphthalate. *Journal of Steroid Biochemistry* **20**, (6B), 1475.

Yalow, S. and Berson, A. (1971). Immunological assays. In *Principles of competitive protein binding assays* (ed. W. Odell and F. Daughady) p. 377. J. B. Lippincott, Philadelphia.

Young, I. N., Leone, R. N., and Silman, R. E. (1985). The mass spectrometric analysis of the urinary metabolites of melatonin and its deuterated analogues confirming their identity as N-acetylserotonin and 6-hydroxymelatonin. *Biomedical Mass Spectrometry* **12**, 319–37.

14. Melatonin—future clinical perspectives

Andrew Miles, David R. S. Philbrick, and Christopher Thompson

The idea of the pineal as a biologically redundant gland, and the equating of this organ with the appendix vermiformis and prostate gland as a source of more potential clinical problems than probable functional relevance is no longer a widely quoted one. The current volume bears witness to the considerable achievements in melatonin research as we celebrate the thirtieth anniversary of its discovery by Lerner and his associates (1958). Despite a great deal of research having been performed since this time (Miles and Philbrick 1987) we remain as ignorant of the definitive human physiological function of melatonin as we were in 1958. The abundance of conclusive animal data contrasts sharply with the comparative paucity of human data in this regard. However, when one considers the profound importance of melatonin in subhuman species (see Reiter, Chapter 1; Vriend and Steiner, Chapter 5; Blask and Hill, Chapter 7) only the naive scientist would conclude that melatonin is devoid of biological effects in man. The paucity of positive data from research aimed at clarifying the role of melatonin in human physiology has, in the opinion of the editors, much to do with the poorly thought out experimental designs that have been employed in physiological investigations of melatonin function. For example, in studies which have examined the endocrine consequences of melatonin administration in humans, timing and dosage of melatonin administered have not been well selected (see Reiter, Chapter 1 for listed examples). Neither has long-term screening for endocrine and non-endocrine correlates of chronic appropriately timed or continuously available melatonin been performed as an essential feature of preliminary physiological studies. Implementation of such designs has produced results in animals and may well yield interesting information if such an approach could be worked out for human studies.

The possible involvement of melatonin in various disease processes has been well covered in the present volume and is an area which is highly likely to be critically evaluated in the next decade of research. No investigators have yet been able to positively identify a clinical syndrome clearly caused by disordered pineal function, although the secretion of melatonin has been demonstrated as altered in many disease processes (Miles and Philbrick 1987, 1988; Miles and Thomas, Chapter 13; Blask and Hill, Chapter 7). It may well be that such alterations in secretion represent purely secondary phenomena of little or no functional relevance. Alternatively, such alterations may be aetiologically involved in the disease process or may come to influence the development and course of the disease, and thus prove of diagnostic and prognostic importance. It is indeed pertinent to note that complete lack of plasma melatonin (idiopathic in some, a result of β-receptor blockade in others) is not necessarily associated with any major abnormalities (Arendt *et al.* 1985). However, an apparent lack of detectable *plasma* melatonin does not necessarily mean that melatonin is not reaching critical receptor sites, perhaps in hypothalamus, in sufficient quantity to elicit effects of physiological importance. Similarly, in patients with a given disease process in which plasma melatonin levels and circadian organization are apparently 'normal', pineal involvement is not necessarily precluded because the secretion of other pineal products may be abnormal or the sensitivity to melatonin of target organs may be altered, possibly to a clinically relevant degree.

Implicit in the quest for possible melatonin action in man is the philosophy that melatonin is a biochemical signal capable of transmitting information of physiological/pathophysiological importance, and that such information might be conveyed by changes in melatonin secretion, levels attained in plasma, duration of availability, and circadian organization of secretion. The time has now come to relate changes in these parameters—which can be discerned by our ever-increasing technical facilities (Miles and Philbrick 1987; Miles and Thomas, Chapter 13) to changes in behavioural (Lieberman, Chapter 6; Arendt, Chapter 2), physiological (Reiter, Chapter 1; Arendt, Chapter 2), and pathophysiological (Miles and Thomas, Chapter 13; Sack and Lewy, Chapter 10; Blask and Hill, Chapter 7) processes. We may then begin properly to understand the role of melatonin in human health and disease.

It has not been our intention to review the constituent chapters of this volume here. The contributions speak for themselves, and the directions that future research should take are advised within each one. It is likely that the next decade of research will be an especially exciting one as present knowledge of melatonin function in animals is applied with our existing analytical facilities to the human in carefully designed and well controlled studies. Goethe has said of mankind 'what we know we do not need, and what we need we do not know'. Perhaps in human melatonin research we will begin in the next decade to 'know what we need to know'.

References

Arendt, J. *et al.* (1985). Some effects of melatonin and the control of its secretion in humans. In *Photoperiodism, melatonin and the pineal,* Ciba Foundation Symposium 117 (ed. D. Evered and S. Clark), pp. 266–83. Pitman, London.

Lerner, A. B., Case, J. D., Takahashi, Y., Lee, T. H., and More, W. (1958). Isolation of melatonin, the pineal factor that lightens melanocytes. *Journal of the American Chemical Society* **80**, 2587.

Miles, A. and Philbrick, D. R. S. (1987). Melatonin: perspectives in laboratory medicine and clinical research. *CRC Critical Reviews in Clinical Laboratory Sciences* **25**, 231–53.

Miles, A. and Philbrick, D. R. S. (1988). Melatonin and psychiatry, a review. *Biological Psychiatry* **23**, 405–25.

Index

283